AF743786

Springer Series in Materials Science

Volume 356

Series Editors

Robert Hull, Center for Materials, Devices, and Integrated Systems, Rensselaer Polytechnic Institute, Troy, USA

Chennupati Jagadish, Research School of Physics and Engineering, Australian National University, Canberra, Australia

Yoshiyuki Kawazoe, Center for Computational Materials, Tohoku University, Sendai, Japan

Jamie Kruzic, School of Mechanical and Manufacturing Engineering, UNSW Sydney, Sydney, Australia

Jürgen Parisi, Universität Oldenburg, Oldenburg, Germany

Udo W. Pohl, Department of Materials Science and Engineering, Technical University of Berlin, Berlin, Germany

Tae-Yeon Seong, Department of Materials Science and Engineering, Korea University, Seoul, Korea (Republic of)

Shin-ichi Uchida, Electronics and Manufacturing, National Institute of Advanced Industrial Science and Technology, Tsukuba, Japan

Zhiming M. Wang, Institute of Fundamental and Frontier Sciences, University of Electronic Science and Technology of China, Chengdu, China

Gang Zhang, Beijing Institute of Technology, Jiaxing, Zhejiang, China

The Springer Series in Materials Science covers the complete spectrum of materials research and technology, including fundamental principles, physical properties, materials theory and design. Recognizing the increasing importance of materials science in future device technologies, the book titles in this series reflect the state-of-the-art in understanding and controlling the structure and properties of all important classes of materials.

Alexei Boulbitch · Alexander Korzhenevskii

Stress-Induced Local Phase Transformations Near Cracks and Dislocations

The Field-Theoretical Approach

Springer

Alexei Boulbitch
Igel, Rheinland-Pfalz, Germany

Alexander Korzhenevskii
Laboratory of Materials Micromechanics
Institute for Problems of Mechanical Engineering
St. Petersburg, Russia

ISSN 0933-033X ISSN 2196-2812 (electronic)
Springer Series in Materials Science
ISBN 978-3-032-00944-9 ISBN 978-3-032-00945-6 (eBook)
https://doi.org/10.1007/978-3-032-00945-6

© The Editor(s) (if applicable) and The Author(s), under exclusive license to Springer Nature Switzerland AG 2026

This work is subject to copyright. All rights are solely and exclusively licensed by the Publisher, whether the whole or part of the material is concerned, specifically the rights of translation, reprinting, reuse of illustrations, recitation, broadcasting, reproduction on microfilms or in any other physical way, and transmission or information storage and retrieval, electronic adaptation, computer software, or by similar or dissimilar methodology now known or hereafter developed.
The use of general descriptive names, registered names, trademarks, service marks, etc. in this publication does not imply, even in the absence of a specific statement, that such names are exempt from the relevant protective laws and regulations and therefore free for general use.
The publisher, the authors and the editors are safe to assume that the advice and information in this book are believed to be true and accurate at the date of publication. Neither the publisher nor the authors or the editors give a warranty, expressed or implied, with respect to the material contained herein or for any errors or omissions that may have been made. The publisher remains neutral with regard to jurisdictional claims in published maps and institutional affiliations.

This Springer imprint is published by the registered company Springer Nature Switzerland AG
The registered company address is: Gewerbestrasse 11, 6330 Cham, Switzerland

If disposing of this product, please recycle the paper.

To the authors of Springer Verlag

Preface

What is This Book About

Addressing the Challenge: Elastic Defects, Phase Transformations, and Process Zones

Solids invariably harbor defects, ranging from dislocations in various forms and packing defects to inclusions, cracks, and their combinations. Even in ideally grown crystals, dislocations of growth persist. These defects give rise to elastic fields in their immediate vicinity, gradually diminishing as one moves away from the isolated defect. Within these pages, our focus is primarily directed towards two noteworthy defects: a dislocation and a mode I crack.

Further, a majority of solids undergo phase transformations. In these transformations, the solid's atomic, electronic, or magnetic structure undergoes a comprehensive change across the entirety of the material while its chemical composition remains constant. These distinct structural states are referred to as *phases*.

The elastic field generated by defects interacts with phase transformations in multifaceted ways. Under specific conditions, stress concentration around a defect can induce the formation of a region characterized by a phase different from the one occupying the rest of the crystal. We refer to this localized phase as a *process zone*.

This endeavor encompasses two intertwined aspects. First, describing the process zone and understanding how its properties depend on its location within the solid's phase diagram presents a formidable challenge.

However, the greatest interest is the study of physical and mechanical properties that arise due to the presence of the process zone rather than the examination of the process zone as such. The ultimate goal of this exploration is to comprehend the process zone's influence on the originating defect's characteristics and, ultimately, on the macroscopic mechanical properties of the solid. We address the latter group of issues through examples of process zones generated by cracks and dislocations.

Zones forming near defects, such as crack tips and dislocations, exhibit several universal features, including the configuration of the phase diagram, the quasi-viscous friction force exerted on a moving defect, and others. However, the shapes of process zones at cracks and dislocations differ significantly, as do some of their non-universal properties. Consequently, both shared and fundamentally distinct qualitative effects may arise near phase transition points due to these zones. Therefore, it is advisable to analyze the universal properties of cracks and dislocations together. The uncommon features, in contrast, should be examined individually for each type of process zone.

* * *

To summarize, *this book examines the formation and behavior of process zones, their influence on the properties of the defects that gave rise to them, and their consequential effects on the macroscopic properties of solid materials.*

On the Importance of Studying the "Dressing" of an Isolated Crack and a Single Dislocation

In the extensive literature on phase transitions in defective crystals, the primary question is typically posed as follows:

"How do defects affect changes in the macroscopic properties of phase transitions?"

The impact of defects on a specific property is typically additive, their effect being generally small. This is particularly true for thermodynamic properties, which are determined by the free energy and its derivatives, such as heat capacity or susceptibility. The universal reason for this negligible impact is straightforward: the contribution is proportional to the ratio of the volume of "defective" regions (i.e., significantly perturbed areas) to the total volume of the sample.

However, for cracks and dislocations, which are the focus of this book, the situation is entirely different. Macroscopic phenomena for which these defects are responsible—such as material fracture and plastic deformation—are primarily governed by local processes occurring in the "defective" regions.

In the case of crystals near phase transition points, these regions are nothing other than process zones. The macroscopic effects of material cracking and plasticity are fundamentally determined by the influence of process zones on localized processes such as the generation of dislocation loops by sources (e.g., Frank–Read sources) and the transformation properties of process zones at the tips of isolated cracks.

The quantitative characteristics in this context are the yield strength (the critical stress value for a Frank–Read source) and the critical stress intensity factor.

* * *

In this book, *we demonstrate how these quantities can be calculated through analytical and numerical evaluation of the parameters of process zones under various external conditions.*

The Field-Theoretical Approach: Its Meaning and Significance

In our quest to describe the process zone, we employ a field associated with a physical quantity known as the *order parameter*. This order parameter aligns with the classical concept used in Landau's theory of phase transformations. We have derived a system of equations governing the order parameter and the strain field. Their combination completely characterizes the process zone. Furthermore, we posit that it is *always possible to eliminate elastic degrees of freedom*, reducing the description of the zone to solving an equation solely for the order parameter field. Consequently, we refer to this approach as *field-theoretical*.

Mathematically, the order parameter equation represents a time-dependent Ginzburg-Landau one (or a system of such equations), with coefficients contingent on the strain field generated by the defect in question. The spatial distribution of the order parameter and its temporal evolution hold comprehensive information about the process zone, including strain and stress distribution. Analysis and solution of these equations is the mathematical framework of this book.

Target Groups

Our work aims to serve multiple audiences.

The first target group comprises experimentalists in physics and mechanics working on plasticity, fracture, or both. We acknowledge that the theoretical and mathematical aspects of this book may seem burdensome to people belonging to this group. However, we've strived to articulate the qualitative outcomes of our theory clearly throughout the text. For a dozen years, one of us served as a "pocket theorist" in an experimental group and appreciated the value of simple analytical results that can be easily applied to experimental data. Accordingly, we have endeavored to provide such insights wherever possible.

The second target group consists of researchers focused on the theoretical mechanics of fracture and plasticity. These scientists may use methodologies that diverge from ours, often ignoring Landau's theory. Our objective is to integrate field-theoretical approaches into the academic discourse of this community, hoping to establish a common mathematical and terminological framework.

Our third audience is students embarking on research in this field. To cater to them, we offer detailed derivations for every new result and statement, no matter how elementary, although seasoned readers may find these sections overly detailed.

The fourth group includes theoretical physicists specializing in phase transitions who are already well-versed in Landau theory. If you fall into this category, please skip the introductory sections.

The fifth target audience includes materials science, metallurgy, and geosciences professionals interested in applied aspects of fracture and plasticity.

We have made every effort to ensure that this book serves the needs of all these diverse groups. The effectiveness of our endeavor ultimately lies in the hands of the reader.

Content of the Book

As authors, we find ourselves in a challenging position. The subject matter is both intricate and voluminous. While we strive for succinctness, focusing directly on the core issues, establishing a strong foundation for the field-theoretical description of the process zone is essential, especially concerning the concept of the order parameter.

We commence the book with chapters that outline well-known facts about dislocations, cracks, and bifurcation theory. These chapters introduce no new information; instead, they serve as a reference and save readers from having to search through extensive literature. Readers are not required to read these chapters in their entirety; we recommend using them as a quick reference to refresh your memory on necessary formulas when needed. However, for those who seek deeper exploration, we provide references to the source books we consulted ourselves.

Chapter 5 summarizes well-known results from the Landau theory of transitions described by a one-component order parameter. Though there are crystals with transitions described by such an order parameter, it is a simplified Landau theory, a kind of "scientific comics," rather than an advanced theory. This section serves practical purposes since, in this book, we study the one-component order parameter case as the most simple one. It also functions as a reference.

An additional aim of these introductory sections is to establish a common language that we use throughout the book.

Thus far, the book comprises introductory material. We have included this introduction to cater to the diverse target audiences outlined earlier. Given this approach, we trust that readers can selectively skip any sections they find unnecessary while still deriving value from the remainder of the content.

Starting from Part I, the book reports our original results. Here, we describe the process zone at a defect without specifying the defect's nature. This approach allows us to discover universal properties of the zone applicable to any extended defect.

Part II covers the geometrical and physical properties of the process zone at the crack tip and dislocation, while Part III contains numerical estimates of key parameters of the zones of both types.

Part IV describes the zone's impact on crack behavior. Here, we address problems such as the origin of transformation toughening, crack propagation accompanied by velocity jumps, and its stick-slip mode of motion.

Part V describes zones at the edge dislocation and studies how the zone influences dislocation properties and macroscopic plasticity.

The book uses a huge array of variables, nearly all of which carry consistent meanings throughout. In the few cases where a variable's meaning changes, we explicitly state this. To ease the reading experience, we have compiled a list of variables with explanations and references to the formulas where each variable is first introduced. Additionally, we often provide page references for objects mentioned far from their initial reference.

Many discussions of a mere technical character have been moved to annexes, allowing us to condense the book's main body while retaining detailed information for those interested. The book is based on our original findings, the old and recent ones. For our old publications, we give references in the footnotes. For recent ones, the footnote states that this material is being published here for the first time.

This book is lengthy, and we acknowledge that it does not need to be read sequentially. Readers can explore sections of interest independently, following the advice of Ya. B. Zeldovich and A.D. Myshkis in "Elements of Applied Mathematics."

We trust you will find this book helpful and as reader-friendly as its complexity allows.

* * *

Igel, Germany — Alexei Boulbitch
St. Petersburg, Russia — Alexander Korzhenevskii

Acknowledgement We are grateful to Prof. Valery Levitas for several fruitful conversations. ALK expresses his deep gratitude to Prof. Richard Bausch for many meaningful discussions and collaboration with him on dynamics of extended defects interacting with phase transitions.

* * *

Contents

Chapter 1
Cracks and Dislocations

Abstract In this chapter, we have compiled selected results pertaining to cracks and dislocations. They will be employed in subsequent sections of this book. This chapter serves as a convenient point of reference for readers, decreasing the necessity to consult additional textbooks on dislocations and cracks.

In this book, we consider dislocations as linear defects in crystals that cause a discontinuity in the strain field. Specifically, when we trace a closed contour enclosing the dislocation line, the elastic displacement vector $\mathfrak{u}$ experiences a finite increment, denoted as $\mathfrak{b}$.[1] This quantity is referred to as the Burgers vector and possesses an absolute value equal to one of the lattice vectors in its direction.

1.1 Mode I Crack

This chapter cites some results from linear elastic fracture mechanics concerning Mode I cracks chosen specifically for their relevance to our subsequent calculations. It serves as a convenient reference on cracking, eliminating the need for readers to consult other fracture textbooks.

***

In this book, we explore cracks that occur in brittle or quasi-brittle solids.[2] Specifically, our focus lies on *open-mode* cracks, often also referred to as *mode I* cracks. Mode I cracks are those loaded by a normal force applied to their surfaces, as schematically depicted in Fig. 1.1.

It's worth noting that in practical cases, the stress is rarely applied to the crack faces. Instead, it is typically applied at various points on the solid's outer boundaries

[1] Hereafter we always have in mind undissociated dislocations.

[2] Quasi-brittle solids are materials characterized by a small plastic zone size, allowing them to be treated similarly to brittle materials. The theory formulated in this book is applicable, provided the plastic zone (if any) is significantly smaller than the transformational one. In the following, we assume this condition to hold and do not discuss it anymore.

© The Author(s), under exclusive license to Springer Nature Switzerland AG 2026
A. Boulbitch and A. Korzhenevskii, *Stress-Induced Local Phase Transformations Near Cracks and Dislocations*, Springer Series in Materials Science 356,
https://doi.org/10.1007/978-3-032-00945-6_1

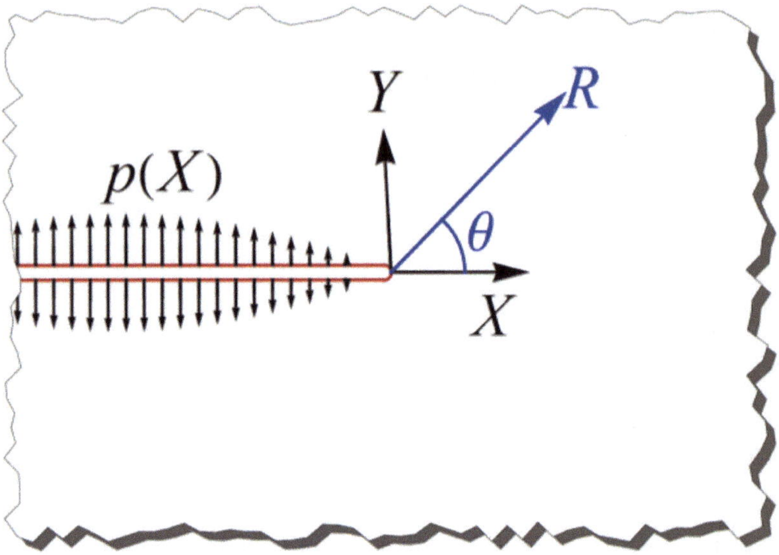

Fig. 1.1 Schematic view of a crack loaded by force, $p(X)$, applied normally to its surfaces. The red contour schematically shows the crack surface

or within its bulk. Following Brueckner's approach [1], the stresses applied to the outer surface, and bulk forces are effectively "translated" into the distribution of force, $p(X)$, exerted on the crack surfaces.[3] For a more comprehensive exploration of this concept, readers can refer to Chap. VI of Muschelishvili's book [2].

1.1.1 Stress at the Crack Tip

The components of the stress tensor at the crack tip can be expressed as follows:

$$\begin{aligned}
\sigma_{11} &= \frac{K_I}{\sqrt{2\pi R}} \cos\frac{\theta}{2}\left(1 - \sin\frac{\theta}{2}\sin\frac{3\theta}{2}\right)\\
\sigma_{22} &= \frac{K_I}{\sqrt{2\pi R}} \cos\frac{\theta}{2}\left(1 + \sin\frac{\theta}{2}\sin\frac{3\theta}{2}\right)\\
\sigma_{33} &= \frac{2\nu K_I}{\sqrt{2\pi R}} \cos\frac{\theta}{2}\\
\sigma_{12} &= \frac{K_I}{\sqrt{2\pi R}} \sin\frac{\theta}{2}\cos\frac{\theta}{2}\cos\frac{3\theta}{2}\\
\sigma_{13} &= \sigma_{23} = 0
\end{aligned} \tag{1.1}$$

Here, K_I is a parameter referred to as the *stress intensity factor, R* and θ are cylindrical coordinates measured from the crack tip (as it is shown in Fig. 1.2).

[3] We always assume that the forces applied to the upper and lower crack surfaces have equal magnitudes but opposite directions, as shown in Fig. 1.2.

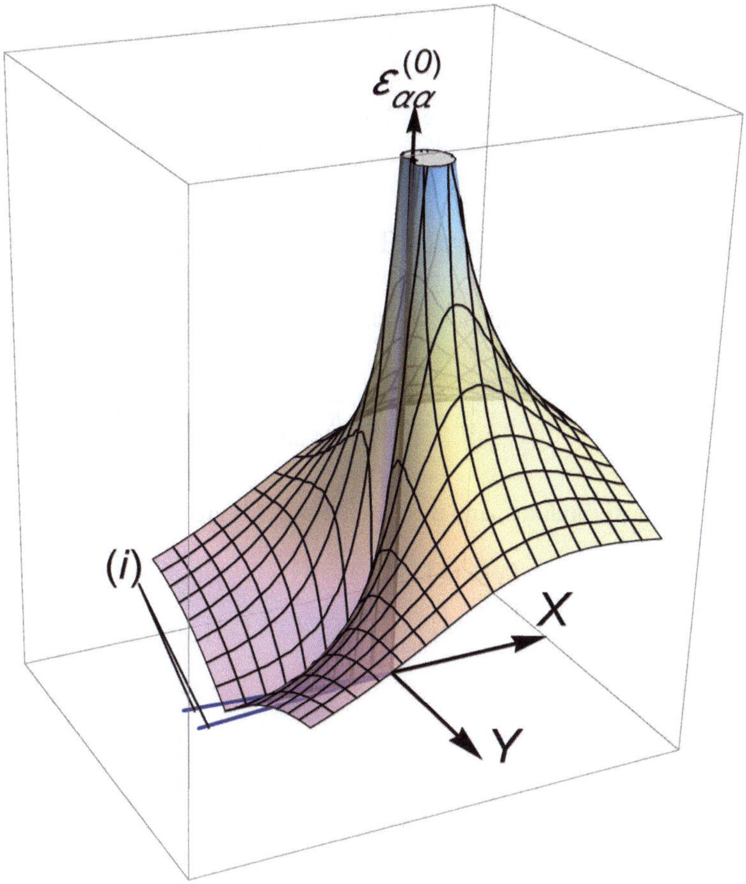

Fig. 1.2 Strain spur distribution at the crack's tip. The solid lines indicated by (i) schematically show the crack contour

1.1.2 Stress Intensity Factor

For a crack with a length of $2l$ and the coordinate origin placed at its midpoint, K_I can be expressed as follows [3–5]:

$$K_I = \frac{1}{\sqrt{\pi l}} \int_{-l}^{l} p(X) \sqrt{\frac{l+X}{l-X}} dX \tag{1.2}$$

For a semi-infinite crack shown in Fig. 1.2, placing the coordinate origin in its tip ($X' = l - X; l + X = 2l - X' \approx 2l$; $p(X) \to p(X')$), one obtains the stress intensity factor (1.2) in the following form:

$$K_I = \sqrt{\frac{2}{\pi}} \int_0^\infty \frac{p(-X')}{\sqrt{X'}} dX' \tag{1.3a}$$

In Sih's textbook [6], one finds stress intensity factors for various configurations exhibiting a wide variety of $K_I = K_I(l)$ dependencies.

1.2 Strain Around the Crack Tip

In linear elastic fracture mechanics, the displacement vector, $\mathrm{u}_\alpha^{(0)}$ $(\alpha = 1, 2)$, at the crack tip can be described as follows [4]:

$$\mathrm{u}_\alpha^{(0)} = \frac{K_I}{\mu} \sqrt{\frac{R}{2\pi}} \varphi_\alpha(\theta) \tag{1.4}$$

where μ is the shear modulus, and the functions $\varphi_\alpha(\theta)$ are defined as:

$$\varphi_\alpha(\theta) = \begin{cases} \cos\left(\frac{\theta}{2}\right)\left[1 - 2\nu + \sin^2\left(\frac{\theta}{2}\right)\right], & \alpha = 1 \\ \sin\left(\frac{\theta}{2}\right)\left[2 - 2\nu - \cos^2\left(\frac{\theta}{2}\right)\right], & \alpha = 2 \end{cases} \tag{1.5}$$

1.2.1 The Strain Spur

The strain spur, $\varepsilon_{\alpha\alpha}^{(0)}$, within linear elastic fracture mechanics[4] can be calculated using (1.1) and is given by:

$$\varepsilon_{\alpha\alpha}^{(0)}(\mathbf{R}) = \frac{(1+\nu)(1-2\nu) K_I}{E(2\pi R)^{1/2}} \cos\left(\frac{\theta}{2}\right) \tag{1.6}$$

Morozov [4] and Rabotnov [5]. Here E and ν are Young's modulus and Poisson's ratio, respectively. The spatial distribution in Cartesian coordinates (X, Y) measured from the crack tip is illustrated in Fig. 1.2.

It's important to note that the stresses σ_{11} and σ_{22} (1.1) represent stretching and diverge like $R^{-1/2}$ as R approaches 0, which is the mechanical origin of fracture.

1.2.1.1 Wedging

In Chap. 19, we will address the cleavage of a solid by a thin wedge. The stress intensity factor at the tip of the crack formed during cleavage takes the following form [7]:

---

[4] We use the superscript "0" here to distinguish the spur (1.6) obtained within linear fracture mechanics from the one we derive later.

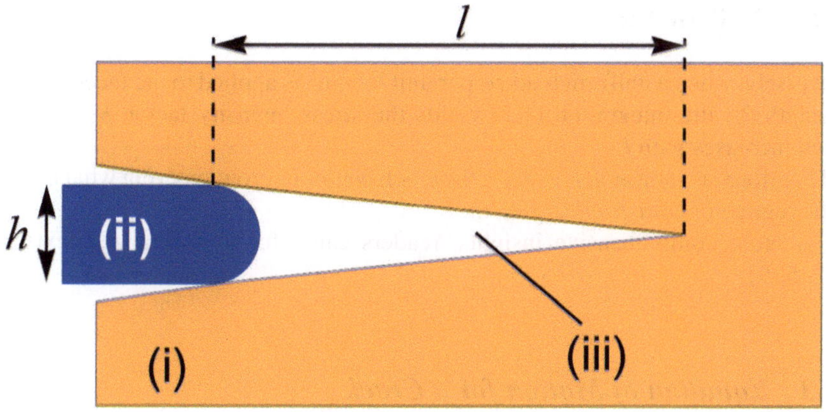

Fig. 1.3 Cleavage of a solid with a thin wedge. A solid elastic body (i) is cleaved by a thin wedge (ii) (thickness h), giving rise to a crack (iii) with the length l

$$K_I = \frac{Eh}{(1-\nu^2)\sqrt{2\pi l}} \tag{1.7}$$

where h is the wedge thickness, and l is the crack length. One defines l as the distance between the crack tip and the extreme point where the wedge contacts the crack faces, as depicted in Fig. 1.3.

1.2.2 Fracture Energy and Fracture Toughness

Each brittle or quasi-brittle material is characterized by the material constant referred to as the fracture toughness, K_{IC}. It is related to the solid's *surface energy* (or *fracture energy*), Σ, as follows:

$$K_{IC}^2 = \frac{2\Sigma E}{1-\nu^2} \tag{1.8}$$

The crack becomes unstable when $K_I = K_{IC}$.

Let us examine two types of crack loading that result in qualitatively different fracture scenarios.

1.2.2.1 Localized Force Loading

If the crack is loaded with a force $p(X) = p_0\delta(X)$ localized at the middle of the crack, the stress intensity factor K_I, as derived from (1.2), is $K_I = p_0/\sqrt{\pi l}$. In this case, the crack length l increases with the rising stress intensity factor, eventually leading to self-arrest when K_I equals K_{IC}.

1.2.2.2 Uniform Load

Conversely, when a uniform load p_0 per unit length[5] is applied to the faces of a crack of length $2l$, the integration (1.2) yields the stress intensity factor $K_I = p_0\sqrt{\pi l}$, which increases with l.

Therefore, as soon as $K_I > K_{IC}$, the crack begins to grow, entering what is known as the regime of *catastrophic growth.*

For more comprehensive insights, readers can refer to specialized texts such as [2–5, 9].

1.2.3 Equation of Motion for a Crack

If the crack in an infinite elastic plane propagates its velocity is described by Freund's equation [9]:

$$\left(1 - \frac{V}{V_{\mathrm{R}}}\right) G = \Gamma \tag{1.9}$$

where V is the crack speed, V_{R} is Rayleigh's velocity,

$$G = \frac{K_I^2(1-\nu^2)}{E}; \quad \Gamma = \frac{K_{IC}^2(1-\nu^2)}{E} \tag{1.10}$$

are the driving force for the crack propagation and its threshold value. Equation (1.9) predicts the crack velocity,

$$V = V_{\mathrm{R}}\left(1 - \frac{\Gamma}{G}\right) \tag{1.11}$$

to monotonously increase from zero to V_R when the driving force, G, increases from Γ to infinity.

***

In summary, *this chapter presents the equations related to the theory of dislocations and cracks, which will be extensively used throughout this book. While readers familiar with the subject may not find new information here, we hope that newcomers to the field will gain an understanding of the physical origins of these results and save time in their reading. For those seeking a deeper exploration of the theory of dislocations, we recommend the books by Friedel* [10] *and Hirth and Lothe* [11]. *For crack theory, one of the books of Mushelishvili* [2], *Morozov* [4], *Savin* [13], *Cherepanov* [3], *Rabotnov* [5], *and Freund* [9] *are suggested.*

---

[5] The case of the uniform load, p_0, of an elastic plane with a cut goes back to Griffith [8].

1.3 Edge Dislocation

This section provides basic information on edge dislocations, intended to serve as a quick reference.

***

Visualize an edge dislocation as follows: imagine the crystal lattice as a series of stacked atomic planes. Now, select one of these planes positioned somewhere within the crystal's interior and remove half of it while the other half remains in place. This remaining part of the atomic plane in the following we refer to as the *half-plane* for shortness. The edge of the remaining semi-plane represents the dislocation line. By design, the Burgers vector of such a dislocation is oriented normally to the dislocation line. Let us align the Z-axis of the coordinate system along this line. In the vicinity of the edge dislocation, atoms in the (X, Y) plane adjust their positions to fill the void, resulting in the creation of strain and stress fields. These fields gradually diminish with increasing distance from the dislocation line.

The strain tensor components, $\varepsilon_{\alpha\beta}$, in the proximity of the edge dislocation are given by:

$$\begin{aligned} \varepsilon_{11} &= -\frac{\mathfrak{b}}{4\pi(1-\nu)}\frac{Y\left[X^2(3-2\nu)+Y^2(1-2\nu)\right]}{(X^2+Y^2)^2} \\ \varepsilon_{22} &= \frac{\mathfrak{b}}{4\pi(1-\nu)}\frac{\left[X^2Y(1+2\nu)-Y^3(1-2\nu)\right]}{(X^2+Y^2)^2} \\ \varepsilon_{12} &= \frac{\mathfrak{b}}{4\pi(1-\nu)}\frac{X(X^2-Y^2)}{(X^2+Y^2)^2} \\ \varepsilon_{13} &= \varepsilon_{23} = \varepsilon_{33} = 0 \end{aligned} \tag{1.12}$$

Here $\mathfrak{b}$ represents the absolute value of the Burgers vector, ν stands for Poisson's ratio, and $\alpha, \beta = 1, 2, 3$.

For our subsequent discussions, we require the trace of the strain tensor, $\varepsilon_{\alpha\alpha}$, which can be computed as $\varepsilon_{11} + \varepsilon_{22} + \varepsilon_{33}$. This yields:

$$\varepsilon_{\alpha\alpha} = -\frac{(1-2\nu)\mathfrak{b}}{2\pi(1-\nu)}\frac{Y}{R^2} = -\frac{(1-2\nu)\mathfrak{b}}{2\pi(1-\nu)}\frac{\sin\theta}{R} \tag{1.13}$$

Here $R = (X^2 + Y^2)^{1/2}$ and $\theta = \arctan(Y/X)$ represent polar coordinates. Figure 1.4 illustrates the relationship between $2\pi(1-\nu)\varepsilon_{\alpha\alpha}(X, Y)/(1-2\nu)\mathfrak{b}$ and the spatial coordinates X and Y.

The elastic energy associated with the edge dislocation, denoted as W, can be expressed as:

$$W^{(0)}_{\text{edge}} = \frac{\mu\mathfrak{b}^2\mathbb{L}}{4\pi(1-\nu)}\ln\left(\frac{\mathbb{R}_{\text{cr}}}{R_0}\right) \tag{1.14}$$

Here, μ represents the shear modulus:

$$\mu = \frac{E}{2(1+\nu)} \tag{1.15}$$

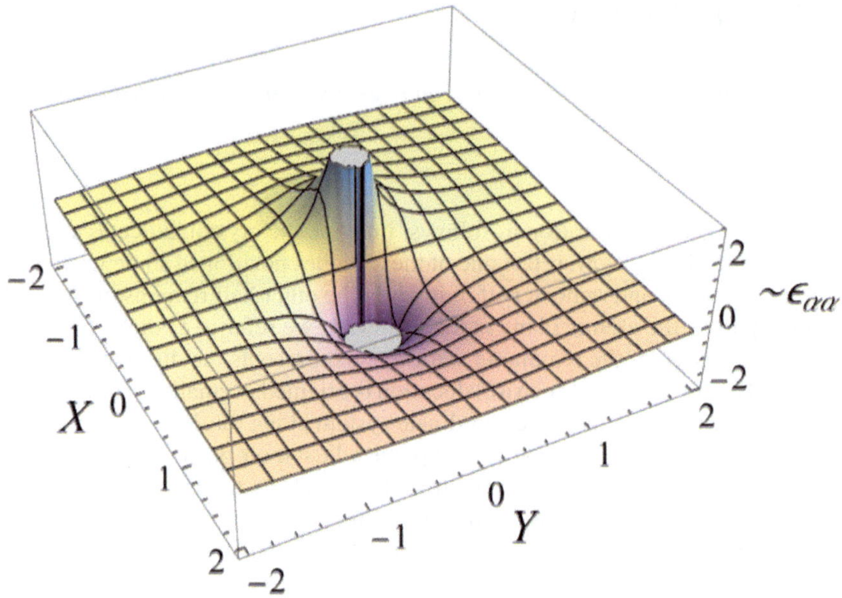

Fig. 1.4 Distribution of the trace of the strain tensor, $\varepsilon_{\alpha\alpha}$, in the vicinity of the edge dislocation

$\mathbb{L}$ stands for the dislocation length, $\mathbb{R}_{cr}$ indicates the crystal size and $R_0 \sim \mathfrak{b}$ represents the cut-off radius, E is Young's modulus and ν is Poisson's ratio.

1.4 Screw Dislocation

This section compiles fundamental facts about screw dislocations, which are utilized in the subsequent text.

***

Consider the crystal lattice as a stack of atomic planes, and envision the screw dislocation as follows: Imagine "cutting" the lattice along a half-plane situated between the atomic planes. Then, shift the lattice parts on either side of the cut in opposite directions, traversing a distance equal to one lattice period parallel to the edge of the cut. This process yields the screw dislocation, where the Burgers vector lies along the dislocation line. The strain field generated by the screw dislocation, which aligns with the Oz axis, exhibits pure shear:

$$\begin{gathered} \varepsilon_{13} = -\frac{\mathfrak{b}}{4\pi}\frac{Y}{X^2+Y^2} \\ \varepsilon_{23} = \frac{\mathfrak{b}}{4\pi}\frac{X}{X^2+Y^2} \\ \varepsilon_{11} = \varepsilon_{22} = \varepsilon_{33} = \varepsilon_{12} = 0 \end{gathered} \tag{1.16}$$

Notably, the strain distribution around the screw dislocation closely resembles the pattern shown in Fig. 1.1, with the exception that the sign differs.

The energy associated with the screw dislocation can be expressed as:

$$W^{(0)}_{\text{screw}} = \frac{\mu \mathfrak{b}^2 \mathbb{L}}{4\pi} \ln\left(\frac{\mathbb{R}_{\text{cr}}}{R_0}\right) \tag{1.17}$$

All parameters—$\mathfrak{b}$, μ, $\mathbb{L}$, $\mathbb{R}_{\text{cr}}$, and R_0—are identical to those for the edge dislocation.

It's worth noting that the Burgers vector remains conserved along the dislocation. In the general case, a dislocation can have both edge and screw components. This implies that it generates a superposition of strains originating from both elements. If the dislocation direction deviates from its original orientation, the values of these components change accordingly. The bending of a straight screw dislocation is accompanied by the formation of the edge component, while bending of the edge one gives rise to the screw component.

For further in-depth information on dislocations, one can refer to the books of Friedel [10], Hirth and Lothe [11], and Sect. 27 of the Theory of Elasticity by Landau and Lifshitz [14].

1.4.1 Dislocation Line Tension

The plastic behavior hinges on the dislocation line tension, τ, as outlined by Friedel [10], and Hirth and Lothe [11]. Traditionally, for an unaltered edge dislocation, its line tension $\tau_{\text{e}}^{(0)} > 0$, is equated to its elastic energy, $W^{(0)}$ (1.14), of the edge dislocation per unit length:

$$\tau_{\text{e}}^{(0)} = \frac{W^{(0)}}{\mathbb{L}} = \frac{E\mathfrak{b}^2}{4\pi(1-\nu)} \ln\left(\frac{\mathbb{R}_{\text{cr}}}{\mathfrak{b}}\right) \tag{1.18}$$

Analogously for the screw dislocation equation (1.17) yields the line tension $\tau_{\text{s}}^{(0)}$:

$$\tau_{\text{s}}^{(0)} = \frac{W^{(0)}}{\mathbb{L}} = \frac{E\mathfrak{b}^2}{4\pi} \ln\left(\frac{\mathbb{R}_{\text{cr}}}{\mathfrak{b}}\right) \tag{1.19}$$

They only differ from one another by the factor $1 - \nu$:

$$\tau_{\text{e}}^{(0)} = \frac{\tau_{\text{s}}^{(0)}}{1-\nu} \tag{1.20}$$

1.5 Dislocation in the Ultrasonic Field

In this section, we revisit the classical theory of dislocations subjected to an ultrasonic elastic field. Following the derivation by Hirth and Lothe [11], *we briefly outline the main aspects of the theory proposed by Granato and Lücke* [12], *including the equation of dislocation motion and the internal friction arising from their oscillations.*

***

1.5.1 Vibrating Dislocation

In this section, we briefly recall the results regarding the oscillations of a dislocation segment subjected to an ultrasonic wave. We follow the derivation provided in the book by Hirth and Lothe [11].

Consider a dislocation segment of length $\mathbb{L}$, aligned along the OZ-axis and clamped at $Z = 0$ and $Z = L$ subjected to a shear stress $\sigma_{yz} = \sigma_{yz}(Z, t) \sim e^{i\omega t}$, where t is the time and ω is the ultrasonic frequency. The largest oscillation amplitude corresponds to the lowest mode:

$$Y_1(Z, t) = \frac{16\mathbb{L}^2(1-\nu)\sigma_0}{\pi^2\mu\mathfrak{b}(1-2\nu)}\sin(\omega t)\sin\left(\frac{\pi Z}{\mathbb{L}}\right) \tag{1.21}$$

Thus, the amplitude $\mathfrak{A}$ of oscillation is:

$$\mathfrak{A} = \frac{16(1-\nu)\sigma_0}{\mu\mathfrak{b}(1-2\nu)}\left(\frac{\mathbb{L}}{\pi}\right)^2 \tag{1.22}$$

A more detailed derivation one finds in the book [11].

1.5.2 Dissipation Mechanism of Granato and Lücke

The energy, ΔW, dissipated by oscillations of a single dislocation segment during the period $2\pi/\omega$, is given by:

$$\Delta W = \frac{\pi\mathfrak{A}^2\mathbb{L}\omega}{6\mathfrak{b}^2 c_t} k_B T \tag{1.23}$$

If the number of vibrating segments is $\mathcal{N}$, it is convenient to define the total active length, L_{act}, of dislocations contributing to energy dissipation per unit volume:

$$L_{\text{act}}=\frac{\mathcal{N}\mathbb{L}}{V} \tag{1.24}$$

Here, V is the crystal volume. Consequently, the total energy loss in the entire crystal during one period becomes $\Delta W_{\text{tot}} = \Delta W \times \mathcal{N}$, leading to:

$$\Delta W_{\text{tot}} = \frac{\pi \mathfrak{A}^2 V L_{\text{act}} \omega}{6 \mathfrak{b}^2 c_{\text{t}}} k_{\text{B}} T \tag{1.25}$$

The elastic energy, $W = \sigma_0^2 V/2\mu$, stored in a crystal of volume V, when expressed in terms of the amplitude $\mathfrak{A}$, takes the form:

$$W = \frac{V\mu}{2}\left[\frac{\pi^2 \mathfrak{A}\mathfrak{b}(1-2\nu)}{16\mathbb{L}^2(1-\nu)}\right]^2 \tag{1.26}$$

Internal friction is defined as the ratio of energy loss to stored energy:

$$Q^{-1} = \Delta W_{\text{tot}}/W \tag{1.27}$$

yielding:

$$Q_{\text{GL}}^{-1} = \frac{\omega L_{\text{act}}}{3\pi^3 \mu c_{\text{t}}}\left[\frac{16(1-\nu)}{1-2\nu}\right]^2 \left(\frac{\mathbb{L}}{\mathfrak{b}}\right)^4 k_{\text{B}} T \tag{1.28}$$

Expression (1.28) represents the energy dissipation mechanism described by Granato and Lücke, with the subscript "GL" denoting "Granato-Lücke."

***

In summary, *dislocations subjected to an ultrasonic wave undergo oscillations. As we will see, these oscillations induce a specific dynamics within the process zone (see Sect.* 29.3). *The key results to be used later are the equation of dislocation motion* (1.21) *and the findings from Sect.* 1.5.2.

References

1. H.F. Brueckner, Trans. ASME **80**, 1225 (1958)
2. N.I. Mushelisvili, *Some Basic Problems of the Mathematical Theory of Elasticity*. Reprint of the Second English edition, translated by J.R.M. Radok (Leyden, The Netherlands, Noordoff International Publishing, 1977)
3. G.P. Cherepanov, *Mechanics of Brittle Fracture* (McGraw-Hill, New York, 1979)
4. N.F. Morozov, *Mathematical Problems of the Theory of Cracks* (Nauka, Moscow, 1984). In Russian
5. Y.N. Rabotnov, *Introduction into Fracture Mechanics* (Nauka, Moscow, 1987). In Russian
6. G.C. Sih, *Handbook of Stress Intensity Factors* (Lehigh University, Bethlehem, Pensylvania, USA, Institute of Fracture and Solid Mechanics, 1973)
7. G.I. Barenblatt, G.P. Cherepanov, J. Appl. Math. Mech. **24**, 993 (1960)

8. A.A. Griffith, *Proceedings of the 1st International Congress for Applied Mechanics, Delft* (1924); G.C. Sih, H. Liebowitz, Int. J. Solids Struct. **3**, 1 (1967)
9. L.B. Freund, *Dynamic Fracture Mechanics* (Cambridge University Press, Cambridge, 1998)
10. J. Friedel, *Dislocations* (Pergamon Press, Oxford, 1964)
11. J.P. Hirth, J. Lothe, *Theory of Dislocations* (Krieger Publishing Company, Malabar, Florida, US, 1991)
12. A. Granato, K.J. Lücke, Appl. Phys. **27**, 583 (1956)
13. G.N. Savin, *Stress Concentration Around Holes* (Pergamon Press, 1961)
14. L.D. Landau, E.M. Lifshitz, *Theory of Elasticity* (Pergamon Press, Oxford, 1986)

Chapter 2
Bifurcation Theory

Abstract A bifurcation represents a qualitative change in the solution of a differential equation or a system of equations that depends on parameters. It occurs when a small, smooth change in a parameter value causes a qualitative change in the behavior of the solution. It's worth noting that the approach described in this chapter, while heavily involving the solution of the linear equation, ultimately yields an asymptotically exact solution of the nonlinear equation. This chapter presents a thorough collection of established mathematical principles related to soft bifurcations in nonlinear differential equations. Central to the text, the theory of soft bifurcations is foundational, enabling this chapter to function as a comprehensive reference and eliminating the need for supplementary textbooks.

In this text, we explore process zones at defects, which exhibit characteristics similar to those of bulk phase transitions. These transitions are categorized into two main types: first-order and second-order. The former is marked by the discontinuity across all physical quantities, whereas the latter shows piecewise-continuous changes in some parameters during transitions. For additional insight into this classification, readers are directed to Landau and Lifshitz's work [1].

The formation of a process zone at a defect can exhibit properties of either first- or second-order transitions, hence termed first- and second-order zones, respectively. The occurrence of a second-order process zone typifies a soft bifurcation, rendering bifurcation theory an effective analytical tool. We will briefly revisit significant findings from this theory, which are applied subsequently. For in-depth discussions and proofs, readers are encouraged to consult other sources, such as [2].

Consider a nonlinear equation in the form:

$$\hat{L}(a)u = \hat{N}(u) \tag{2.1}$$

where $\hat{L}(a)$ is a linear operator depending on parameter a, $u = u(R)$ is a dependent variable, and $\hat{N}(u)$ is a nonlinear operator satisfying the following condition:

$$\hat{N}(0) = 0$$

The boundary conditions are defined as:

© The Author(s), under exclusive license to Springer Nature Switzerland AG 2026

A. Boulbitch and A. Korzhenevskii, *Stress-Induced Local Phase Transformations Near Cracks and Dislocations*, Springer Series in Materials Science 356,
https://doi.org/10.1007/978-3-032-00945-6_2

$$u_{|\partial\Omega} = 0 \tag{2.2}$$

where Ω is the domain of the (2.1) and $\partial\Omega$ is its boundary.

While our discussion applies to differential, integral, or integro-differential operators $\hat{L}(a)$ and $\hat{N}(u)$, our focus is on $\hat{L}(a)$ as a differential operator, and $\hat{N}$ as either a polynomial or an integral operator.

Given $\hat{N}(0) = 0$, (2.1) and (2.2) admit a trivial solution $u = 0$, assumed that it is stable for some $a > 0$.

We introduce an *auxiliary equation*:

$$\hat{L}(a_n)\psi_n(\mathbf{R}) = 0 \tag{2.3}$$

subjected to the same boundary conditions: $\psi_{n|\partial\Omega} = 0$. Here a_n denotes the discrete spectrum's eigenvalues (if any), and $\psi_n(R)$ are the corresponding eigenfunctions, indexed by natural numbers $n = 1, 2, \ldots$. We arrange a_n in ascending order and regard a_1 and $\psi_1(R)$ as the *ground state* eigenvalue and eigenfunction.

Theorem 2.0.1 *If the linear operator, $\hat{L}(a_n)$, possesses at least one discrete spectrum eigenvalue, a_1, the trivial solution to (2.1) becomes unstable when a reaches a_1. Near the bifurcation point, the primary term of the overcritical solution takes the form:*

$$u(\mathbf{R}) \approx \xi_1\psi_1(\mathbf{R}) \tag{2.4}$$

where ξ_1 is the amplitude, indicating an asymptotically precise solution.

The established overcritical solution form (2.4) leaves ξ_1 as the sole unknown, derived by substituting (2.4) into (2.1) and reformulating it to the following identical form:

$$\hat{L}(a_1)u = f \tag{2.5}$$

where f is defined as:

$$f = [\hat{L}(a_1) - \hat{L}(a)]u + \hat{N}(u) \tag{2.6}$$

Since the linear part, $\hat{L}(a_1)u$, of (2.5) is on the spectrum, $a = a_1$, Fredholm's alternative applies.[1] The latter states that a nontrivial solution to (2.5) exists if and only if f is orthogonal to $\psi_1(R)$, resulting in the following relation:

$$\int \psi_1^*(\mathbf{R})\left[\hat{L}(a_1) - \hat{L}(a)\right]u(\mathbf{R})d\Omega + \int \psi_1^*(\mathbf{R})\hat{N}\left[u(\mathbf{R})\right]d\Omega = 0$$

[1] Here, we formulate it in a simplified form adequate to equations met later in this book. The rigorous formulation one finds in the book [2].

where $\psi_1^*(R)$ is the conjugate function to $\psi_1(R)$. By substituting the solution (2.4) and taking into account that $\hat{L}(a_1)\psi_1(R) = 0$ according to (2.3), one obtains the following relation:

$$-\xi_1 \int \psi_1^*(\mathbf{R})\hat{L}(a)\psi_1(\mathbf{R})d\Omega + \int \psi_1^*(\mathbf{R})\hat{N}\left[\xi_1\psi_1(\mathbf{R})\right]d\Omega = 0 \tag{2.7}$$

In all cases we will face in this book, $\hat{N}(u)$ is a polynomial. Therefore, integrating (2.7) yields a nonlinear algebraic equation for ξ_1 known as the *ramification equation* [2] providing ξ_1 and finalizing the solution. This method is effective when $(a - a_1)/a_1 \ll 1$ and ξ_1 remains small.

Throughout our discussions, we frequently encounter cases where the original equation has been derived from a functional $\mathcal{F} = \mathcal{F}\{u(R)\}$ in such a way that the condition for its minimum is expressed as:

$$\delta\mathcal{F}/\delta u = 0$$

yields (2.1). Here δ denotes the variational derivative. By substituting the solution (2.4) into this functional:

$$\mathcal{F} = \mathcal{F}\{\xi\psi_1(\mathbf{R})\} \tag{2.8}$$

and performing the integration, $\mathcal{F}$ transforms into a nonlinear function of ξ: $\mathcal{F} = \mathcal{F}(\xi)$. The condition for its minimum concerning the amplitude ξ:

$$\frac{\partial\mathcal{F}}{\partial\xi} = 0 \tag{2.9}$$

yields the ramification equation, which is identical to (2.7). This approach provides a simpler and more transparent method for deriving the ramification equation compared to the previous one.[2]

***

To summarize, *it's worth noting that this approach, while heavily involving the solution of the linear equation, ultimately yields an asymptotically exact solution of the nonlinear equation. For complete details, theorems, and proofs, please refer to the book by Vainberg and Trenogin* [2].

---

[2] In this book, we will consistently utilize this approach, particularly when the original equation has been derived from a functional. In our case, this functional invariably represents the free energy.

References

1. L.D. Landau, E.M. Lifshitz, *Statistical Physics*, 3 edn. (Pergamon Press, Oxford, 1980)
2. M.M. Vainberg, V.A. Trenogin, *Theory of Branching of Solutions of Non-Linear Equations* (Noordhoff, Leyden, 1974)

Chapter 3
Landau's Theory with a One-Component Order Parameter

Abstract Landau's theory employs an order parameter to describe phase transitions. Generally, it is an n-component entity characterized by specific symmetry properties. However, this book focuses on cases involving a one-component order parameter for several reasons. First and foremost, this represents the most technically simple scenario. Given the complexities surrounding the problem of process zones, it is pragmatic to start with a technically less demanding case. Moreover, many characteristics of process zones are observed to be invariant, irrespective of the number of order parameter components. Thus, investigating these fundamental properties through simple examples becomes important. Additionally, in the case of a multicomponent order parameter, there are often low-symmetry phases where its different nonzero components are equal to one another. One adequately describes such phases by a single order parameter component. With certain caveats, one can model the corresponding process zones using a one-component order parameter framework. Last, we want to get the reader to the point as soon as possible rather than engage in a voluminous discussion of the classical Landau's theory for transitions with multicomponent order parameters. Consequently, our analysis centers on the characteristics of zones governed by a one-component order parameter. Landau's theory of phase transitions is fundamentally based on two key concepts: (i) the relationship between phase transformation and changes in crystal symmetry, and (ii) the bifurcation phenomenon. The former is expressed using the language of group theory, while the latter is described by the time-dependent Ginzburg-Landau equation or a system of such equations. These equations can be interpreted as equations of motion or state for the atoms involved in the phase transformation, averaged over their ensemble. Together, they provide a universal framework for understanding the internal mechanisms governing phase transformations.

3.1 Landau's Potential

In this section, we derive Landau's potential from the perspective of statistical mechanics using a straightforward example. This approach demystifies the origin of the order parameter and Landau's potential. We then discuss the general frame-

© The Author(s), under exclusive license to Springer Nature Switzerland AG 2026
A. Boulbitch and A. Korzhenevskii, *Stress-Induced Local Phase Transformations Near Cracks and Dislocations*, Springer Series in Materials Science 356,
https://doi.org/10.1007/978-3-032-00945-6_3

work proposed by Landau for constructing the free energy density and the free energy functional as functions of the order parameter. While these results are well-known and available in the literature on phase transformations, they are included here for reference and the reader's convenience.

***

3.1.1 Origin and Significance

The discourse begins with the genesis and physical justification for Landau's potential. Let us recall that in classical statistical mechanics, the thermodynamics of a crystal is encapsulated by the partition function, $\mathcal{Z}$, formulated as follows:

$$\mathcal{Z} = \frac{1}{N!} \int \exp\left[-\frac{E(\mathbf{p}, \mathbf{q})}{k_B T} \right] \frac{d\Gamma}{(2\pi\hbar)^s} \tag{3.1}$$

where $\mathbf{p}, \mathbf{q}$ denotes the sets of momenta, $\mathbf{p}^{(i)}$, and coordinates, $\mathbf{q}^{(i)}$, for the atoms indexed by $i = 1, 2, ..., N$. Here, $E(\mathbf{p}, \mathbf{q})$ represents the total energy of N atoms in a crystal, T is the temperature, k_B is Boltzmann's constant, $\hbar$ is Planck's constant, and s signifies the number of degrees of freedom. The phase space volume element is given by:

$$d\Gamma = \prod_{i=1}^{N} d^3 p^{(i)} d^3 q^{(i)}$$

The conventional definition of the crystal's free energy is then expressed as:

$$\mathcal{F} = -k_B T \ln(\mathcal{Z}) \tag{3.2}$$

Details can be found in Sect. 31 of the book by Landau and Lifshitz [2].

Let us explore an example of a phase *transformation of the displacive type*.[1] In the high-temperature phase (which we also refer to as the *mother phase*), atoms oscillate around their equilibrium positions within the crystal lattice. These oscillations group into what are known as *normal modes of oscillations* or *phonon modes*. These modes represent displacements of single atoms or linear combinations of several atoms' displacements, each oscillating around its equilibrium position.[2]

[1] The second large group of structural transformations are those of the order-disorder type. In these cases, one or a few atoms are distributed over multiple positions in the crystal lattice, whose number exceeds the number of the distributed atoms. In the high-symmetry phase, the occupation probabilities of all those positions are equal; the situation changes with the transition to the low-symmetry phases.

[2] For more details, refer to well-known textbooks such as [2, 4, 5] to name a few.

The displacive transition to a low-symmetry phase termed the *daughter phase*, occurs when the equilibrium position of at least one mode shifts from its original location, involving the displacement of one or several atoms. Let $\boldsymbol{\eta} = (\eta^1, \eta^2, ..., \eta^m)$ be such a shift of an m-component normal mode from its equilibrium position in the mother phase, where, by definition, $\boldsymbol{\eta} = 0$. In the daughter phase, $\boldsymbol{\eta}$ assumes a non-zero value.

First, let us perform a linear transformation of the atomic coordinates, $(\mathbf{q}_1, \mathbf{q}_2, ..., \mathbf{q}_N) \rightarrow (\mathbf{q}'_1, \mathbf{q}'_2, ..., \mathbf{q}'_{N-m}, \boldsymbol{\eta})$, where the latter set represents the normal modes. Next, we introduce the incomplete partition function, $\mathcal{Z}(\boldsymbol{\eta})$, integrated over all momenta and all normal coordinates $\mathbf{q}'$, except $\boldsymbol{\eta}$:

$$\mathcal{Z}(\boldsymbol{\eta}) = \frac{1}{N!} \int \exp\left[-\frac{E\left(\mathbf{p}, \mathbf{q}', \boldsymbol{\eta}\right)}{k_\mathrm{B} T} \right] J \frac{d^{3N} p d^{3N-m} q'}{(2\pi\hbar)^s} \tag{3.3}$$

Here, J is the Jacobian of the transformation $(\mathbf{q}_1, \mathbf{q}_2, ..., \mathbf{q}_N) \rightarrow (\mathbf{q}'_1, \mathbf{q}'_2, ..., \mathbf{q}'_{N-m}, \boldsymbol{\eta})$ representing a numerical value. By multiplying the logarithm of the partition function, $\mathcal{Z}(\boldsymbol{\eta})$, by $-k_\mathrm{B}T$, we obtain Landau's potential:

$$\mathcal{F}(\boldsymbol{\eta}) = -k_\mathrm{B} T \ln\left[\mathcal{Z}(\boldsymbol{\eta})\right] \tag{3.4}$$

$\mathcal{Z}(\boldsymbol{\eta})$ and $\mathcal{F}(\boldsymbol{\eta})$ depend on $\boldsymbol{\eta}$ referred to as the *order parameter*.

We now highlight several critical insights following from the above derivation.

1. The order parameter generally consists of multiple components, though there are one-component normal modes giving rise to one-component order parameters. In this book, we focus on this latter case.
2. In the mother phase, the order parameter is zero. It acquires a non-zero value in potential daughter phases, thereby distinguishing between parent and daughter phases, as well as among various daughter phases.
3. The free energy, as defined in (3.3), is averaged over the remainder of the ensemble. This averaging grants the order parameter the status of a *macroscopic thermodynamic quantity*.
4. The partition function reaches a sharp maximum at the equilibrium value of the order parameter, η_eq. By definition (3.4), Landau's potential, $\mathcal{F}(\eta_\mathrm{eq})$, attains a minimum at this juncture.
5. At equilibrium, $\mathcal{F}(\eta_\mathrm{eq})$, as delineated by (3.4) deviates from the free energy defined by the expression (3.2). This deviation stems from (3.3) not accounting for fluctuations of η around η_eq. Nevertheless, when fluctuations are minor, (3.4) provides a sufficiently accurate description, hence the classification of Landau's approach as a mean-field theory. Should the value of η diverge from its equilibrium, $\mathcal{F}(\eta)$ characterizes the free energy of a non-equilibrium state.

***

In conclusion, *the present arguments establish the statistical basis for Landau's potential. Theoretically, if interatomic interactions were known, the equations in* (3.3) *and* (3.4) *would provide a method for deriving Landau's potential from a microscopic perspective. However, direct measurement of interatomic interactions remains unfeasible; they are often undetermined or hypothesized without experimental validation. Additionally, the integration in* (3.3) *is highly complex, requiring significant simplifications. Consequently, instead of a microscopic approach, we follow an alternative strategy, which combines symmetry principles with a phenomenological framework put forward by Landau.*

3.2 Landau's Free Energy Expansion

In this section, we derive the Landau potential as a functional of the order parameter and the strain tensor.

***

3.2.1 The One-Component Order Parameter

In general, the order parameter is a multi-component object, though, in some cases, it can be one-component. Delving into the one-component order parameter presents several advantages. Primarily, its technical simplicity makes it an appealing starting point. Given the uncharted territory of process zones, initiating our investigation with the most basic scenarios seems prudent. Additionally, within the spectrum of daughter phases identified by a multicomponent order parameter, $\boldsymbol{\eta} = (\eta^1, \eta^2, ..., \eta^m) \neq 0$, we encounter phases described by forms like $(\eta, 0, ..., 0)$, $(\eta, \eta, 0, ..., 0)$ or $(\eta, \eta, ..., \eta)$ and so forth with $\eta \neq 0$.[3] The latter suggests that in certain daughter phases, the multi-component order parameter depends on a single independent variable. The characteristics of process zones corresponding to such phases closely resemble those defined by the one-component order parameter. Thus, the one-component order parameter enables one to reveal fundamental properties of the process zones.

Consequently, we will concentrate on the one-component order parameter theory in the subsequent chapters.[4]

The crystal symmetry group plays a critical role in constructing the free energy by imposing constraints on its form. Specifically, the free energy per unit volume

[3] For example, in $BaTiO_3$, the order parameter has three components, (η^1, η^2, η^3). $BaTiO_3$ exhibits a cubic mother phase with $\eta^1 = \eta^2 = \eta^3 = 0$; a tetragonal daughter phase with $\eta^1 \neq 0$ and $\eta^2 = \eta^3 = 0$; an orthorhombic daughter phase with $\eta^1 = \eta^2 \neq 0, \eta^3 = 0$; and a rhombohedral daughter phase: $\eta^1 = \eta^2 = \eta^3 \neq 0$. All these daughter phases are controlled by a single independent component of the order parameter.

[4] Exploring the multicomponent scenario requires a considerable volume, which cannot be included in this book. Instead, we refer the readers to our paper [1].

(referred to as the free energy density, Φ) must remain invariant under the operations of the crystal symmetry group. For a multicomponent order parameter, this invariance requirement results in a complex structure for the free energy density as a function of the order parameter components.

For a one-component order parameter, there are only two possible scenarios:

1. All operations of the crystal symmetry group transform the order parameter into itself: $(\eta \to \eta)$.
2. Some symmetry group operations leave the order parameter invariant $(\eta \to \eta)$, while others reverse its sign $(\eta \to -\eta)$.

The first case describes transitions that do not involve symmetry or structural changes, such as vapor-liquid transitions, phase separation in fluids and polymers, coil-globule transitions in polymers, and isostructural transitions in SmS crystals. These types of transitions are beyond the scope of this book.

The second case, involving a one-component order parameter that changes sign under some symmetry operations is the primary focus of this book. It is commonly observed in various materials, including:

- Strongly anisotropic magnetics such as K_2CoF_4 and Rb_2CoF_4 [6].
- Para to ferroelectric transitions in uniaxial dielectric ferroelectrics, e.g., $LiTiO_3$, $LiNbO_3$, K_2PO_4, Pb_5GeO_{11}, and triglycine sulfate $(NH_2CH_2COOH)_3 \cdot H_2SO_4$ [7].
- Transitions in ferroelectrics-semiconductors like SbSJ [7].
- $\alpha - \beta$ transition in quartz [8].
- The transition between normal and superconducting phases in conventional superconductors at low temperatures, described by the wave function of the superconducting condensate. Although the wave function is a two-component complex scalar, it can be reduced to a one-component order parameter by eliminating the Goldstone mode [9]. Examples include superconducting materials such as Sn, Pb, Fe, [10], Nb and Ta [11], superconductive compounds, such as C_6Ca [12] and $C_{60}K_3$ [13], among many others.

3.2.1.1 The Free Energy Part Dependent on the Order Parameter

Given that $\Phi = \Phi(\eta)$, the requirement of invariance under the transformation $\eta \to -\eta$ implies that Φ must be an even function of η, expressed as $\Phi = \Phi(\eta^2)$.

Pursuant to Landau's approach, we can expand the free energy density as a series in terms of η, assuming its magnitude to be minor:

$$\Phi_{\text{pt}} = \frac{1}{2}\alpha\eta^2 + \ldots \tag{3.5}$$

Here, α signifies a phenomenological coefficient, and Φ_{pt} represents the free energy density portion solely contingent on the order parameter. In equilibrium, the order parameter assumes a value that minimizes the free energy, as established in Sect. 3.1.1. Should the free energy density not be positively defined, it could erroneously suggest a nonphysical prediction at $\eta_{eq} = \pm\infty$. Thus, an essential requirement is that the free energy density be positive-definite, so that the model describes a physically realistic system, which necessitates $\alpha > 0$.

However, α is intricately shaped by interatomic interactions and entropic factors, exhibiting a characteristic dependency on thermodynamic variables like temperature (T) and pressure (p), such that $\alpha = \alpha(T, p)$. Under certain conditions, α may become zero or negative, breaching the positive definiteness. To rectify this, the next positive term is incorporated:

$$\Phi_{\text{pt}} = \frac{1}{2}\alpha\eta^2 + \frac{1}{4}\beta_0\eta^4 \tag{3.6}$$

assuming the phenomenological parameter β_0 is positive. If it turns out negative, the expansion continues to the subsequent positively defined term:

$$\Phi_{\text{pt}} = \frac{1}{2}\alpha\eta^2 + \frac{1}{4}\beta_0\eta^4 + \frac{1}{6}\gamma\eta^6 \tag{3.7}$$

For large values of η, the Landau potential asymptotically behaves as $\Phi_{\text{pt}} \sim \gamma\eta^6$, enforcing the requirement of global positive definiteness, provided $\gamma > 0$. It is worth noting that the fractional coefficients in front of the terms are chosen to ease subsequent calculations.

Furthermore, the order parameter may vary spatially, necessitating the addition of a term $\sim g\,(\nabla\eta)^2$ to account for increasing free energy due to such spatial variations. Here, $g > 0$ is a constant, and $\nabla\eta$ symbolizes the gradient of the order parameter: $\nabla\eta = (\partial\eta/\partial X, \partial\eta/\partial Y, \partial\eta/\partial Z)$.

Thus, the Landau potential that governs the phase transformation is articulated as:

$$\Phi_{\text{pt}} = \frac{1}{2}g\,(\nabla\eta)^2 + \frac{1}{2}\alpha\eta^2 + \frac{1}{4}\beta_0\eta^4 + \frac{1}{6}\gamma\eta^6 \tag{3.8}$$

where g, β_0, and γ are constants, with γ and g necessarily positive, while β_0 could be either positive, or negative. Fundamentally, they exhibit a dependency on thermodynamic parameters, with α showing the most significant variation:

$$\alpha = \alpha_0(T - T_c) \tag{3.9}$$

Here, $\alpha_0 > 0$ represents a constant, T is the temperature, and T_c denotes the Curie temperature at zero pressure. Our exploration is confined to a sixth-order polynomial, the simplest form that permits analysis of both second and first-order transitions.

3.2.1.2 The Strain-Dependent Part of the Free Energy Density

The free energy density encompasses a component that is contingent on the strain $\varepsilon_{\alpha\beta}$:

$$\varepsilon_{\alpha\beta} = \frac{1}{2}\left(\frac{\partial u_\alpha}{\partial X_\beta} + \frac{\partial u_\beta}{\partial X_\alpha}\right) \tag{3.10}$$

Here, u_α denotes the displacement vector components, X_α is a coordinate, and the Greek indices $\alpha, \beta, ... = 1, 2, 3$ represent tensor indices in three-dimensional (3D) Euclidean space, with the Einstein summation convention in effect.

Assuming the crystal exhibits elastic isotropy—a reasonable approximation for many cubic metals according to Eshelby [14]—the strain-dependent portion of the free energy density is expressed as:

$$\Phi_{\mathrm{el}} = \frac{E\nu}{2(1-2\nu)(1+\nu)}\varepsilon_{\alpha\alpha}^2 + \frac{E}{2(1+\nu)}\varepsilon_{\alpha\beta}^2 \tag{3.11}$$

where E symbolizes Young's modulus, and ν denotes Poisson's ratio. It is vital to distinguish between the Greek characters used for constants such as α, β_0, γ, ν, and others within the free energy expressions from those serving as tensor indices.

3.2.1.3 Interaction Between Strain and Order Parameter

The free energy component that encapsulates the interaction between the elastic degrees of freedom and the order parameter is defined as:

$$\Phi_{\mathrm{int}} = A\varepsilon_{\alpha\alpha}\eta^2, \tag{3.12}$$

where A symbolizes the interaction constant referred to as the *striction constant*.

Note 3.2.1 *In scenarios involving a one-component order parameter, no mixed invariants linear in terms of the strain tensor and squared in terms of the order parameter exist, with the sole exception being the expression (3.12).*

3.2.1.4 Total Free Energy

The aggregate free energy is determined by integrating the free energy density across the crystal volume Ω:

$$\mathcal{F} = \mathcal{F}_0 + \int \left(\Phi_{\mathrm{pt}} + \Phi_{\mathrm{el}} + \Phi_{\mathrm{int}}\right) d\Omega,$$

Here, $\mathcal{F}_0$ represents the free energy of the crystal in the mother phase ($\eta \equiv 0$), and the subsequent term delineates its perturbation due to the phase transition. Inserting the expressions (3.8), (3.11) and (3.12) yields the following free energy density:

$$\Phi = \frac{1}{2}g\,(\nabla\eta)^2 + \frac{1}{2}\alpha\eta^2 + \frac{1}{4}\beta_0\eta^4 + \frac{1}{6}\gamma\eta^6 +$$

$$+ \frac{E\nu}{2(1-2\nu)(1+\nu)}\varepsilon_{\alpha\alpha}^2 + \frac{E}{2(1+\nu)}\varepsilon_{\alpha\beta}^2 + A\varepsilon_{\alpha\alpha}\eta^2 \tag{3.13}$$

and the free energy functional:

$$\mathcal{F} = \mathcal{F}_0 + \int\left[\frac{1}{2}g\,(\nabla\eta)^2 + \frac{1}{2}\alpha\eta^2 + \frac{1}{4}\beta_0\eta^4 + \frac{1}{6}\gamma\eta^6 +\right.$$

$$\left. + \frac{E\nu}{2(1-2\nu)(1+\nu)}\varepsilon_{\alpha\alpha}^2 + \frac{E}{2(1+\nu)}\varepsilon_{\alpha\beta}^2 + A\varepsilon_{\alpha\alpha}\eta^2\right]d\Omega \tag{3.14}$$

***

To summarize, *the Landau potential, representing the free energy density associated with phase transition, is expressed as a polynomial of the order parameter, its gradients, and strain tensor components, constrained by crystal symmetry. The coefficients of this polynomial—such as* g, α_0, β_0, T_c *and* A*—along with elastic moduli* E *and* ν*, serve as key material constants. These constants are critical in determining the material's response to plastic deformation and fracture.*

3.3 Homogeneous Case

In the subsequent discussion, we will draw comparisons between local phase transitions near a crack tip or dislocation and those occurring within the bulk of a defectless crystal. To facilitate this, we will revisit fundamental findings concerning the equation of state of a crystal exhibiting a phase transition and its associated phase diagram. These results have been extensively documented in numerous books and research papers. We provide references to these sources below for our readers' convenience.

***

3.3.1 Elimination of Elastic Variables in the Homogeneous Case

Let us consider an infinite, homogeneous crystal. In such a crystal, the phase transformation is also uniform throughout. Consequently, the order parameter η becomes independent of coordinates, and we have $\nabla\eta \equiv 0$. Our focus now turns to the equilibrium state of this crystal.

In this scenario, we can study the density of the free energy, Φ, instead of the functional representation (3.14). Φ is given by:

$$\Phi = \frac{1}{2}\alpha\eta^2 + \frac{1}{4}\beta_0\eta^4 + \frac{1}{6}\gamma\eta^6 + \\ + \frac{E\nu}{2(1-2\nu)(1+\nu)}\varepsilon_{\alpha\alpha}^2 + \frac{E}{2(1+\nu)}\varepsilon_{\alpha\beta}^2 + A\varepsilon_{\alpha\alpha}\eta^2. \tag{3.15}$$

The stress tensor is defined as

$$\sigma_{\alpha\beta} = \frac{\partial\Phi}{\partial\varepsilon_{\alpha\beta}}.$$

It adopts the following form:

$$\sigma_{\alpha\beta} = \frac{E\nu}{(1-2\nu)(1+\nu)}\varepsilon_{\gamma\gamma}\delta_{\alpha\beta} + \frac{E}{(1+\nu)}\varepsilon_{\alpha\beta} + A\eta^2\delta_{\alpha\beta} \tag{3.16}$$

For its trace, $\sigma_{\gamma\gamma}$, we derive the expression:

$$\sigma_{\gamma\gamma} = \frac{E}{1-2\nu}\varepsilon_{\gamma\gamma} + 3A\eta^2 \tag{3.17}$$

Assuming a *stress-free crystal*, $\sigma_{\gamma\gamma} = 0$, we can deduce the strain tensor: $\varepsilon_{\gamma\gamma} = -3A\eta^2(1-2\nu)/E$. Substituting this result into (3.16), where $\sigma_{\alpha\beta} = 0$, we finally express the strain tensor as:

$$\varepsilon_{\alpha\beta} = -\frac{(1-2\nu)A\eta^2}{E}\delta_{\alpha\beta} \tag{3.18}$$

The above expression characterizes what is known as *spontaneous strain*. It is the strain generated within a free, homogeneous crystal due to the phase transformation. While it is absent in the *mother phase* ($\eta = 0$), it becomes nonzero in the *daughter phase* ($\eta \neq 0$).

Substitution (3.18) into (3.15) yields the potential:

$$\Phi_{\mathrm{pt}}(\eta) = \frac{1}{2}\alpha\eta^2 + \frac{1}{4}\beta_1\eta^4 + \frac{1}{6}\gamma\eta^6 \tag{3.19}$$

with

$$\beta_1 = \beta_0 \left[1 - \frac{3A^2(1-2\nu)}{2E\beta_0} \right] \tag{3.20}$$

It is important to note that in (3.19) $\gamma > 0$, while α and β_1 can have either sign.

Let us address two opposite scenarios for the mechanical constraints on a crystal: a clamped crystal and a free one.

Case 3.3.1 Consider the situation of a clamped crystal. In this case, the displacements at the boundary are zero, leading to the vanishing of all components of the strain tensor $\varepsilon_{\alpha\beta}$. The crystal can be clamped uniaxially, biaxially, or triaxially. We will focus on the tri-axial case for the reason that will become clear below. Here, there is no need to eliminate the elastic variables as all strain tensor components automatically turn to zero. The potential Φ (3.15), thus, simplifies to

$$\Phi = \frac{1}{2}\alpha\eta^2 + \frac{1}{4}\beta_0\eta^4 + \frac{1}{6}\gamma\eta^6.$$

We can conclude that the constant β_0 applies to this triaxially clamped geometry. However, achieving such a setup in practice is challenging. A close approximation could be a composite material where grains of one substance are embedded in a more rigid matrix of another. Let us also note that uniaxial and biaxial clamping types give different values of the factor in front of η^4.

Case 3.3.2 For the second scenario, consider a free crystal. Here, the stress is zero, which aligns with the conditions discussed in this section. Upon eliminating the elastic variables, the potential reduces to Φ_{pt} as described by the (3.19). Therefore, the constant β_1 describes the coefficient in front of η^4 in the free crystal case. This free-crystal setup is the most accessible from an experimental standpoint. All measurements determining the constants of Landau's potential are conducted in this geometry. This is also true for the material parameters listed in Tables 34.0.1 and 34.0.2.

3.3.2 Transformation Phase Diagram for a Homogeneous Crystal

3.3.2.1 Phase Diagrams: The Introduction

Phase transformations are characterized by phase diagrams, which serve as a material's "fingerprint." The phase diagram generally represents the division of the space of thermodynamic parameters, such as temperature, pressure, concentration, and electric and magnetic fields, into domains of existence of various phases. In practice, however, experimental analysis of phase diagrams is limited by the availability of instrumentation, enabling the study of phase transformations. The latter typically

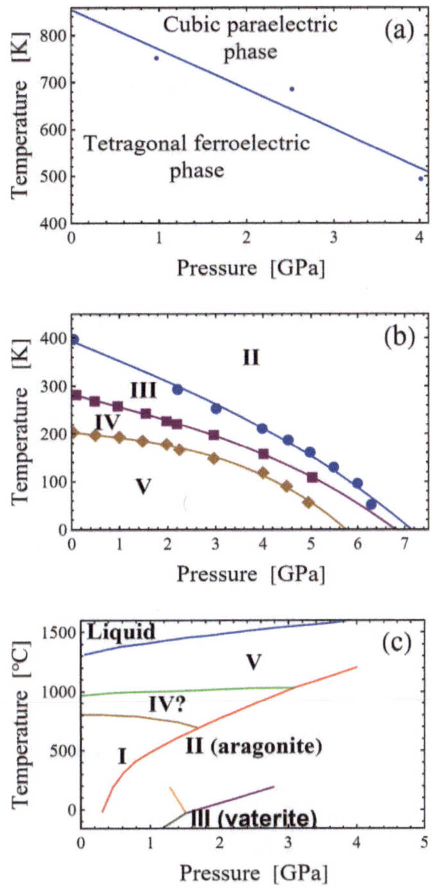

Fig. 3.1 A few examples of $p - T$ phase diagrams. **a** Phase diagram of $PbTiO_3$ [15]. Filled circles are the experimental data. **b** Phase diagram of $BaTiO_3$. The markers indicate the experimental points [16]. The cubic II, tetragonal III, orthorhombic IV, and rhombohedral phase V are shown. The enumeration of the phases is according to [10]. **c** Phase diagram of calcite. Below its liquidus transition line calcite exhibits the phase I (often referred to as "calcite"), phase II (referred to as the "aragonite") stable at high pressure, the high-temperature calcite V, and the high-pressure calcite III (vaterite) and a few other phases [17]

yields the phase diagram regarding two thermodynamic parameters, thus projecting the multidimensional phase diagram onto a specific thermodynamic plane. The most often met are the 2D phase diagrams regarding the temperature and concentration and those in coordinates with the hydrostatic pressure, $p = -\sigma_{\alpha\alpha}$, and temperature, T. In this book, we focus on the latter phase diagram type. They divide the $(p - T)$-plane into regions of existence of different phases exhibited by the solid. Figure 3.1 provides examples of such phase diagrams for lead titanate ($PbTiO_3$), barium titanate ($BaTiO_3$), and calcite ($CaCO_3$). Many more phase diagrams can be found, e.g., in the books of Tonkov and Ponyatovsky [10].

A cursory examination of these phase diagrams [10] reveals that regions of existence of different phases can vary greatly in size, ranging from minuscule to substantial. However, the typical width of these domains along the temperature axis is of the order of 100 K.

Notably, the vast majority of solids exhibit first-order phase transformations rather than second-order ones. The boundaries between these regions are defined by the lines of first-order phase transitions, $T_0 = T_0(p)$, also known as *binodals*, surrounded by hysteresis domains where the merging phases coexist. The latter are typically ~ 1 to 10 K wide, with some exceptions like martensite-austenite steels, nitinol, or ZrO_2 [10], which can reach widths of $\sim 50 \div 100$ K. The boundaries of the hysteresis regions are referred to as the *spinodals*.

3.3.2.2 Rescaling

By rescaling the order parameter and introducing rescaled control parameters a and b:

$$\eta \to u\gamma^{-1/6}; \quad a = \alpha\gamma^{-1/3}; \quad b_1 = -\beta_1\gamma^{-2/3} \tag{3.21}$$

we simplify the free energy density (3.7) to:

$$\Phi_{\mathrm{pt}} = \frac{1}{2}au^2 - \frac{1}{4}b_1u^4 + \frac{1}{6}u^6 \tag{3.22}$$

By analogy with (3.9) we introduce

$$a = a_0(T - T_{\mathrm{c}}) \tag{3.23}$$

where the relation between a_0 and α_0 is like the one between a and α. Note that while a is dimensionless, the dimension of a_0 is K^{-1}.

This simplified formulation lays the groundwork for our analysis, making it easier to compare with more complex cases.

3.3.2.3 Analysis of Landau's Potential

As previously discussed in Sect. 3.1.1, the equilibrium value of the order parameter is determined by minimizing the Landau's potential (3.22) with respect to u:

$$\begin{cases} d\Phi/du = 0 \\ d^2\Phi/du^2 > 0 \end{cases} \tag{3.24}$$

From the first equation under (3.24), the equation of state for the rescaled order parameter is derived as:

$$u(a - b_1u^2 + u^4) = 0 \tag{3.25}$$

This yields the straightforward solution:

$$u = 0, \tag{3.26}$$

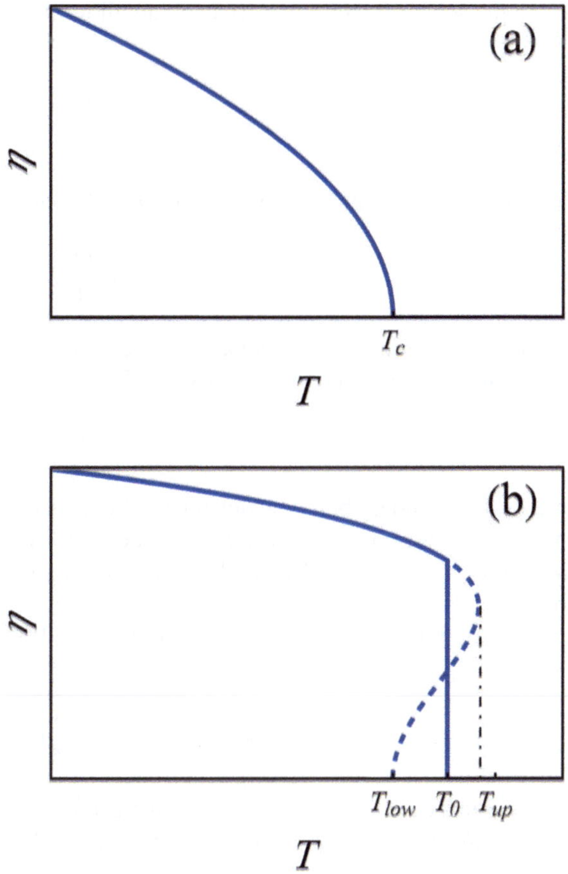

Fig. 3.2 Dependence of the order parameter on the temperature in Landau's theory. **a** In the case of the second-order phase transition, the order parameter dependence on the temperature, T, is non-smooth but continuous. The first derivative discontinuity takes place at $T = T_c$. **b** In the case of $b_1 > 0$, the first-order phase transformation occurs. The order parameter exhibits a discontinuous behavior with the jump at the binodal point, T_0 (the solid line). The dashed line shows the unstable states of the order parameter. The temperatures T_{low} and T_{up} correspond to the lower and upper spinodals correspondingly

which indicates the high-symmetry (mother) phase, occurring when $a > 0$.

For non-trivial solutions where $u \neq 0$, indicating the low-symmetry (daughter) phase, we consider two cases: $b_1 < 0$ and $b_1 > 0$. When $b_1 < 0$, ignoring the small u^4 term within the parentheses simplifies the equation to:

$$a - b_1 u^2 = 0$$

with the solution:

$$u = \pm\sqrt{-\frac{a}{b_1}} \tag{3.27}$$

This solution applies when $a < 0$ and is characteristic of second-order phase transitions, where the order parameter changes continuously but not smoothly at the Curie temperature T_c, as illustrated in Fig. 3.2a.

In the scenario where $b > 0$, keeping all terms gives:

$$u = u_{\mathrm{d}} = u_0 = \sqrt{\frac{b_1 + \sqrt{b_1^2 - 4a}}{2}} \tag{3.28}$$

This solution identifies a minimum for (3.22), valid when $a \leq a_{\mathrm{up}}$, where:

$$a_{\mathrm{up}} = \frac{b_1^2}{4} \tag{3.29}$$

corresponds to the *upper spinodal* in the phase diagram, delineating the threshold below which the daughter phase exists. The *lower spinodal*, a_{low}:

$$a_{\mathrm{low}} = 0 \tag{3.30}$$

marks the boundary below which the mother phase is unstable, and only the daughter phase exists for both $b_1 > 0$ and $b_1 < 0$ conditions. The range defined by $a_{\mathrm{low}} \leq a \leq a_{\mathrm{up}}$ for $b_1 > 0$ is known as the hysteresis domain. See a more thorough discussion below.

Equating the free energies of the mother ($\Phi(0) \equiv 0$) and daughter ($\Phi(u)$) phases leads to:

$$\frac{1}{2}au^2 - \frac{1}{4}b_1u^4 + \frac{1}{6}u^6 = 0 \tag{3.31}$$

Solving this equation concerning a while using the solution for u from (3.28) defines the *binodal* line, $a_{\mathrm{b}} = a_{\mathrm{b}}(b_1)$, indicating the condition for a first-order transition:

$$a_{\mathrm{b}} = \frac{3b_1^2}{16} \equiv a_0(T_0 - T_{\mathrm{c}}) \tag{3.32}$$

where T_0 is the *binodal temperature*.

The interplay (3.9) between α and T as well as the relations (3.29), (3.30), and (3.32) enable one the delineation of the *upper spinodal* (T_{up}), binodal, and *lower spinodal* (T_{low}) temperatures in terms of the dimensional variables:

$$T_{\mathrm{up}} = T_{\mathrm{c}} + \frac{\beta_1^2}{4\alpha_0 g}; \quad T_0 = T_{\mathrm{c}} + \frac{3\beta_1^2}{16\alpha_0\gamma}; \quad T_{\mathrm{low}} \equiv T_{\mathrm{c}} \tag{3.33}$$

Figure 3.2b depicts the dependence of the order parameter on temperature during a first-order phase transition, where the transition occurs abruptly at the binodal temperature ($T = T_0$). At this point, the free energies of the mother and daughter phases equalize. This temperature is flanked by two critical temperatures, the lower spinodal T_{low} and the upper spinodal T_{up}, which define the stability limits for each phase. Beyond T_{up}, the mother phase remains stable, while the daughter phase becomes

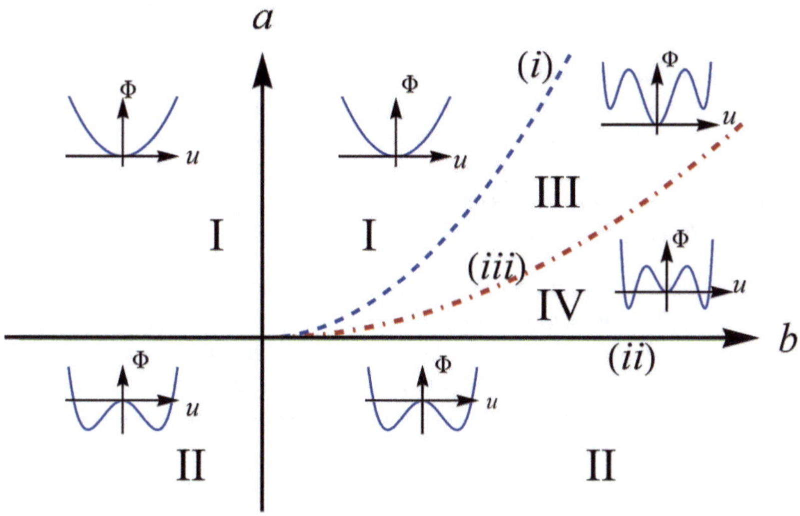

Fig. 3.3 Phase diagram of the phase transformation described by the potential (3.22) with the one-component order parameter. (i), (ii), and (iii) indicate the upper and lower spinodals and the binodal correspondingly. The mother phase exists in the domain indicated by I, and the daughter phase is in the one marked by II. Domains III and IV show the hysteresis region, where the mother and daughter phases coexist, one stable and the other metastable. The insets schematically show the dependence of $\Phi = \Phi(u)$ in each domain

unstable. Conversely, below T_{low}, the mother phase is no longer viable, giving way to the daughter phase. The interval $T_{\text{low}} \leq T \leq T_{\text{up}}$ constitutes a hysteresis region in terms of temperature, within which both phases are possible, albeit in different stability states. The phase with the lower free energy density minimum is considered stable, whereas the other is deemed metastable. One can see the behavior of the free energy density minimums in the insets of Fig. 3.3.

Figure 3.3 presents a detailed phase diagram driven by a one-component order parameter with the potential (3.22), dividing the (b_1, a) plane into four distinct areas. Region I houses the mother phase, above the boundary $a = 0$ at $b_1 < 0$ and beyond $a = b_1^2/4$ at $b_1 > 0$. The latter line corresponds to the upper spinodal. Region II, below the $a = 0$ line, is occupied solely by the daughter phase. The hysteresis Regions III and IV, denote the coexistence area for both phases. In Region III, the mother phase is stable, while the daughter phase is metastable. The situation reverses in Region IV, with the daughter phase becoming stable. The binodal line (iii, dot-dashed), determined by (3.32), traces the boundary of the first-order transition across these coexistence areas. Each diagram inset visually represents the $\Phi = \Phi(u)$ relationship in the respective region.

Note 3.3.1 *It's important to keep in mind that the parameters a and b_1 are functions of thermodynamic variables, namely temperature (T) and pressure (p), expressed as $a = a(T, p)$ and $b_1 = b_1(T, p)$, respectively. In principle, this relationship allows* *for the translation of the (b_1, a) plane into the (p, T) plane, effectively transforming it into a phase diagram. This diagram is segmented into various regions, each corresponding to distinct phases of matter. Thus, the (b_1, a) plane serves as a proxy for the (p, T) phase diagram, providing a detailed representation of phase transitions and stability domains based on thermodynamic conditions. For this reason, in the following we refer to the (b_1, a) plane as to the phase diagram. A few examples of the (p, T) phase diagrams of various materials on finds in Fig. 3.3. Much more examples the reader finds, e.g., in the handbook [10].*

Note 3.3.2 *Figure 3.3 illustrates the phase diagram corresponding to the polynomial potential (3.22). It's important to note that the transformation leading from (3.8) to this particular potential is not unique. As we progress, our exploration will encompass more intricate potentials and transformations, all converging towards the form equal or close to that expressed by (3.22). This ensures that once we apply a transformation culminating in the potential (3.22), its phase diagram adheres to the layout presented in Fig. 3.3. This consistency provides a foundation for comparing various phase behaviors.*

3.3.3 Striction Constant and Binodal Slope

In this study, we are particularly interested in the relationship between the striction constant A and the slope k_0 of the binodal in the (p, T)-phase diagram. If none of the Landau potential coefficients depend on the pressure p, this relationship is straightforward.

The stress tensor $\sigma_{\alpha\beta}$ under hydrostatic pressure satisfies the relation:

$$\sigma_{\alpha\beta} = -p\delta_{\alpha\beta} \tag{3.34}$$

where $\delta_{\alpha\beta}$ is the Kronecker delta (see Sect. 4 of [18]).

For the mother phase ($\eta = 0$), assuming $\varepsilon_{11} = \varepsilon_{22} = \varepsilon_{33} = \varepsilon_0$ and substituting (3.34) into (3.17), the strain ε_0 due to hydrostatic pressure is given by:

$$\varepsilon_0 = -\frac{(1-2\nu)}{E}p \tag{3.35}$$

Now, consider the strain tensor $\varepsilon^{(0)}_{\alpha\beta}$ produced by an undressed defect in the crystal within the mother phase under hydrostatic pressure. This strain accounts for:

$$\varepsilon^{(0)}_{\alpha\beta}(\mathbf{r}) = \varepsilon_0\delta_{\alpha\beta} + \varepsilon^{(1)}_{\alpha\beta}(\mathbf{r}) \tag{3.36}$$

The first term represents the hydrostatic strain, while the second arises from the elastic defect. Substituting this relation and (3.35) into the equation of state (4.12) yields:

$$g\Delta\eta = \left[\alpha_0(T - T_{c0}) - \frac{6Ap(1-2\nu)}{E} + 2A\varepsilon_{\alpha\alpha}^{(1)}(\mathbf{r})\right]\eta + \beta\eta^3 + \gamma\eta^5 \tag{3.37}$$

where T_{c0} is the Curie temperature at zero pressure. The Curie temperature under pressure, $T_c(p)$, is defined by:

$$\alpha_0(T_c - T_{c0}) - \frac{6Ap(1-2\nu)}{E} = 0 \tag{3.38}$$

from which:

$$T_c(p) = T_{c0} + \frac{6A(1-2\nu)}{\alpha_0 E}p \tag{3.39}$$

The slope of the spinodal line is then:

$$k_c = \frac{dT_c(p)}{dp} = \frac{6A(1-2\nu)}{\alpha_0 E} \tag{3.40}$$

With this definition, (3.37) takes its classical form:

$$g\Delta\eta = \left\{\alpha_0\left[T - T_c(p)\right] + 2A\varepsilon_{\alpha\alpha}^{(1)}(\mathbf{r})\right\}\eta + \beta\eta^3 + \gamma\eta^5 \tag{3.41}$$

which only differs from (4.12) by replacing T_{c0} with $T_c(p)$.

Similarly, from (3.41), one obtains the binodal temperature, $T_0(p)$, expressed as:

$$T_0(p) = T_c + \frac{3\beta^2}{16\gamma\alpha_0} + \frac{6A(1-2\nu)}{\alpha_0 E}p \tag{3.42}$$

where

$$T_c + \frac{3\beta^2}{16\gamma\alpha_0} = T_0^{(0)},$$

$T_0^{(0)}$ being the binodal temperature at zero pressure. Since α_0, β, and γ are independent of p, we find the derivative $dT_0(p)/dp = k_0$, leading to:

$$k_0 = \frac{6A(1-2\nu)}{\alpha_0 E} \tag{3.43}$$

Equation (3.43) describes the slope of the binodal line in the (p, T)-phase diagram.

However, in principle, Landau's potential coefficients β and γ may or may not depend on p. If such a dependence exists, it can be either linear or nonlinear, particularly at large p. For instance, if $\beta \approx \beta^{(0)} + \beta^{(1)}p$, substituting this expression into (3.43) gives the binodal slope:

$$k_0 \approx \frac{6A(1-2\nu)}{\alpha_0 E} + \frac{3\beta^{(0)}\beta^{(1)}}{8\alpha_0\gamma} + \frac{3\beta^{(1)2}}{8\alpha_0\gamma}p$$

This shows that k_0 is not necessarily directly proportional to A in general, especially if multiple coefficients depend on p, as in multicomponent order parameter systems.

Nevertheless, in many cases where the dependence of Landau coefficients on p is weak, the sign of A coincides with the sign of k_0.

Note 3.3.3 *In the following, we assume that the sign of the binodal slope k_0 coincides with the sign of the striction constant A.*

3.3.4 Conjugate Field-Driven Phase Transition

Assume that there is a field, h, with the same symmetry as the order parameter.[5] In other words, h transforms according to the same irreducible representation of the crystal symmetry group as u. Specifically, this means that h remains invariant under symmetry operations that leave u unchanged, and it transforms to $-h$ under operations that reverse the sign of u. Under these conditions, the product $h \times u$ is invariant with respect to this group, and the field, h, is referred to as conjugated to the order parameter. The free energy density should then be supplemented as follows:

$$\Phi = \Phi_{\text{pt}} - hu, \tag{3.44}$$

Here Φ_{pt} is given by the expression (3.22).

Let us briefly examine the impact of the conjugated field on the solution for the one-component order parameter, which describes a second-order phase transition in a homogeneous crystal. In the case of the second-order transition ($b_1 < 0$), we omit the term $\sim u^6$ and arrive at the following equation of state, $\partial\Phi/\partial u = 0$:

$$au - b_1 u^3 = h \tag{3.45}$$

An exact solution to this equation is easy to obtain. However, for our future discussion (see Sect. 22.2.2), we need its approximation at small h.

For $a > 0$, the presence of a small field h induces a correspondingly small order parameter, allowing for the simplification by neglecting the term proportional to u^3, leading to the approximation:

$$u \approx \frac{h}{a} \tag{3.46}$$

For $a < 0$, expressing the solution as:

[5] The solution to this problem can be found in Sect. 144 of Landau and Lifshitz's book [2].

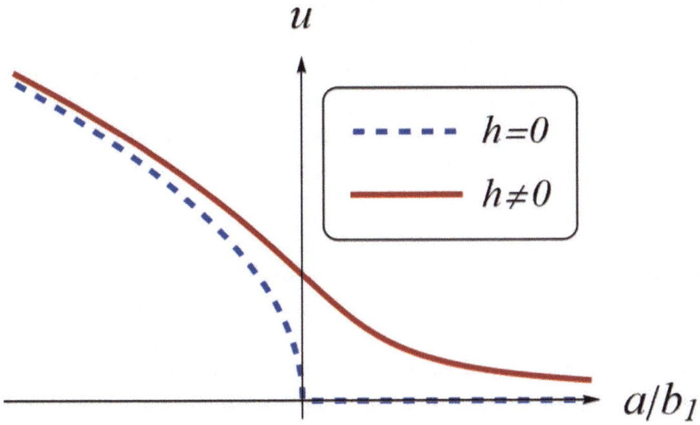

Fig. 3.4 Behavior of the order parameter in the conjugated field. The order parameter in the absence of a field, shown in dashed blue, exhibits standard second-order behavior: $u = 0$ at $a \geq 0$ and $u \neq 0$ at $a < 0$. In contrast, the solid red line represents the behavior of the order parameter in the presence of the conjugated field. In this case, the order parameter never reaches zero

$$u = \sqrt{-\frac{a}{b_1}} + v$$

and transforming the (3.45):

$$-2av + 3\sqrt{-ab_1}v^2 + bv^3 = h$$

simplifies the analysis. This transformation effectively reduces the problem to a previously discussed case, where the solution for v, induced by a small field h, is minor. By disregarding terms proportional to v^2 and v^3, the approximation yields: $v \approx -h/2a$. Thus, we obtain:

$$u \approx \sqrt{-\frac{a}{b_1}} - \frac{h}{2a} \tag{3.47}$$

Figure 3.4 illustrates the behavior of the order parameter as a function of a/b_1 at two different values of h/b_1: $h/b_1 = 0$ and $h/b_1 = 0.4$:

At $h = 0$, the order parameter remains nonzero only for $a < 0$, and it transitions to zero when $a \geq 0$. This is indicative of the characteristic behavior taking place during soft bifurcations.

Note 3.3.4 *Unlike the scenario where $h = 0$, the order parameter persists without vanishing for $h \neq 0$.*

Without an external field ($h = 0$), bifurcation occurs precisely at $a = 0$. However, with a nonzero field, the sharpness of the bifurcation point diminishes, transforming the precise bifurcation point into a broader interval. Moreover, introducing a conjugate field reduces the symmetry of the high-temperature phase to match that of the low-temperature one, distinguishing the phases primarily through the magnitude of the order parameter, which is lesser in the high-temperature phase.

***

In summary, *this section examines the phase diagram for a transformation governed by the potential* (3.7) *within a homogeneous crystal. While the fundamentals of such transformations are well-documented in numerous publications and widely understood in the scientific community, we briefly revisit these concepts to establish a reference point for subsequent discussions. For a more detailed exploration, readers are encouraged to consult Landau's original contributions* [19] *and comprehensive works on phase transformations, such as* [7, 20, 21].

Additionally, we briefly describe the phase transformation when the crystal is subjected to a field conjugate to the order parameter. Details on this topic can be found in Sect. 144 of Landau and Lifshitz's book [2]. *Although this may initially appear unrelated to the content of the following sections, this methodology will similarly be applied to analyze the zone of the mother phase encapsulated within the bulk daughter phase in Sect. 22.2.2.*

3.4 Dynamics of Phase Transformation

This section derives equations of motion and state for a one-component order parameter and, based on these, discusses its characteristic scale. The chapter concludes with analyzing the friction force exerted on propagated defects, attributing it to energy dissipation during the order parameter evolution.

***

3.4.1 Equations of Motion and State

In classical mechanics, the equations of motion are derived as the conditions for minimizing the action, $\mathcal{A}$, which is defined as the integral of the system's Lagrangian, $\mathcal{L}$, over time, t:

$$\mathcal{A} = \int \mathcal{L} dt \tag{3.48}$$

The Lagrangian is constructed by subtracting the system's potential energy from its kinetic energy [3]. To transition to thermodynamics, the system's potential energy is replaced by its free energy, resulting in the following form of the Lagrangian:

$$\mathcal{L} = \int \left\{ \frac{1}{2}\rho \dot{\mathfrak{u}}_\alpha^2 + \frac{\mathfrak{m}}{2}\dot{\eta}^2 - \Phi(\eta, \varepsilon) \right\} d\Omega \tag{3.49}$$

Here, ρ represents the mass density of the solid, $\dot{\mathfrak{u}}_\alpha = \partial \mathfrak{u}_\alpha / \partial t$ denotes the time derivative of the displacement vector components, Φ is the density of the Landau free energy, as defined elsewhere (3.13), Ω denotes the spatial domain, typically a three-dimensional domain, with $d\Omega = dXdYdZ$.

This book focuses on cracks and straight dislocations, where the stress and order parameter distributions depend only on X and Y, independent of Z. Consequently, Ω is treated as a two-dimensional domain in the (X, Y)-plane, with $d\Omega = \mathbb{L}dXdY$, where $\mathbb{L}$ is the length of the solid in the Z-direction. Greek indices α, β, γ,... take the values of 1 and 2 in this context.

Further, $\mathfrak{m}$ represents the mass density associated with the order parameter. For the transitions of the order-disorder type $\mathfrak{m} = 0$. For displacive transitions, $\mathfrak{m} > 0$, and it is closely related to the frequency ω of the *soft mode*. Soft mode frequencies typically range from $\omega \sim 10^{12}$ Hz to 10^{13} Hz. The mass $\mathfrak{m}$ can be determined using the relation imposed on the soft mode: $\mathfrak{m}\omega^2 = \alpha_0(T - T_c)$. From this, the mass is derived as follows:

$$\mathfrak{m} = \alpha_0 \left(\frac{\partial \omega^2}{\partial T} \right)^{-1} \tag{3.50}$$

The derivative, $\partial\omega^2/\partial T$ can be determined from measurements of the high-frequency dielectric response (see, e.g., [22, 23]). Using typical values in the CGS unit system, the estimated range for $\mathfrak{m}$ is:

$$\mathfrak{m} \sim 10^{-25}\,\mathrm{s}^2 \text{ to } 10^{-26}\,\mathrm{s}^2$$

The dissipation function is expressed as a quadratic form of the order parameter rate $\dot{\eta} \equiv \partial\eta/\partial t$:

$$\mathcal{D} = \frac{\kappa}{2} \int \dot{\eta}^2 d\Omega \tag{3.51}$$

Here, κ denotes the kinetic constant of the order parameter.

It is important to note that the dissipation function (3.51) does not include terms related to energy dissipation during plastic deformation. The current approach excludes the study of plasticity, which lies beyond the scope of this book.

The procedure for deriving the dissipative equations of motion from the Lagrangian and dissipation function follows a variational principle[6]:

$$\begin{cases} \frac{\delta \mathcal{L}}{\delta \eta} = \frac{\delta \mathcal{D}}{\delta \dot{\eta}} \\ \frac{\delta \mathcal{L}}{\delta \mathrm{u}_\alpha} = \frac{\delta \mathcal{D}}{\delta \dot{\mathrm{u}}_\alpha} \end{cases} \tag{3.52}$$

where δ denotes the variational derivative [2]. Applying this principle results in the following system of equations:

$$\begin{cases} \mathrm{m}\ddot{\eta} + \kappa\dot{\eta} = g\Delta\eta - [\alpha + 2A\varepsilon_{\gamma\gamma}(\mathbf{R})]\eta - \beta_0\eta^3 - \gamma\eta^5 \\ \rho\frac{\partial^2 \mathrm{u}}{\partial t^2} = \partial\sigma_{\alpha\beta}/\partial X_\beta \end{cases} \tag{3.53}$$

where $\dot{\eta} = \partial\eta/\partial t$ and $\ddot{\eta} = \partial^2\eta/\partial t^2$, and Δ is the two-dimensional Laplace operator:

$$\Delta\eta = \frac{\partial^2\eta}{\partial X^2} + \frac{\partial^2\eta}{\partial Y^2}$$

In the following, we study defects (cracks and dislocations) propagating with velocity V. The joint propagation of the defect-process zone complex along the X-axis is described using an automodel solution:

$$\eta(\mathbf{R}, t) = \eta(X - Vt, y)$$

For this solution, the inertial term in (3.53) becomes $\mathrm{m}V^2\partial^2\eta/\partial X^2$. Comparing it to $g\partial^2\eta/\partial X^2$ gives the ratio $\mathrm{m}V^2/g$. Assuming $V \ll c$, where c is the sound velocity, and using typical values $V \sim 10^{-1}c$, $c \sim 10^5$ cm/ s, and $g \sim 10^{-16}\,\mathrm{cm}^2$ (refer to Tables 34.0.1 and 34.0.2), we find:

$$\frac{\mathrm{m}V^2}{g} \sim 10^{-1} \div 10^{-2} \ll 1$$

This suggests that at $V \ll c$, the kinetic term can be neglected. Similarly, the term $\rho\dot{\mathrm{u}}_\alpha^2$ in (3.49) is omitted. Consequently, the Lagrangian simplifies to $\mathcal{L} = -\mathcal{F}$, reducing the equation to a form describing relaxational dynamics:

$$\begin{cases} \kappa\dot{\eta} = g\Delta\eta - [\alpha + 2A\varepsilon_{\gamma\gamma}(\mathbf{R})]\eta - \beta_0\eta^3 - \gamma\eta^5 \\ \frac{\partial\sigma_{\alpha\beta}}{\partial X_\beta} = 0 \end{cases} \tag{3.54}$$

Finally, in the stationary case $\partial\eta/\partial t = 0$, we obtain the equation of state:

[6] Refer to Sect. 121, (121.8) in the book of Landau and Lifshitz [2].

$$\begin{cases} g\Delta\eta = [\alpha + 2A\varepsilon_{\gamma\gamma}(\mathbf{R})]\eta + \beta_0\eta^3 + \gamma\eta^5 \\ \frac{\partial\sigma_{\alpha\beta}}{\partial X_\beta} = 0 \end{cases} \tag{3.55}$$

3.4.2 Latent Heat Release

The system of (3.55) fully describes the motionless defect-zone complex. In contrast, the system (3.54) is incomplete. During motion, the defect-zone complex generates latent heat at the zone's leading edge and absorbs it at the trailing edge. Additionally, part of this heat diffuses away, sinks into the substrate (if present), and radiates into the surrounding space. To account for these phenomena, a heat transfer equation must be added to the system (3.54).

The heat rate per unit volume, $\dot{Q} = \partial Q/\partial t$, associated with phase transformations is expressed as $\dot{Q} = T\dot{S}$, where S is the entropy per unit volume $\dot{Q} = T\dot{S}$. With $S = -\partial\Phi/\partial T$, substituting the expression for Φ (3.19) and using the relation (3.9) yields:

$$\dot{Q} = -\frac{\alpha_0 T}{2}\frac{\partial\eta^2}{\partial t}$$

This results in the heat transfer equation that accounts for latent heat generation and absorption:

$$\dot{T} = \chi\Delta T - \frac{\alpha_0 T}{2c}\frac{\partial\eta^2}{\partial t} \tag{3.56}$$

where c is the specific heat, χ is the thermal diffusivity of the crystal, and $\dot{T} = \partial T/\partial t$ represents the rate of temperature change. Together, (3.54) and (3.56) form the complete system of equations describing the propagating defect-zone complex.[7]

In certain scenarios, however, thermal effects can be neglected for the following reasons.

[7] Equation (3.56) disregards other heat source, such as those generated at the crack tip during propagation, which are beyond the scope of this study.

3.4.2.1 Minor Temperature Changes

In some cases, the latent heat has minimal influence on zone formation because the temperature at the zone boundary only slightly varies. For example, Gollerthan and co-authors measured the temperature distribution around the process zone at a crack tip in NiTi alloy [24]. The observed temperature variations, ranging from 0.1 to 0.5 K, are negligible compared to the typical difference between the Curie temperature and the temperature at which the zone emerges (see Table 16.1).

3.4.2.2 Heat Evacuation to the Substrate

When a thin crystal slab lies on a substrate with high thermal conductivity, the substrate absorbs most of the heat, rendering it insignificant to zone motion. This effect was observed in the stick-slip propagation of phase fronts[8] in $PbTiO_3$ [26].

3.4.2.3 Model Transparency

Including the heat transfer equation increases the mathematical complexity of the problem. To maintain clarity, and given the minimal impact of latent heat in the cases described above, this book neglects its effects in most of the dynamic problems analyzed.

3.4.3 *Energy Dissipation Within the Zone*

3.4.3.1 Free Energy Dissipation Rate

In this section, we establish a relationship between the rate of change of free energy, denoted as $d\mathcal{F}/dt$, and the dissipation function $\mathcal{D}$. To derive the free energy rate, we start with the straightforward calculating of its time derivative:

$$\frac{d\mathcal{F}}{dt} = \int \left(\frac{\partial \Phi}{\partial \eta} - \nabla \frac{\partial \Phi}{\partial \nabla \eta} \right) \times \frac{\partial \eta}{\partial t} d\Omega$$

Now, we can simplify this expression further. By noting that

$$\frac{\partial \Phi}{\partial \eta} - \nabla \frac{\partial \Phi}{\partial \nabla \eta} = [\alpha + 2A\varepsilon_{\gamma\gamma}(\mathbf{R})]\eta + \beta_0 \eta^3 + \gamma \eta^5 - g\Delta\eta,$$

and utilizing the equation of motion (3.54), we arrive at:

---

[8] Let us note that the release of the latent heat represents the origin also of the stick-slip phenomenon during the phase front motion [25].

$$\frac{d\mathcal{F}}{dt} = -\kappa \int \left(\frac{\partial \eta}{\partial t}\right)^2 d\Omega \equiv -2\mathcal{D} \tag{3.57}$$

Let us emphasize that this relationship is exact. It directly stems from the structure of the dissipation function, $\mathcal{D}$, which is a quadratic function of the order parameter rate. A derivation of the relation (3.57) within a different problem one finds in Sect. 121 of the book [2]. We reiterate it here in the context of this work.

3.4.3.2 Friction Force

We analyze the behavior of a crack-zone or dislocation-zone complex as it propagates through a crystal. Consider a point P located far ahead of the defect-zone complex. Initially, the order parameter at P is zero. As the complex approaches P, the order parameter gradually increases, reaching its maximum when P lies within the zone. Subsequently, as the defect-zone complex moves past P, the order parameter decreases and eventually returns to zero. During this evolution, the time derivative of the order parameter, $\dot{\eta}$, remains nonzero, while the dissipation function $\mathcal{D}$ (3.51) is always positive. The latter generates a friction force, f, exerted on the defect. The power associated with this friction force is $f \times V$. Consequently, we can establish a fundamental relationship between friction force and the dissipation function by considering the rate of change of free energy, $d\mathcal{F}/dt$:

$$\frac{d\mathcal{F}}{dt} = -fV$$

This relationship reveals a fundamental link between f and the dissipation function:

$$f = \frac{2\mathcal{D}}{V} \tag{3.58}$$

When the defect propagates with the constant velocity, V, along X-axis, the zone follows the defect. One can express $\eta(X, Y, t) = \eta(X - Vt, Y)$. By substituting it into the dissipation function (3.51) one finds:

$$\mathcal{D} = \frac{\kappa V^2}{2} \int \left(\frac{\partial \eta}{\partial X}\right)^2 d\Omega \tag{3.59}$$

Using (3.58) one finds the expression for the friction force:

$$f = \kappa V \int \left(\frac{\partial \eta}{\partial X}\right)^2 d\Omega \tag{3.60}$$

***

In essence, *we derived the equation of motion for the order parameter and a simple expression for the friction force exerted on the propagating defect carrying the process zone.*

3.5 Phase Front Propagation: Gordon's Velocity

In this section, for reference purposes, we outline the theory of phase front propagation [27]. Later, we use its results to derive properties of the process zone during its propagation.

***

3.5.1 Statement of Problem

Let us consider a scenario where a crystal initially resides in the mother phase and undergoes rapid cooling, resulting in a homogeneous decrease in temperature throughout the entire bulk. Additionally, we assume that the mother phase becomes metastable while the daughter phase remains stable. Consequently, the region occupied by the mother phase gradually shrinks due to the advancement of its interface, making way for the daughter phase to emerge and grow.

In addressing this problem, we made several assumptions:

- The crystal has transformed from its initial metastable state (the mother phase) to a homogeneous overcooled state corresponding to the daughter phase.
- The process of switching into the overcooled state should be much faster than the motion of the front.

To quantify this latter condition, let us consider a crystal in the form of a slab with length $\mathbb{L}$ and thickness $\mathbb{H}$, characterized by temperature diffusivity χ. The time required to cool the crystal down ($\sim \mathbb{H}^2/\chi$) should be significantly shorter than the time ($\sim \mathbb{L}/V_G$) needed for the phase front to traverse the distance $\mathbb{L}$. Here, V_G represents the velocity of the phase interface. Therefore, the condition:

$$\frac{\mathbb{H}^2 V_G}{\mathbb{L}\chi} \ll 1 \tag{3.61}$$

must be met.[9] It's worth noting that the condition mentioned above places limits on the applicability of Gordon's solution.

While this may initially seem unrelated to the properties of process zones, it will prove useful for deriving significant results later in our study. These results can

[9] In cases where this condition is not satisfied, heat slowly removes from the crystal, resulting in an uneven transient temperature distribution and complicating the kinetics of the transition.

also be applied in cases involving propagating spatial inhomogeneities generated by moving defects. In such scenarios, overcooling or overheating occurs due to changes in the stressed state rather than temperature penetration. These changes can happen much more rapidly than the heat evacuation. As a result, the relation (3.61) becomes irrelevant. This allows us to use the results on front propagation listed below to describe the properties of a propagating zone.

Here we describe the phenomenon of the front propagation provided the corresponding conditions are fulfilled.

In line with the approach adopted throughout this book, we describe the first-order phase transformation using the time-dependent Ginzburg-Landau equation (4.10). As we are considering a defect-free crystal, the term $\sim \varepsilon^{(0)}_{\alpha\alpha}(\mathbf{R},t)$ vanishes. Specifically, we focus on a planar phase front perpendicular to the X-axis. Consequently, the order parameter is independent of the Y-coordinate, and we can express it as $\eta = \eta(X, t)$. The equation of motion takes the following form:

$$\kappa\dot{\eta} = g\frac{\partial^2\eta}{\partial X^2} - \alpha\eta - \beta_1\eta^3 - \gamma\eta^5 \tag{3.62}$$

where $\dot{\eta} = \partial\eta/\partial t$ represents the time derivative of the order parameter. Let V_{G} denote the velocity of the phase interface propagation, referred to as Gordon's velocity. By switching to a co-moving coordinate system:

$$X' = X - V_{\mathrm{G}}t$$

the equation transforms into:

$$g\frac{\partial^2\eta}{\partial X'^2} + \kappa V_{\mathrm{G}}\frac{\partial\eta}{\partial X'} - \alpha\eta - \beta_1\eta^3 - \gamma\eta^5 = 0 \tag{3.63}$$

The phase front exists only within the hysteresis region. We focus on the upper portion of this region, specifically between the binodal and the upper spinodal (refer to Sect. 3.3.2.3):

$$\frac{3\beta_1^2}{16\gamma} \leq \alpha \leq \frac{\beta_1^2}{4\gamma}; \quad \beta_1 < 0 \tag{3.64}$$

Within this domain, we seek a solution that satisfies the boundary conditions at infinity:

$$\eta(-\infty) = \eta_{\mathrm{d}}; \quad \eta(\infty) = 0 \tag{3.65}$$

where

$$\eta_{\mathrm{d}} = \left(\frac{-\beta_1 + \sqrt{\beta_1^2 - 4\alpha\gamma}}{2\gamma}\right)^{1/2} \tag{3.66}$$

is the solution of the equation:

$$\alpha\eta_d + \beta_1\eta_d^3 + \gamma\eta_d^5 = 0$$

This corresponds to the non-rescaled version of (3.25).

3.5.2 Gordon's Solution

3.5.2.1 Velocity of a Plane Phase Front

Equation (3.63), subject to the boundary conditions (3.65) and within the hysteresis domain specified in (3.64), describes the progression of the daughter phase from the left, displacing the mother phase to the right. Its solution:

$$\eta(X') = \frac{\eta_d}{\sqrt{1 + \exp(X'/\mathbb{D})}} \quad (3.67)$$

was originally demonstrated by Gordon [27]. Here, $\mathbb{D}$ represents the thickness of the phase front and is expressed as:

$$\mathbb{D} = \sqrt{\frac{3}{8\alpha^2}\left(\beta_1^2 - 2\alpha - \beta_1\sqrt{\beta_1^2 - 4\alpha}\right)} \quad (3.68)$$

Figure 3.5 illustrates this solution.

This solution represents a kink.

The velocity V_G represents a nonlinear eigenvalue of (3.63) and is expressed as:

$$V_G = \sqrt{\frac{g}{6\gamma}} \frac{3\beta_1^2 - 8\alpha\gamma + 3\beta_1\sqrt{\beta_1^2 - 4\alpha\gamma}}{\kappa\left|\beta_1^2 - 2\alpha\gamma + \beta_1\sqrt{\beta_1^2 - 4\alpha\gamma}\right|^{1/2}} \quad (3.69)$$

Within the upper part of the hysteresis region:

$$\frac{3\beta_1^2}{16\gamma} \leq \alpha \leq \frac{\beta_1^2}{4\gamma} \quad (3.70)$$

the velocity V_G (3.69) describes phase front propagation in an overheated crystal. Conversely, for the lower part:

$$0 \leq \alpha < \frac{3\beta_1^2}{16\gamma} \quad (3.71)$$

it represents front motion in an overcooled crystal.

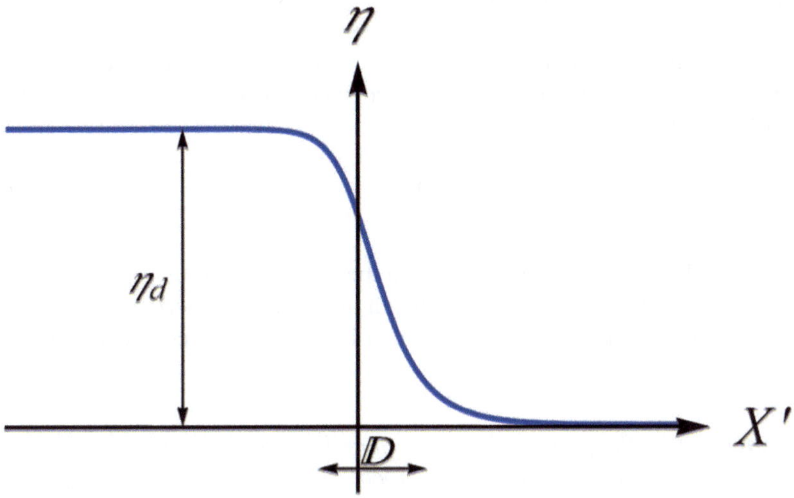

Fig. 3.5 Profile of the phase front propagating in a crystal

Verification of relations (3.67)–(3.69) can be achieved by direct substitution into (3.62). A rigorous derivation is detailed in Gordon's paper [28].

Velocity Analysis

Overcooled Case Within the interval (3.71) , $V_G > 0$, indicating propagation from left to right. This scenario corresponds to the overcooled mother phase, causing the daughter phase to expand in this direction.

Binodal Point At $\alpha = 3\beta_1^2/16\gamma$ the phase boundary is stationary. Near this point, V_G can be approximated in terms of temperature. Substituting $\alpha = \alpha_0(T - T_c)$ and $\alpha_0(T_0 - T_c) = 3\beta_1^2/16\gamma$ into (3.69), and expanding around $T - T_0$, we obtain:

$$V_G^{(bin)} \approx \frac{2\sqrt{\alpha_0 g}}{\kappa\sqrt{T_0 - T_c}}(T - T_0) \tag{3.72}$$

Overheated Case For the hysteresis region (3.70), $V_G < 0$, indicating propagation from right to left. This corresponds to an overheated daughter phase, forcing the mother phase to expand in this direction.

Extreme Negative Value of Velocity Gordon's velocity achieves its extreme negative value, $V_G^{(extr)}$, at the upper spinodal $\alpha\gamma/\beta_1^2 = 1/4$:

$$V_G^{(extr)} = -\frac{1}{\sqrt{3}}\frac{\beta_1}{\kappa}\sqrt{\frac{g}{\gamma}} \tag{3.73}$$

3.5.2.2 Velocity of a Curved Front

The interface curvature, represented by $\mathbb{R}_{\text{curv}}^{-1}$, significantly affects the front velocity by either decreasing or increasing it as demonstrated by Bakaleynikov, Felsteiner and Gordon [29]:

$$V_{\text{G}}^{(\pm)} = V_{\text{G}} \pm \frac{g}{\kappa \mathbb{R}_{\text{curv}}} \tag{3.74}$$

Here, $V_{\text{G}}^{(\pm)}$ is the velocity of the front with the curvature radius $\mathbb{R}_{\text{curv}}$. The negative sign should be used when the surface tension of the curved interface slows down the front. The positive sign applies in the case of its accelerating configuration.

***

To summarize, *the planar phase interface in a homogeneously overcooled or overheated crystal moves at Gordon's velocity* (3.69), *whereas an interface with a curvature radius* $\mathbb{R}_{\text{curv}}$ *propagates at a higher or lower velocity* (3.74).

3.6 Cornerstones of Landau's Theory

In this section, we formulate two fundamental ideas forming basis for the Landau theory.

***

3.6.1 Interpreting the Derived Equations Physically

As previously emphasized, the order parameter represents an intrinsic degree of freedom within a crystal. In the case of displacive transitions, it corresponds to the position of a specific atom in the crystal lattice or, more broadly, to a multi-component combination of several atoms' coordinates. In the order-disorder case, it may signify a combination of occupancy probabilities at sites within the crystal lattice.

To illustrate, let us consider the first case where the order parameter, η, is viewed as the displacement of an atom from its equilibrium position in the crystalline lattice cell. This atom moves within a field created by neighboring atoms, known as the crystalline field. The potential of this field aligns with the free energy. By finding its variation with the sign minus with respect to the atom's displacement η, we obtain the force, f_{cryst}, exerted on the atom from the crystalline field:

$$f_{\text{cryst}} = g\Delta\eta - \alpha\eta - \beta\eta^3 - \gamma\eta^5 - 2A\varepsilon_{\alpha\alpha}\eta \tag{3.75}$$

This formulation allows us to recast the first equation of motion (3.53) as:

$$\mathfrak{m}\ddot{\eta} + \kappa\dot{\eta} = f_{\text{cryst}} \tag{3.76}$$

Equation (3.76) represents Newton's second law applied to an atom within the crystalline field. Moreover, the state equation, the first line in the system (3.55), articulates as:

$$f_{\text{cryst}} = 0 \tag{3.77}$$

indicating the equilibrium condition for the atom in the crystalline field.

It's important to note that both the atomic displacement η and the free energy are averaged over a small yet macroscopic volume encompassing numerous lattice cells. Post-averaging, both the order parameter and the equations of motion or of state acquire a macroscopic character and become dependent on thermodynamic parameters, such as temperature, pressure.

3.6.2 *First Cornerstone: Phase Transformations, Symmetry, and Order Parameter*

Landau's theory fundamentally addresses crystalline structure and symmetry changes during phase transformations. This is achieved by defining the order parameter, representing the internal degrees of freedom driving the phase change. Identifying the order parameter is complex, as the elementary cell of a crystal may have many internal degrees of freedom, each potentially influencing the phase transformation. Landau's approach offers a universal method to pinpoint the order parameter, which is key in capturing critical structural changes and distinguishing different phases based on their symmetry.

Order parameter serves as a macroscopic degree of freedom or a set of such degrees of freedom, distinguishing between mother and daughter phases and between different daughter phases,[10] encapsulating structural and symmetry data. They represent the average values of these degrees of freedom over a small but macroscopic volume.

Contrasting with the phase field, the order parameter is not merely an abstract mathematical construct but a quantifiable degree of freedom supported by empirical evidence. In static situations, it is observable through atomic displacements or redistributions within the crystal lattice cell. These changes can be precisely measured using techniques such as X-ray analysis, high-resolution electron microscopy, and neutron scattering. In dynamic contexts, the movements of these atoms can be examined using optical and neutron experimentation.

The role of the equations of motion (3.53) in phase transitions is analogous to that of the Navier–Stokes equations in hydrodynamics. Both sets of equations are

[10] The latter—in the case of a multicomponent order parameter.

grounded in Newton's second law and deal with quantities averaged over small but macroscopic scales. However, while the former describes averaged atomic displacement, the latter is concerned with the averaged velocity of fluid atoms. While the movement of liquid particles is visually observable, clearly necessitating hydrodynamic equations for fluid dynamics, the involvement of the order parameter in a crystal's evolution is less immediately apparent, requiring advanced investigative techniques for its study.

Therefore, a comprehensive account of phase transformation must include the equations of motion (3.53) in dynamic scenarios or the equation of state (3.55) in equilibrium situations.

3.6.3 Second Cornerstone: Bifurcation

Changes in interatomic interactions can significantly alter the effective crystalline field, particularly when some thermodynamic parameters, like temperature, exceed critical thresholds. These changes instigate instability, leading to atomic displacement, redistribution of occupancy probabilities, or both. This forms the essence of a phase transition. The restructuring of the crystal field, expressed through the order parameter, is precisely what the Landau potential describes. Mathematically, this restructuring is represented by a bifurcation in solutions of equations of state or motion.

Thus, a pivotal principle in Landau's theory is *depicting phase transformation as a bifurcation* in nonlinear equations of state or motion.

In statics and dynamics, the order parameter obeys the generalized Ginzburg-Landau equations. It's worth noting that these equations describe a wide range of phenomena in various scientific domains. Its popularity arises from its ability to offer a universal description of the most common type of bifurcation in nonlinear equations. It is the same bifurcation that also lies at the core of process zone formation. This concept underscores the link between the second cornerstone of Landau's theory and the appropriate mathematical framework based on Ginzburg-Landau equations.

***

To summarize, *Landau's theory of phase transitions is fundamentally based on two key concepts: (i) the relationship between phase transformation and changes in crystal symmetry, and (ii) the bifurcation phenomenon. The former is expressed using the language of group theory, while the latter is described by the time-dependent Ginzburg-Landau equation or a system of such equations. These equations can be interpreted as equations of motion or state for the atoms involved in the phase transformation, averaged over their ensemble. Together, they provide a universal framework for understanding the internal mechanisms governing phase transformations.*

References

1. A. Boulbitch, Y.M. Gufan, A.L. Korzhenevskii, Phys. Rev. E **96**, 013005 (2017)
2. L.D. Landau, E.M. Lifshitz, *Statistical Physics*, 3th edn. (Pergamon Press, Oxford, 1980)
3. L.D. Landau, E.M. Lifshitz, *Mechanics* 3th edn. (Butterworth-Heinemann, 1982)
4. W.A. Wooster, *Tensors and Group Theory for the Physical Properties of Crystals* (Clarendon Press, Oxford, 1973)
5. C. Kittel, *Introduction to Solid State Physics*, 4th edn. (John Wiley and Sons Inc, New York, 2005)
6. A.Z. Patashinskij, V.L. Pokrovskij, *Fluctuation Theory of Phase Transitions* (Pergamon Press, Oxford, 1979)
7. M.E. Lines, A.M. Glass, *Principles and Applications of Ferroelectrics and Related Materials* (Clarendon Press, Oxford, 1977)
8. T.A. Aslanian, A.P. Levanyuk, Solid State Commun. **31**, 547 (1979)
9. L.P. Pitaevskii, E.M. Lifshitz, L.D. Landau, *Statistical Physics. Part 2. Theory of Condensed State* (Butterworth-Heinemann Ltd, 1995)
10. E.Y. Tonkov, *High Pressure Phase Transformations: a Handbook* (Gordon and Breach, Amsterdam, 1992), vol. 1–3; E.Y. Tonkov, E.G. Ponyatovsky, *Phase Transformations of Elements Under High Pressure*, vol. 4 (CRC Press, Boca Raton, FL, 2005)
11. T.H. Geballe, V.B. Compton, Rev. Mod. Phys. **35**, 1 (1963)
12. N. Emery, C. Hérold, J.F.O. Marêché, P. Lagrange, Sci. Technol. Adv. Mater. **9**, 044102 (2008)
13. T. Rachi et al., Sci. Technol. Adv. Mater. **7**, S88 (2006)
14. J.D. Eshelby, *Elastic inclusions and inhomogeneities* (North Holland Publishing Co., Amsterdam, 1961)
15. S.G. Jabarov, D.P. Kozlenko, S.E. Kichanov et al., Phys. Solid State **53**, 2300–2304 (2011)
16. T. Ishidate, S. Abe, H. Takahashi et al., Phys. Rev. Lett. **78**, 2397–2400 (1997)
17. N.S. Bagdassarov, A.B. Slutskii, Phase Transit. **76**, 1015–1028 (2003); M.J. Harris, M.T. Dove, I.P. Swainson, M.E.J. Hagen, Phys Cond. Matter **10**, L423–L429 (1998); M.T. Dove, I.P. Swainson, B.M. Powell, D.C. Tennant, Phys. Chem. Min. **32**, 493–503 (2005); M.T. Dove, B. Winkler, M. Leslie, M.J. Harris, E.K.H.A. Salje, Am. Mineral. **77**, 244–250 (1992); R.C. Garvi, Hannink, R.T. Pascoe, Nature **258**, 703 (1975); D.L. Porter, A. H. Heuer, J. Am. Ceram. Soc. **60**, 183 (1977); T.K. Gupta, F.F. Lange, J.H. Behtold. J. Mater. Sci. **13**, 1464 (1978); L.K. Lenz, A.H. Heuer, J. Am. Ceram. Soc. **65**, 192 (1982); R.P. Indel, R.W. Rice, D. Lewis, J. Am. Ceram. Soc. **65**, 108 (1982); N. Claussen, Z. Werkstofftechnik **13**, 138 (1982); F.F. Lange, J. Mater. Sci. **17**, 240 (1982); P.F. Becher, V.J. Tennery, in *Fracture Mechanics of Ceramics*, vol. 6, ed. by R.C. Bradt (Plenum, New York, 1983), p. 383
18. L.D. Landau, E.M. Lifshitz, *Theory of Elasticity* (Pergamon Press, Oxford, 1986)
19. L.D. Landau, *Collected Papers* (Pergamon Press, Oxford, 1965)
20. Y.M. Gufan, *Structural Phase Transitions* (Nauka, Moscow, 1983). ([in Russian])
21. J.C. Toledano, P. Toledano, *The Landau Theory of Phase Transitions* (World Scientific, Singapore, 1987)
22. J. Hlinka, B. Hehlen, A. Kania, I. Gregora, Phys. Rev. B **87**, 064101 (2013)
23. J. Hlinka, T. Ostapchuk, D. Nuzhnyy, J. Petzelt, P. Kuzel, C. Kadlec, P. Vanek, I. Ponomareva, L. Bellaiche, Phys. Rev. Lett. **101**, 167402 (2008)
24. S. Gollerthan, M.L. Young, K. Neuking, U. Ramamurty, G. Eggeler, Acta Mater. **57**, 5892 (2009)
25. A. Boulbitch, A.L. Korzhenevskii, Phys. Rev. E **108**, 014114 (2023)
26. E.G. Fesenko, V.G. Gavrilyachenko, A.F. Semenchev, in *Domain Structure of Multiaxial Ferroelectric Crystals* (Rostov University, Rostov-on-Don, 1990) [in Russian]; T.E. Hooper, A.J. Bell, J. Appl. Phys. **127**, 104102 (2020)
27. A. Gordon, Phys. Lett. A **99**, 329 (1983)
28. A. Gordon, Physica B **138**, 239 (1986)
29. L. Bakaleynikov, J. Felsteiner, A. Gordon, Physica **B369**, 104 (2005); L. Bakaleinikov, A. Gordon, Physica **B388**, 359 (2007)

Part I
Universal Properties of the Process Zone Formation

Until now, our discussion has primarily focused on defects, and bifurcation theory as well as on the well-established aspects of Landau's theory, encompassing phase transformations. Starting with this part, we shift our attention to our original results regarding the formation of a process zone, characterized by a one-component order parameter. This part explores properties of the process zone independent of the nature of the generating defect, which we categorize as universal properties.

Chapter 4
Inhomogeneous Case: Elimination of Elastic Degrees of Freedom

Abstract We demonstrate that one can eliminate the variables related to the displacement from the system of equations of state or motion. The latter condenses into a single equation solely affecting the order parameter. The elimination of elastic variables is fundamental. It lies in the foundation for all subsequent discussions in this book. We show that if a process zone exists, the order parameter induces a localized spontaneous strain within the zone. This, in turn, generates a strain field throughout the solid. Two primary elastic fields influence the daughter phase embedded within the mother phase matrix. The first originates from stress concentrators such as cracks or dislocations. The daughter phase itself generates the second. We also prove that one can eliminate the strain induced by the process zone from the equations of motion and the equations of state. As a result, one describes the process zone by a single nonlinear equation with the solitary dependent variable, the order parameter. This equation depends only on the trace of the strain tensor of the undressed defect. The latter is known from linear elastic mechanics. The solution to this equation provides complete information on the dynamics and statics of the process zone. Henceforth, in the following, we focus exclusively on these equations to describe process zones (Boulbitch 1992).

Note 4.0.1 Unless specified otherwise, we assume that the crystal remains in its mother phase, except for the vicinity of an inhomogeneity, such as a crack tip or dislocation. The region at the inhomogeneity may or may not contain the daughter phase. In the former case, we refer to it as the *dressed defect*, and in the latter, as the *undressed defect*.

This book primarily focuses on cases where the inhomogeneity source is motionless or propagates at a velocity, V, significantly less than the speed of sound, denoted as c. Under this condition, equation of motion (3.54) applies.

© The Author(s), under exclusive license to Springer Nature Switzerland AG 2026

A. Boulbitch and A. Korzhenevskii, *Stress-Induced Local Phase Transformations Near Cracks and Dislocations*, Springer Series in Materials Science 356,
https://doi.org/10.1007/978-3-032-00945-6_4

4.1 Strain Generated by the Order Parameter Distribution

In this section, we eliminate the elastic variables from the system of equations describing a propagating or stationary defect. This transformation simplifies the problem to a single dependent variable: the order parameter that obeys a single nonlinear differential equation. The solution of this equation enables one to describe all aspects of the process zone exhaustively. This simplification is pivotal for advancing in the field theory of the process zone, and therefore, we provide a detailed derivation.

***

4.1.1 Particular Solution of the Mechanical Equation

Elimination of variables[1] we begin by substituting the expression for stress (3.16) into the second equation (3.54). Utilizing the standard definition of the strain field (3.10), we rewrite this equation in terms of the displacement vector u_α:

$$\frac{E}{2(1-2\nu)(1+\nu)}\frac{\partial^2 u_\beta}{\partial X_\alpha \partial X_\beta}+\frac{E}{2(1+\nu)}\frac{\partial^2 u_\alpha}{\partial X_\beta^2}+A\frac{\partial \eta^2}{\partial X_\alpha}=0 \tag{4.1}$$

Here, the vector indices α and β range from 1 to 3. We then introduce the Fourier transforms of the displacement vector and the square of the order parameter:

$$u_\alpha(\mathbf{R})=\int u_\alpha(\mathbf{q})\exp(i\mathbf{qR})\frac{d^3q}{(2\pi)^3};$$

$$\eta^2(\mathbf{R})=\int Q(\mathbf{q})\exp(i\mathbf{qR})\frac{d^3q}{(2\pi)^3} \tag{4.2}$$

The Fourier transform, $u_\alpha(\mathbf{q})$, should not be confused with the displacement vector, $u_\alpha(\boldsymbol{R})$, in the spatial domain. By performing a Fourier transform on (4.1), we obtain the following equation:

$$-\frac{E\nu}{(1-2\nu)(1+\nu)}q_\alpha q_\beta u_\beta(\mathbf{q})-\frac{E}{(1+\nu)}q_\alpha^2 u_\alpha(\mathbf{q})+Aiq_\alpha Q(\mathbf{q})=0 \tag{4.3}$$

---

[1] Eliminating elastic variables in a simplified case describing the formation of a spherical nucleus of a daughter phase has been first reported by Bar'jakhtar, Vitebsky and Jablonsky [1]. This section outlines another, general and universal approach put forward by Boulbitch, 1992 [2].

Further, we multiply (4.3) by q_α and consider the Einstein summation convention. This enables us to derive the expression for the scalar product, $q_\alpha \mathrm{u}_\alpha(\mathbf{q})$:

$$q_\alpha \mathrm{u}_\alpha(\mathbf{q}) = i\frac{(1-2\nu)(1+\nu)A}{E(1-\nu)}Q(\mathbf{q}) \tag{4.4}$$

where i is the imaginary unit. By substituting (4.4) into (4.3), we obtain the Fourier-transformed particular solution of the equation (4.1) for the displacement vector:

$$\mathrm{u}_\alpha(\mathbf{q}) = i\frac{(1-2\nu)(1+\nu)A}{E(1-\nu)}\frac{q_\alpha}{q^2}Q(\mathbf{q}). \tag{4.5}$$

Here $q^2 = q_1^2 + q_2^2 + q_3^2$.

The general solution in the spatial domain takes the following form:

$$\mathrm{u}_\alpha(\mathbf{R}) = \mathrm{u}_\alpha^{(0)}(\mathbf{R}) + i\frac{(1-2\nu)(1+\nu)A}{E(1-\nu)}\int \frac{q_\alpha}{q^2}Q(\mathbf{q})\exp(i\mathbf{qR})\frac{d^3q}{(2\pi)^3}. \tag{4.6}$$

Here, $\mathrm{u}_\alpha^{(0)}(R)$ represents the solution of the homogeneous equation, which describes the displacement around an undressed defect (where $\eta = 0$). In contrast, the second contribution in (4.6) describes the displacement field resulting from the distribution of the order parameter. In other words, *the second term is generated by the process zone.*

By substituting (4.6) into the definition of the strain tensor (3.10), we obtain:

$$\varepsilon_{\alpha\beta}(\mathbf{R}) = \varepsilon_{\alpha\beta}^{(0)}(\mathbf{R}) - \frac{(1-2\nu)(1+\nu)A}{E(1-\nu)}\int \frac{q_\alpha q_\beta}{q^2}Q(\mathbf{q})\exp(i\mathbf{qR})\frac{d^3q}{(2\pi)^3}, \tag{4.7}$$

where $\varepsilon_{\alpha\beta}^{(0)}(R)$ is the strain tensor generated by the undressed defect (where $\eta = 0$), and $\varepsilon_{\alpha\beta}(R)$ is the total strain accounting for the defect-zone contribution.

It is worth highlighting that the integral in (4.7) is convergent. This can be verified by noting that the term $q_\alpha q_\beta/q^2$ can be expressed as a product of cosines for a range of values of the tensor indices α and β. Each of these cosine terms is bounded above by 1 and below by -1. Therefore, by replacing $q_\alpha q_\beta/q^2$ with ± 1, we arrive at an integral that acts as a majorant (i.e., an upper bound) to the original met in (4.7):

$$\left| \int \frac{q_\alpha q_\beta}{q^2}Q(\mathbf{q})\exp(i\mathbf{qR})\frac{d^3q}{(2\pi)^3}\right| \leq \int Q(\mathbf{q})\exp(i\mathbf{qR})\frac{d^3q}{(2\pi)^3}$$

The latter integral is already known to converge according to (4.2), thereby confirming the convergence of the original integral.

Next, we turn our attention to the trace of the strain tensor, denoted as $\varepsilon_{\alpha\alpha}$. Utilizing (4.7), it is found that:

$$\varepsilon_{\alpha\alpha}(\mathbf{R}) = \varepsilon_{\alpha\alpha}^{(0)}(\mathbf{R}) - \frac{(1-2\nu)(1+\nu)A}{E(1-\nu)}\int Q(\mathbf{q})\exp(i\mathbf{qR})\frac{d^3q}{(2\pi)^3}$$

Incorporating (4.2), we arrive at:

$$\varepsilon_{\alpha\alpha}(\mathbf{R}) = \varepsilon_{\alpha\alpha}^{(0)}(\mathbf{R}) - \frac{A}{E}\frac{(1-2\nu)(1+\nu)}{1-\nu}\eta^2(\mathbf{R}) \tag{4.8}$$

Until this point, we have assumed that the functions $\eta = \eta(R)$, $u_\alpha(R)$, and $\varepsilon_{\alpha\beta}(R)$ are time-independent. Nonetheless, when the velocity of defect propagation is significantly smaller than the sound speed ($V \ll c$), the mechanical equation of motion (3.53) can be replaced by (3.54) accurately describing the system's mechanical behavior. Hence, we extend this equation to time-dependent scenarios:

$$\varepsilon_{\alpha\alpha}(\mathbf{R},t) = \varepsilon_{\alpha\alpha}^{(0)}(\mathbf{R},t) - \frac{A}{E}\frac{(1-2\nu)(1+\nu)}{1-\nu}\eta^2(\mathbf{R},t). \tag{4.9}$$

***

To elaborate, *the initial terms,* $\varepsilon_{\alpha\beta}^{(0)}(R)$ *and* $\varepsilon_{\alpha\alpha}^{(0)}(R,t)$, *in (4.7) and (4.9) outline the strain field associated with an undressed defect, whereas the subsequent terms describe contributions arising from the process zone. This delineation is particularly evident in (4.9), where the distribution* $\eta^2(R,t)$ *explicitly appears.*

4.1.2 Equations of Motion

Upon substituting the trace of the strain tensor, as given by (4.9), into the governing equation of motion (3.54), a modified equation of motion is derived:

$$\kappa\frac{\partial\eta}{\partial t} = g\Delta\eta - [\alpha + U_0(\mathbf{R},t)]\eta - \beta\eta^3 - \gamma\eta^5 \tag{4.10}$$

where we introduced a following notation:

$$U_0(\mathbf{R},t) = 2A\varepsilon_{\alpha\alpha}^{(0)}(\mathbf{R},t) \tag{4.11}$$

Let us emphasize that (4.10) depends only on the strain, $\varepsilon_{\alpha\alpha}^{(0)}(\mathbf{R},t)$, of the bare defect, which was described within the framework of linear elastic mechanics.

In the case of the motionless defect ($\varepsilon_{\alpha\alpha}^{(0)} = \varepsilon_{\alpha\alpha}^{(0)}(\mathbf{R}) \neq \varepsilon_{\alpha\alpha}^{(0)}(\mathbf{R},t)$), $\dot{\eta}$ becomes zero, leading to the equation of state:

$$g\Delta\eta - [\alpha + U_0(\mathbf{R})]\eta - \beta\eta^3 - \gamma\eta^5 = 0 \tag{4.12}$$

Here, the parameter β is defined in terms of β_0 as follows:

$$\beta = \beta_0\left[1 - \frac{2A^2}{E\beta_0}\frac{(1-2\nu)(1+\nu)}{1-\nu}\right] \tag{4.13}$$

Equation (4.10) completely describes the dynamics, while (4.12) describes the statics of the process zone.

4.1.3 Free Energy and Lagrangian in Terms of the Bare Defect Strain

The generating functional, $\mathcal{F}_{\text{pt}}$, can be straightforwardly derived as:

$$\mathcal{F}_{\text{pt}} = \int \left\{ \frac{1}{2} g\,(\nabla\eta)^2 + \frac{1}{2}\left[\alpha + U_0(\mathbf{R}, t)\right]\eta^2 + \frac{1}{4}\beta\eta^4 + \frac{1}{6}\gamma\eta^6 \right\} d\Omega \tag{4.14}$$

This functional is a part of the total free energy functional, $\mathcal{F}_{\text{tot}}$. Its variation with respect to the order parameter produces the equation of state (4.12), as expected. Additionally, varying $\mathcal{F}_{\text{tot}}$ with respect to the displacement vector u_α must yield the mechanical equilibrium equation (4.1), which can be expressed in the form of the second equation (3.55) when considering the relation (3.16).

Accounting for these conditions we obtain the free energy density:

$$\begin{aligned}\Phi_{\text{tot}} &= \frac{E\nu}{2(1-2\nu)(1+\nu)}\varepsilon_{\alpha\alpha}^{(0)2} + \frac{E}{2(1+\nu)}\varepsilon_{\alpha\beta}^{(0)2} \\ &+ \frac{1}{2} g\,(\nabla\eta)^2 + \frac{1}{2}\left[\alpha + U_0(\mathbf{R}, T)\right]\eta^2 + \frac{1}{4}\beta\eta^4 + \frac{1}{6}\gamma\eta^6\end{aligned} \tag{4.15}$$

while the total free energy of the crystal is its integral over the whole domain:

$$\mathcal{F}_{\text{tot}} = \int \Phi_{\text{tot}} d\Omega \tag{4.16}$$

and the Lagrangian is expressed as:

$$\mathcal{L}_{\text{tot}} = \frac{1}{2}\int \left[\rho\left(\frac{\partial \mathrm{u}_\alpha^{(0)}}{\partial t}\right)^2\right] d\Omega - \mathcal{F}_{\text{tot}} \tag{4.17}$$

It is worth emphasizing that the total free energy density (4.15) relies only on the strain $\mathrm{u}_\alpha^{(0)}$ produced by the bare defect, unlike the expressions given in (3.8), (3.11), (3.12). The same is true for (4.16) and the Lagrangian (4.17).

Equation of motion for the order parameter one obtains using the Lagrangian (4.17) and the total free energy (4.16) according to the standard approach employing the dissipation function $\mathcal{D}$ (3.52):

$$\frac{\delta \mathcal{L}_{\text{tot}}}{\delta \eta} = \frac{\delta \mathcal{D}}{\delta \dot{\eta}} \tag{4.18}$$

This equation again brings us to (4.10), as one expects.

***

In summary, *if a process zone exists ($\eta(\mathbf{R}, t) \neq 0$), the order parameter induces a localized spontaneous strain within the zone. This, in turn, generates a strain field throughout the solid. Two primary elastic fields influence the daughter phase embedded within the mother phase matrix. The first originates from stress concentrators such as cracks or dislocations. The daughter phase itself generates the second. Equation* (4.10) *incorporates both contributions.*

We have shown that one can eliminate the strain induced by the process zone from the equations of motion and the equations of state. As a result, one describes the process zone by a single nonlinear equation (4.10) *or* (4.12) *with the solitary dependent variable, η. This equation depends only on the trace of the strain tensor of the undressed defect. The latter is known from linear elastic mechanics. The solution to this equation provides complete information on the dynamics and statics of the process zone, such as its configuration and energy. After the equation of motion (or of state) has been solved, one can determine the zone-generated strain using* (4.7).

Henceforth, we will focus exclusively on these equations to describe process zones. In deriving these equations, we made three key simplifications:

1. *We only considered the one-component order parameter.*
2. *By assuming, $V \ll c$, we simplified the mechanical equation of motion to a quasistatic form. This is inapplicable for $V \sim c$ or in situations involving acoustic waves.*
3. *For simplification, we only assumed the spherically symmetric elastic crystal.*
4. *We considered only the interaction term $A\eta^2\varepsilon_{\alpha\alpha}$ involving pure dilatational strains, thus simplifying the problem mathematically. The latter made it possible to obtain the expression* (4.9). *Consequently, the equation of motion was reduced to a partial differential equation that depends exclusively on the order parameter. This equation provides a comprehensive description of the process zone and serves as the foundational mathematical framework for our subsequent analyses.*

These simplifications raised questions about their applicability in more general settings. The answer is positive. One can eliminate the elastic variables without the simplifications 1–4. Analogously to the conclusion derived in this book, solely the order parameter distribution exhaustively describes the process zone in this case. In contrast to the case described here, an integro-differential equation (or a system of such equations) describes one- or multicomponent order parameter instead of a differential equation. We address these questions in paper [3] *to which we refer the interested reader.*

References

1. V.G. Bar'jakhtar, I.M. Vitebsky, D.A. Jablonsky, Fizika Tverdogo Tela **19**(2), 347–356 (1977); ibid, idem **23**(5), 1448–1455 (1982)
2. A.A. Bulbich, J. Mater. Sci. **27**, 1070 (1992)
3. A. Boulbitch, Y.M. Gufan, A.L. Korzhenevskii, Phys. Rev. E **96**, 013005 (2017)

Chapter 5
Rescaling of Equations Describing the Process Zone

Abstract In this chapter, we modify the spatial, temporal, and order parameter scales. Consequently, we transition to equations expressed in dimensionless coordinates and dimensionless dependent variables. Additionally, we significantly reduce the number of control parameters, which are also rendered dimensionless. In this format, we discuss the equations of state and motion throughout the book. This constitutes the final preparatory technical section before introducing the original context.

In this chapter, we rescale the equations of state and motion that describe the process zone through the following substitutions:

$$X \to \mathbb{R}_0 x; \quad Y \to \mathbb{R}_0 y; \quad t \to \mathbb{T}_0 \tau; \quad \eta(X, Y, t) \to \eta_0 u(x, y, \tau) \tag{5.1}$$

We select the scaling factors $\mathbb{R}_0$, $\mathbb{T}_0$, and η_0 such that the new coordinates, x, y, time, τ, and order parameter, u, become dimensionless. Thus, we introduce a natural spatial scale $\mathbb{R}_0$, measured in centimeters, and a natural temporal scale $\mathbb{T}_0$, measured in seconds. Additionally, η_0 represents the scaling factor with the dimension of the order parameter.

The introduced rescaling serves three primary purposes:

1. It reduces the number of independent control parameters, thus simplifying the search for an analytical solution and making numerical analysis more manageable.
2. It ensures that the equation depends solely on dimensionless coordinates and parameters, enabling the consolidation of various solution families into a unified framework.
3. It enables comparisons with results from the homogeneous case discussed in Sect. 3.3.2 and allows for mapping the bifurcation phase diagram onto the bulk one.

© The Author(s), under exclusive license to Springer Nature Switzerland AG 2026
A. Boulbitch and A. Korzhenevskii, *Stress-Induced Local Phase Transformations Near Cracks and Dislocations*, Springer Series in Materials Science 356,
https://doi.org/10.1007/978-3-032-00945-6_5

5.1 Second-Order Process Zone: Rescaling

Previously, we derived the equation of motion for the second-order transition, which is repeated here for convenience:

$$\kappa\dot{\eta} = g\Delta\eta - [\alpha + U_0(\mathbf{R},t)]\eta - \beta\eta^3 \tag{5.2}$$

In our subsequent analyses, $U_0(R)$ consistently adopts the following expression:

$$U_0(\mathbf{R}) = \frac{B}{R^z} f(\varphi) \tag{5.3}$$

where B is a positive constant, φ is the polar angle $\varphi = \arctan(Y/X)$, and $f(\varphi)$ is a function dependent solely on a combination of $\sin\varphi$, $\cos\varphi$, and potentially a constant. Additionally, z is a positive number that ensures $U(\infty) = 0$ as outlined in (6.2). We posit that the potential $U(R)$ is negative, at least within one region. It is due to the function $f(\varphi)$, which we assume to be negative at some intervals of φ.

In the case of the defect propagating with the velocity considerably below the sound speed, the elastic field is accurately described by passing to the co-moving system:

$$U_0(\mathbf{R},t) = \frac{B}{\left[(X - Vt)^2 + Y^2\right]^{z/2}} f(\varphi) \tag{5.4}$$

Note 5.1.1 *In the following, we focus on the cases of the edge dislocation and mode I crack. For the dislocation, $z = 1$ with $f(\varphi) = \sin\varphi$ and B as defined in (20.2). For a crack, $z = 1/2$ with $f(\varphi) = \cos(\varphi/2)$, and B outlined in (9.2).*

Regardless of the chosen rescaling factors $\mathbb{R}_0$, $\mathbb{T}_0$, and η_0 in (5.1), the functions $\sin\varphi$ and $\cos\varphi$ and $f(\varphi)$ remain invariant. In the context of the second-order process zone, we select the scaling factors[1] as follows:

$$\mathbb{R}_0 = \left(\frac{g}{B}\right)^{\frac{1}{2-z}}; \quad \mathbb{T}_0 = \frac{\kappa}{2\sqrt{a_d}} \left(g^z B^{-2}\right)^{\frac{1}{2-z}}; \quad \eta_0 = \beta^{-1/2} \left(B g^{-z/2}\right)^{\frac{1}{2-z}} \tag{5.5}$$

Except a_d, all parameters entering (5.5), (5.6) are coefficients of Landau's potential. We will determine a_d later on.[2]

Applying the transformation (5.5) to (4.12) we introduce the dimensionless control parameter a:

$$a = B^{\frac{2}{z-2}} g^{-\frac{z}{z-2}} \alpha \tag{5.6}$$

[1] A. Boulbitch and A.L. Korzhenevskii 2016 [1].

[2] See Note 6.1.3 in Sect. 6.1.1.

The equation of state has the form:

$$\frac{\partial^2 u}{\partial x^2}+\frac{\partial^2 u}{\partial y^2}-[a+U(\mathbf{r})]\,u-u^3=0 \tag{5.7}$$

Note 5.1.2 *Let us stress that* x *and* y *are the new dimensionless coordinates,* $\mathbf{r}=(x,y)$ *is the dimensionless 2D radius-vector,* τ *is the dimensionless time, and* $U(\mathbf{r})$ *represents the rescaled potential:*

$$U(\mathbf{r})=\frac{f(\varphi)}{r^z} \tag{5.8}$$

Substitution (5.5), (5.6) into (5.2) leads to the rescaled equation:

$$2\sqrt{a_d}\frac{\partial u}{\partial \tau}=\frac{\partial^2 u}{\partial x^2}+\frac{\partial^2 u}{\partial y^2}-[a+U(\mathbf{r})]\,u-u^3 \tag{5.9}$$

Regarding the propagating defect-zone complex, we transition in (5.9) to the co-moving system: $x'=x-v\tau$, $y'=y$, $r'=(x',y')$, and $u(x,y,\tau)=u(x-v\tau,y)\equiv u(x',y')$. This transition leads to the automodel equation:

$$\frac{\partial^2 u}{\partial x'^2}+\frac{\partial^2 u}{\partial y'^2}+2\sqrt{a_d}v\frac{\partial u}{\partial x'}-\left[a+U(\mathbf{r}')\right]u-u^3=0 \tag{5.10}$$

Note 5.1.3 *Henceforth, we omit the prime signs and use* x, y, $\mathbf{r}$ *in place of* x', y', *and* $\mathbf{r}'$.

Here, the *dimensionless velocity* v is defined as $v=x/\tau$, while the velocity of the defect-zone complex in the original coordinates is $V=X/t$ yielding the following relation between them:

$$V=\frac{\mathbb{R}_0}{\mathbb{T}_0}v \tag{5.11}$$

whence it follows:

$$v=\frac{\kappa B^{\frac{1}{z-2}}g^{\frac{1-z}{z-2}}}{2\sqrt{a_d}}V \tag{5.12}$$

By substitution of (5.1), (5.5), and (5.12) into the free energy formula[3] stated in (4.14) we derive the following expression:

$$\mathcal{F}=\frac{B^{\frac{2}{2-z}}g^{\frac{2(1-z)}{2-z}}}{\beta}F \tag{5.13}$$

[3] Considering the specific case for second-order transitions, one should put $\gamma=0$.

where the factor F represents the integral:

$$F = \int \left\{ \frac{1}{2}\left(\frac{\partial u}{\partial x}\right)^2 + \frac{1}{2}\left(\frac{\partial u}{\partial y}\right)^2 + \frac{1}{2}\left[a + U(\mathbf{r})\right]u^2 + \frac{1}{4}u^4 \right\} dxdy \tag{5.14}$$

The integral in (5.14) is dimensionless, while its pre-factor in (5.13) has the dimension of energy.

We then consider the dissipation function originated from (3.59):

$$\mathcal{D} = \frac{2a_d}{\beta\kappa} B^{\frac{4}{2-z}} g^{\frac{2-3z}{2-z}} v^2 \int \left(\frac{\partial u}{\partial x}\right)^2 dxdy \tag{5.15}$$

Furthermore, the friction force, as detailed in (3.60), transforms into:

$$f = \frac{2\sqrt{a_d}v}{\beta} B^{\frac{3}{2-z}} g^{\frac{2z-1}{z-2}} \int \left(\frac{\partial u}{\partial x}\right)^2 dxdy \tag{5.16}$$

The rescaled equation of state (5.7) depends solely on the control parameter a, while the automodel equation of motion (5.10) is influenced by both control parameters a and v. These equations will be explored in further detail in the upcoming analysis focusing on the second-order transition zone at the defect.

In contrast, the study of the first-order zone requires a slightly different rescaling approach, which will be discussed in the subsequent section.

***

To summarize, *in this book, we mainly regard two defects: the crack and the dislocation. Both of them create the potential of the form (5.3), differing by the power z value. The latter enables one to apply a universal form of the rescaling, giving rise to the equation of state (5.7) and equation of motion (5.10). The latter's advantage is that they only depend on a limited number of parameters (one or two), significantly simplifying their analysis. After solving these equations, one obtains the original parameters: η, R_0, T_0, α, β, F, and f by using the rescaling formulae (5.5).*

Reference

1. A. Boulbitch, A.L. Korzhenevskii, Eur. Phys. J. B **89**, 261 (2016)

Chapter 6
Formation of the Second-Order Process Zone

Abstract This chapter examines the formation of a second-order process zone around a defect. The nature of this defect we do not specify in advance. The aim is to understand the phenomenon's most universal and general aspects. The focus is particularly on a zone where the daughter phase is embedded in the matrix of the bulk mother phase, as well as the reverse situation where the bulk daughter phase hosts the mother-phase process zone (Boulbitch 1992, Boulbitch and Korzhenevskii 2016).

6.1 Soft Daughter-Phase Process Zone: The Stationary Case

In this section, we examine the formation of the daughter-phase process zone within the stable bulk mother phase, focusing on the domain $a > 0$ and the second-order phase transformations.

***

Note 6.1.1 To distinguish between variables associated with the daughter and mother phases, we use subscripts "d" (for "daughter") and "m" (for "mother"). For example, $\mathbb{R}_{\mathrm{d}}, a_{\mathrm{d}}, \xi_{\mathrm{d}}$ refer to the daughter-phase process zone, while $\mathbb{R}_{\mathrm{m}}, a_{\mathrm{m}}, \xi_{\mathrm{m}}$ pertain to the mother-phase one.

In the stationary case, the strain field is time-independent, $\varepsilon_{\alpha\beta} = \varepsilon_{\alpha\beta}(\mathbf{R})$, and $\partial u/\partial\tau = 0$.

6.1.1 Equation of State

As described above, to study the second-order process zone one uses equation of state (5.7):

$$\left(\frac{\partial^2 u}{\partial x^2} + \frac{\partial^2 u}{\partial y^2}\right) - [a + U(\mathbf{r})]\, u - u^3 = 0 \tag{6.1}$$

© The Author(s), under exclusive license to Springer Nature Switzerland AG 2026

A. Boulbitch and A. Korzhenevskii, *Stress-Induced Local Phase Transformations Near Cracks and Dislocations*, Springer Series in Materials Science 356,
https://doi.org/10.1007/978-3-032-00945-6_6

In the following, we refer to $U(\mathbf{r})$ as the *potential* for a reason that will become clear later.

Since we only consider a solitary defect far removed from the sample boundaries, the potential, $U(\mathbf{r})$, exhibits the asymptotics:

$$U(\infty) = 0 \tag{6.2}$$

In this section, we focus on zone formation properties independent of the particular defect. Since the defect manifests itself only in the potential's spatial form, we discuss the properties that are independent of $U(\mathbf{r})$.

We discuss:

1. The condition for forming instability,
2. The bifurcation point, $a_{\mathrm{d}} > 0$,
3. The resulting non-trivial solution, $u(\mathbf{r}) \neq 0$ at $a \leq a_{\mathrm{d}}$.
4. A brief account of the physical implications of such instability.

To examine the bifurcation, we use the principles of bifurcation theory as outlined in Chap. 2. This theory posits that bifurcation is related to the ground eigenvalue of the following auxiliary equation:

$$\Delta\psi_n(\mathbf{r}) - [a_n + U(\mathbf{r})]\psi_n(\mathbf{r}) = 0 \tag{6.3}$$

Here,

$$\Delta = \frac{\partial^2}{\partial x^2} + \frac{\partial^2}{\partial y^2}$$

is the 2D Laplacian in the dimensionless coordinates x and y, a_n $(n = 1, 2, 3, ...)$ are the eigenvalues belonging to the discrete spectrum of (6.3), and $\psi_n(\mathbf{r})$ $(n = 1, 2, 3...)$ are the corresponding eigenfunctions.

The process zone is localized near the defect and fades away with distance:

$$\psi_n(\infty) = 0; \tag{6.4}$$

Additionally, the order parameter has a finite upper limit because it is related to atom displacements or the probability of occupying specific positions. This condition is expressed as: $\psi_n(\mathbf{r}) < \infty$.

Note 6.1.2 Let us note that (6.3) with the boundary conditions (6.4) has the form of Schrödinger equation describing the finite motion of a quantum particle in a potential well $U(\mathbf{r})$. This analogy we consistently use below.

According to bifurcation theory (Chap. 2), bifurcation occurs when a becomes equal to the ground state eigenvalue a_1. In the following, we denote this ground state eigenvalue $a_1 \equiv a_{\mathrm{d}}$ to stress that the daughter-phase zone emerges at this point.

Note 6.1.3 It is this ground-state eigenvalue a_d representing a dimensionless real number that we used in the rescalings (5.5) and (8.2). After the value a_d has been determined, one relates a_d to the corresponding value α_d using (5.6):

$$\alpha_d = a_d \frac{B^{2/(2-z)}}{g^{z/(2-z)}} \tag{6.5}$$

It's important to note that (6.3) has a universal form that applies to a wide range of defects, including cracks, dislocations, and other structural anomalies. The only term that reflects the unique properties of a defect is the potential, $U(\mathbf{r})$.

6.1.2 Solution

Comparing equation (6.3) with Schrödinger's equation:

$$\frac{\hbar^2}{2m}\Delta\psi_n + [E_n - U(\mathbf{r})]\,\psi_n = 0$$

(see Sect. 17 of the book [3]) we conclude that (6.3) is essentially a modified version of Schrödinger's equation where the sign of a_n is inverted: $E_n = -a_n$ $(n = 1, 2, ...)$. Unlike in quantum mechanics, *the eigenvalues of bound states in the potential well* $U(\mathbf{r}) < 0$ *described by* (6.3) and (6.4) *are positive.*

The formation of a process zone at $a > 0$—as described by (6.1)—is contingent upon the contribution $U(\mathbf{r})$ being negative at some points in space. This leads us to a necessary condition for daughter-phase zone formation: there must exist a localized region near the defect where

$$U(\mathbf{r}) < 0 \tag{6.6}$$

Drawing parallels to quantum mechanics, a_1 and $\psi_1(\mathbf{r})$ serve as the ground-state eigenvalue and eigenfunction, respectively, and are central to the subsequent analysis.

As outlined in Chap. 2, a bifurcation from the bulk mother phase to a mother phase with an embedded daughter-phase process zone occurs at $a = a_1 \equiv a_d$.

Below this critical point, $(0 < a \leq a_d)$, the defect appears dressed with a process zone, containing the island of the daughter phase embedded in the matrix of the mother phase.

According to the general principles of bifurcation theory (2.4), in the proximity of $a = a_d$, the leading term of the solution for (6.1) that describes the daughter phase is:

$$u(\mathbf{r}) \approx \xi_d\psi_d(\mathbf{r}) \tag{6.7}$$

As emphasized in Chap. 2, this solution is *asymptotically exact* near the bifurcation point.

Note 6.1.4 Since $\psi_d(\mathbf{r})$ describes the ground state wave function of Schrödinger's equation, it nowhere turns into zero or, as one says, has no nodes. Besides, $\psi_d(\mathbf{r})$ is real. Without any loss of generality, we can assume that $\psi_d(\mathbf{r}) > 0$ everywhere.

The reader finds detailed proof of the above statements in Sect. 20 of the book of L.D. Landau and E.M. Lifshitz's "Quantum Mechanics" [3].

Specific and Universal Features. We should note that the specific nature of the defect manifests itself solely through the potential $U(\mathbf{r})$. One knows from quantum mechanics that this potential controls:

1. The existence of a discrete spectrum,
2. The particular eigenvalue of interest, namely $a_1 \equiv a_d$, and
3. The structure of the corresponding eigenfunction, $\psi_1(\mathbf{r}) \equiv \psi_d(\mathbf{r})$.

These characteristics are specific to the defect. Conversely, the form of the near-overcritical solution (6.7) is universally applicable.

Conjecture 6.1.1 In the following discourse, we operate under the presumption that at least one discrete eigenvalue $a_1 > 0$ of Eq. (6.3) exists. We further assume that both the value of $a_1 > 0$ and the corresponding eigenfunction $\psi_1(\mathbf{r}) \equiv \psi_d(\mathbf{r})$ are explicitly known.

Note 6.1.5 The requirement: "at least one discrete eigenvalue $a_1 > 0$ exists" is a mathematical formulation of the condition that the defect generates a process zone.

6.1.3 Results

Bifurcation Temperature. The bifurcation point is defined as $a = a_d$. We expressed a in terms of α (5.6), while In Landau's theory, $\alpha = \alpha_0(T - T_c)$ (refer to Sect. 3.1). One can introduce

$$a_0 = B^{\frac{2}{z-2}} g^{-\frac{z}{z-2}} \alpha_0 \tag{6.8}$$

and

$$a = a_0(T - T_c) \tag{6.9}$$

Though a is dimensionless, a_0 has the dimension of K^{-1}.

The emergence of the zone containing the daughter phase occurs at temperature $T = T_d$:

$$T_d = T_c + \frac{a_d}{a_0} \tag{6.10}$$

Given that both $a_d > 0$ and $a_0 > 0$, it follows that $T_d > T_c$. Hence, the daughter-phase process zone of the second order *always* forms at temperatures higher than the Curie temperature, where the bulk mother phase remains stable.

The absolute value of a_d is defect-specific. One concludes, therefore, that *the value of the zone emerging temperature, T_d, and the temperature shift, $T_d - T_c$, are defect-specific*. They depend on the power and configuration of the strain field generated by the defect.

Zone Size. Consider the auxiliary equation (6.3) far from the defect. In this scenario, the potential $U(\mathbf{r})$ can be disregarded. The equation for the ground state then simplifies to:

$$\begin{cases} \Delta\psi_d(\mathbf{r}) \approx a_d\psi_d(\mathbf{r}) \\ \psi_d(\infty) = 0 \end{cases} \tag{6.11}$$

Consequently, far from the defect, the solution exhibits the following asymptotic behavior: $\psi_d(\mathbf{r}) \sim \exp\left(-r\sqrt{a_d}\right)$. This expression defines the dimensionless characteristic size, R_d, of the order parameter distribution (6.7):

$$R_d = \frac{1}{\sqrt{a_d}} \tag{6.12}$$

At significant distances, the order parameter can be approximated by:

$$u(\mathbf{r}) \sim \xi_d \exp\left(-\frac{r}{R_d}\right) \tag{6.13}$$

This analysis suggests that the process zone becomes observable in the temperature range $T_c < T \le T_d$ and exhibits a finite dimensionless spatial size, denoted by R_d. According to the rescaling relation outlined in (5.1), the physical size of the process zone, $\mathbb{R}_d$, can be calculated by scaling the dimensionless size with $\mathbb{R}_0$ as $\mathbb{R}_d = \mathbb{R}_0 R_d$. This yields:

$$\mathbb{R}_d = \frac{1}{\sqrt{a_d}}\left(\frac{g}{B}\right)^{\frac{1}{2-z}} \tag{6.14}$$

In the vicinity of the bifurcation point, the second-order zone exhibits the single spatial dimension $\mathbb{R}_d$.

Note 6.1.6 The magnitudes of $\mathbb{R}_d$ and a_d are specific to the defect under consideration. In contrast, *the expression for size in* (6.12) *and the asymptotic behavior described in* (6.13) *are universal.*

***

In summary, *the universal form of the potential* (5.3) *enables one to simultaneously study the zones of the motionless crack and dislocation by solving* (6.1)*. The corresponding auxiliary equation has the form of Schrödinger equation, enabling one to use its properties already established in quantum mechanics. The bifurcation point, at which the process tone emerges, is related to the eigenvalue of the latter, enabling one to obtain the bifurcation temperature, T_d. It also determines the spatial dimension of the zone, R_d, close to the bifurcation point.*

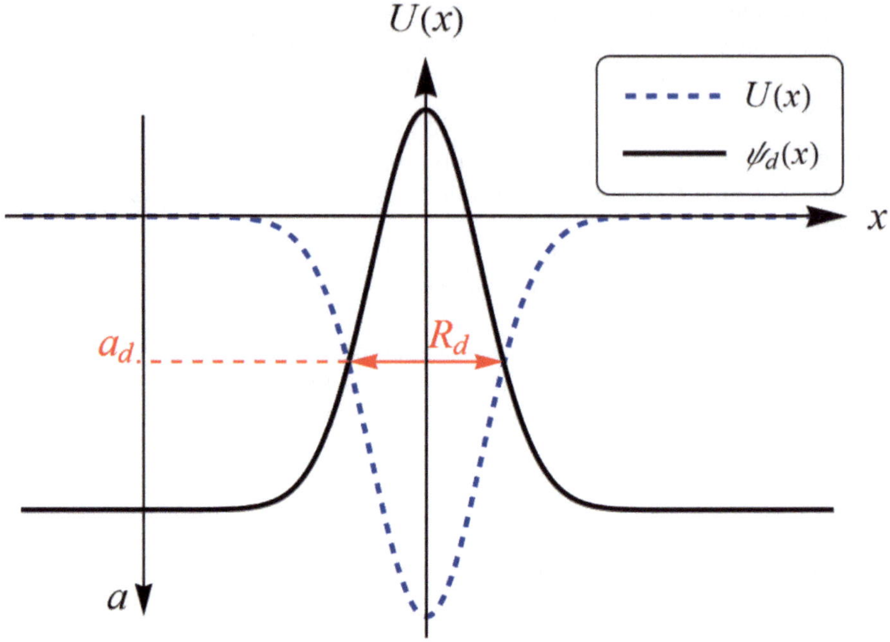

Fig. 6.1 Illustration to the "Old Quantum Theory" approach. See the explanation in the text

6.2 "Old Quantum Theory" Estimates of the Ground State Eigenvalue a_d

In straightforward cases, the magnitude of the ground state eigenvalue a_d can be estimated through a method akin to the "Old Quantum Theory".[1]

Definition 6.2.1 The core principle is as follows: a_d should be determined in a manner that allows the spatial distribution $\psi_\mathrm{d}(\mathbf{r})$ with width R_d (6.12) have a room within the potential well, $U(\mathbf{r})$.

The "Old Quantum Theory" approach is graphically depicted in Fig. 6.1, utilizing a hypothetical one-dimensional (1D) potential, $U(x)$. The dashed blue line in the illustration represents the potential well. As elaborated in Sect. 6.1.2, it is established

[1] "The old quantum theory" refers to this obsolete science domain described in Sect. 48 of "Quantum Mechanics" by Landau and Lifshitz [3]. There is a difference, however: for our purposes, it is enough to find the eigenvalue approximately. The latter contrasts with the old quantum theory, which looked for exact eigenvalues.

that $a_n = -E_n$. Consequently, the a axis is oriented downwards, in contrast to the U axis, and is displayed on the left side of the figure. The eigenfunction, $\psi_d(x)$ (indicated by the solid black line), possesses a width denoted as R_d.

From a qualitative perspective, the eigenfunction is considered to have sufficient space within the potential well if its width can be entirely accommodated within the well, conforming to the criteria outlined in Definition 6.2.1. In Fig. 6.1, the eigenfunction was vertically adjusted to align its width with that of the potential well (as indicated by the double-headed arrow). The energetic level (alternatively referred to as the a level) at which both widths match approximates the value of a_d.

One-Dimensional Case. In a 1D scenario, this method approximates the eigenvalue as:

$$a_d \approx -U(R_d)$$

Multi-Dimensional Cases (2D and 3D). In 2D and 3D, the potential, $U(\mathbf{r})$, is angle-dependent, which complicates substituting a scalar value like R_d into a function relying on radial and angular coordinates. One obtains a more reasonable estimate by averaging $U(\mathbf{r})$ over the angular variables, but only in the region where it is negative. This angular averaging yields an effective, potential, $\overline{U}(r)$, only dependent on $r = |\mathbf{r}|$, where the overline signifies the angular average. One expresses the ground state eigenvalue approximation as follows:

$$a_d \sim -\overline{U}\left(a_d^{-1/2}\right) \tag{6.15}$$

This formula provides an estimated a_d value, contingent on the spatial profile of the potential $U(\mathbf{r})$. Examples concerning specific defects will be elaborated in the next section.

For a precise calculation of a_d, one must solve (6.3) subject to the boundary condition (6.4). Example of such solutions can be found in Sect. 10.2 and Chap. 11.

Note 6.2.1 It is pivotal to note that the equation labeled as (6.15) is *universally applicable for estimating* a_d within this framework.

***

In summary, *a simple approach enables one to rather accurately estimate the eigenvalue of the auxiliary equation, thus obtaining the approximate value of the emergence temperature,* T_d. *In the following, we apply it to cases of crack and dislocation.*

6.3 The Dimensionless Free Energy and Ramification Equation

In this section, we derive the expression for the dimensionless free energy, F, in terms of the amplitude, ξ, of the overcritical solution. We further obtain the ramification equation from the latter expression for F and solve it.

***

Let us apply Gauss's theorem to the gradient term in the expression for effective free energy F (5.14). Taking into account that at infinitely remote boundary u turns into zero, we rewrite F as follows:

$$F = \frac{1}{2} \int u \left[-\Delta_1 u + [a + U(\mathbf{r})]\, u + \frac{1}{2} u^3 \right] dxdy \tag{6.16}$$

Substituting the solution (6.7) into (6.16) and eliminating the terms $-\Delta_1 \psi_d + U(\mathbf{r})\psi_d$ by utilizing the (6.3), the dimensionless free energy associated with a defect zone can be expressed as:

$$F = \Omega_{pz} \left[\frac{1}{2} j_2 (a - a_d)\, \xi^2 + \frac{1}{4} j_4 \xi^4 \right] \tag{6.17}$$

Here j_n represent the dimensionless integrals, defined as:

$$j_n = \frac{1}{\Omega_{pz}} \int \psi_d^n (\mathbf{r})\, dxdy \tag{6.18}$$

Here n is a natural number, and the variable Ω_{pz} denotes the dimensionless volume of the process zone.

Note 6.3.1 *The structure of the effective free energy* (6.17) *is universal. It is valid for a zone forming at any defect. Only the absolute values of the integrals j_2 and j_4, and the dimensionless zone size, Ω_{pz}, are defect-specific.*

Examples of the Zone Size, Ω_{pz}.

Example 6.3.1 Plane Defects: In the case of plane defects like domain and twin walls, the inhomogeneity is one-dimensional. The volume of the process zone is given by $\Omega_{pz} = AR_d$, where A represents the area of the domain (or twin) wall , and $R_d = a_d^{-1/2}$ is the characteristic zone dimension (6.12). Both of them are expressed in terms of the dimensionless coordinates.

Example 6.3.2 Cracks and Dislocations: For cracks and straight dislocations causing plane strain in the (x, y)-plane, the volume of the process zone is $\Omega_{pz} = R_d^2 L$, where L signifies the dimensionless sample length along the crack tip or dislocation.

Example 6.3.3 Three-Dimensional Defects: For defects that are three-dimensional, Ω_{pz} is R_d^3.

Note 6.3.2 *The volume of the zone*, Ω_{pz}, is contingent upon the configuration of the defect and its characteristic dimension, R_d, making it *defect-specific*.

6.3.1 Ramification Equation and Its Solution

Equation (6.17) defines the free energy of the process zone, where the daughter phase is embedded in the mother phase matrix. As described in Chap. 2, minimizing this zone free energy, i.e., setting $\partial F/\partial \xi = 0$, leads to the ramification equation:

$$j_2 (a - a_d) \xi + j_4 \xi^3 = 0 \tag{6.19}$$

It is worth mentioning that the same equation can be derived via (2.7), although the approach employed here seems more transparent to us.

The solution to the ramification equation, $\xi = \xi_d$, is as follows:

$$\xi_d = \begin{cases} 0, & a > a_d \\ \pm\sqrt{\frac{a_d j_2}{j_4}} \left(1 - \frac{a}{a_d}\right)^{1/2}, & a \le a_d \end{cases} \tag{6.20}$$

By substituting these solutions into the dimensionless factor of the free energy equation (6.17), we obtain:

$$F = \begin{cases} 0, & a > a_d \\ -\frac{j_2^2 (a-a_d)^2}{4 j_4} \Omega_{pz}, & a \le a_d \end{cases} \tag{6.21}$$

Property 6.3.1 *The result (6.21) shows that the formation of a transformational process zone always reduces the energy of the defect-zone complex—a topic we will explore in subsequent discussions.*

Note 6.3.3 Both *the form of the ramification equation* (6.19), *the qualitative behaviors of the amplitude* ξ_d (6.20) *and the effective free energy* (6.21) *are universal.*

Note 6.3.4 At this juncture, it is important to debunk a common misconception that the second-order transition is so subtle that one can safely ignore its effects on the defect's behavior. While it is true that near the bifurcation points a decreases, the zone's contribution, $\Delta F = j_2^2 (a - a_d)^2 \Omega_{pz}/4j_4$, can become substantial. Therefore, the transition's second-order nature doesn't necessarily render the zone effect negligible.

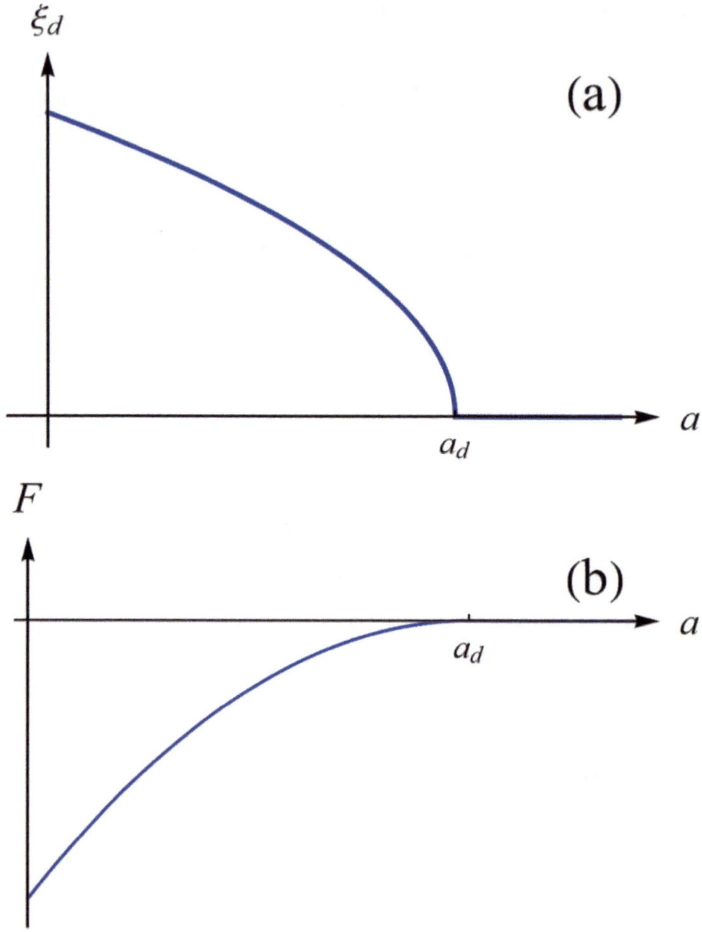

Fig. 6.2 The order parameter amplitude and free energy vs. a. Schematic view of the behavior of the order parameter amplitude, $\xi_d = \xi_d(a)$, described by (6.20) (**a**), and the free energy, $F_{eff} = F_{eff}(a)$, of the defect as described by (6.21). The former increases as $\xi_d \sim (a_d - a)^{1/2}$, while the latter decreases as $F_{eff} \sim -(a_d - a)^2$ as soon as a becomes smaller than a_d

Figure 6.2 graphically portrays the behavior of the order parameter amplitude (panel a) and the defect's effective energy (panel b). It's important to note that panel b of Fig. 6.2 displays the negative values achieved by $F(a)$ when $a < a_d$.

***

In summary, *our discussion began with the derivation and analysis of a condition governing the bifurcation of the equation of state, which leads to the formation of the process zone. This condition is derived from the so-called 'auxiliary equation,' which resembles Schrödinger's equation with an angular-dependent potential. The* *eigenvalue a_d of its ground state determines the bifurcation temperature T_d, while the corresponding eigenfunction $\psi_d(r)$ defines the spatial distribution of the order parameter near the bifurcation point. The auxiliary equation also provides a means to determine the size of the process zone.*

To maintain generality regarding the nature of the defect, this chapter proceeded under the assumption that the bifurcation indeed occurs, that the bifurcation equation has already been solved, and that its ground-state eigenvalue and eigenfunction are explicitly known. Based on this assumption, we developed an approximation for accurately estimating a_d for angular-dependent potentials. This approximation was then applied to calculate a_d for dislocations and cracks. Subsequently, we introduced and solved the ramification equation, which serves as the state equation for the process zone. We demonstrated that the daughter-phase zone is confined to a restricted region of the phase diagram, $0 < a \leq a_d$. Beyond this region, the daughter-phase zone vanishes. Finally, we computed the free energy of the zone, showing that it becomes negative below the bifurcation point. These findings are universally applicable to any defect that forms a process zone.

6.4 Daughter Phase Process Zone at a Propagating Defect

In this section, we focus on the properties of a process zone forming at a defect that propagates at a constant velocity, V. We reduce the auxiliary equation for the zone at the propagating defect to that for a stationary one. As a result, if the eigenvalue and eigenfunction for a motionless defect are already known, those for a propagating defect can be straightforwardly derived. We further explicitly express the order parameter distribution within the propagating process zone in terms of the motionless one.[2]

***

6.4.1 Equation of Motion and Its Bifurcation

Let us address a defect propagating along the x-axis with a constant velocity which is much smaller than the sound speed. It gives rise to the process zone propagating together with the defect described by the automodel dimensionless equation (5.10). Let us restate it here:

$$\Delta_1 u + 2\sqrt{a_d}\, v \frac{\partial u}{\partial x} - [a + U(\mathbf{r})]u - u^3 = 0 \tag{6.22}$$

[2] A. Boulbitch and P.E. Pumpyan 1990 [4]; A. Boulbitch and A.L. Korzhenevskii [2].

where Δ_1 and ∇_1 are the Laplace operator and gradient in terms of the dimensionless coordinates x and y:

$$\Delta_1 = \frac{\partial^2}{\partial x^2} + \frac{\partial^2}{\partial y^2}; \quad \nabla_1 = \left(\frac{\partial}{\partial x}, \frac{\partial}{\partial y}\right)$$

They should not be confused with the Laplace operator, Δ, and gradient, ∇, in terms of the physical coordinates X and Y. Equation (6.22) captures the joint propagation of the defect-process zone complex.

The auxiliary equation (2.3), can be formulated as follows:

$$\Delta_1 \Psi_n(\mathbf{r}) + 2\sqrt{a_\mathrm{d}}\,v \frac{\partial \Psi_n(\mathbf{r})}{\partial x} = [a_n(v) + U(\mathbf{r})]\Psi_n(\mathbf{r}) \tag{6.23}$$

with the standard boundary conditions:

$$\Psi_n(\infty) = 0; \quad \Psi_n(\mathbf{r}) < \infty \tag{6.24}$$

Here, $\Psi_n(\mathbf{r})$ represents the eigenfunctions, and $a_n(v)$ $(n = 1, 2, ...)$ are the corresponding eigenvalues of the dynamic auxiliary equation (6.23).

Note 6.4.1 One should not confuse a_n introduced in (6.3) as the eigenvalue of the problem describing the motionless defect with $a_n(v)$ introduced in (6.23) for the propagating defect.

Let us proceed by making the following substitution:

$$\Psi_n(\mathbf{r}) = \exp\left(-vx\sqrt{a_\mathrm{d}}\right) \times \psi_n(\mathbf{r}) \tag{6.25}$$

where $\psi_n(\mathbf{r})$ is a new dependent variable. This substitution leads to the equation imposed on $\psi_n(\mathbf{r})$:

$$\Delta_1 \psi_n(\mathbf{r}) - \left[a_n(v) + a_\mathrm{d} v^2 + U(\mathbf{r})\right] \psi_n(\mathbf{r}) = 0 \tag{6.26}$$

with the following boundary conditions:

$$\psi_n(\infty) = 0; \ \psi_n < \infty$$

Therefore, we have reduced equation (6.23) to the already examined problem (6.3, 6.4). This allows us to establish the relation between the eigenvalues of (6.26) and those of (6.3):

$$a_n = a_n(v) + a_\mathrm{d} v^2$$

In the dynamic scenario, the ground state eigenvalue $a_\mathrm{d}(v)$ is:

$$a_\mathrm{d}(v) = a_\mathrm{d}\left(1 - v^2\right) \tag{6.27}$$

With the corresponding eigenfunction $\Psi_d(\mathbf{r})$:

$$\Psi_d(\mathbf{r}) = \exp\left(-vx\sqrt{a_d}\right)\psi_d(\mathbf{r}) \tag{6.28}$$

We continue under the agreed assumption that the eigenvalue a_d and eigenfunction $\psi_d(\mathbf{r})$ of (6.3) are known in advance. With this in place, the dynamic problem is solved at this point.

Note 6.4.2 The relationships of $a_d(v)$ to v and of the dynamic function $\Psi_d(\mathbf{r})$ to the static one $\psi_d(\mathbf{r})$ are universally applicable.

One can see that the factor $\exp[-vx/2]$ is everywhere positive. Therefore, its presence cannot change the sign of the expression (6.28).

Note 6.4.3 Building upon this argument and Note 6.1.4, we formulate the following statement: The eigenfunction, $\Psi_d(\mathbf{r})$, has no nodes. In other words, it is everywhere positive.

6.4.2 Analysis of the Results

6.4.2.1 Characteristic Velocity

As elaborated in Sect. 6.1.3, the temperature of the process zone formation, $T_d(V)$, exceeds T_c whenever the eigenvalue $a_d(v)$ is positive. The inequality, $a_d(v) > 0$, stipulates that the daughter-phase zone, embedded within the mother-phase matrix, forms at the propagating defect before the bulk transition occurs: $T_d > T_c$. Observing that $a_d(v) = a_0\,[T_d(v) - T_c]$ according to (6.9). As soon as the whole crystal transforms into the daughter phase, the notion of a "daughter-phase process zone embedded into a mother-phase matrix" becomes senseless. Thus, the condition of vanishing of the process zone is $T_d(v) \leq T_c$, or equivalently,

$$a_d(v) \leq 0 \tag{6.29}$$

Regarding (6.27), one concludes that the zone can only exist within a specific range of the dimensionless velocity:

$$0 \leq v < 1 \tag{6.30}$$

As soon as $v \geq 1$, $T_d(v)$ falls below T_c, indicating that the zone ceases to exist beyond $v = 1$. The dimensionless velocity $v = 1$, thus, represents its critical value.

Switching to the original parameters with (5.11) and (5.6), one finds the physical critical velocity value:

$$V_d = \frac{2\sqrt{g\alpha_d}}{\kappa} \tag{6.31}$$

Let us also give a useful relation between v and V:

$$v = \frac{V}{V_d} \tag{6.32}$$

Note 6.4.4 The existence of the critical velocity, V_d (6.31), is a *universal qualitative characteristic of the process zone*. To elaborate further, this is its characteristic dynamic feature, applicable both in the context of the second-order process zone and the first-order one. It holds true for zones characterized by a single-component order parameter as well as those defined by a multicomponent order parameter. In contrast, the absolute value of V_d is contingent upon α_d, making it *specific to the type of defect involved*.

Note 6.4.5 To return to physical parameters, such as the Landau's coefficient α_d corresponding to the bifurcation point and the dimensional order parameter function $\eta(X, Y)$ and others, one applies the rescaling relations (5.1) and (5.5) yielding:

$$\alpha_d = a_d \frac{B^{2/(2-z)}}{g^{z/(2-z)}}; \quad \eta(X, Y, t) = \frac{B^{\frac{1}{2-z}}}{\beta^{1/2} g^{\frac{z}{2(2-z)}}} u(x, y, \tau)$$

Note 6.4.6 *We established the existence of a critical velocity, V_d, such that a process zone will only form if $0 \le V < V_d$; no zone will emerge for $V \ge V_d$. We also reveal that the temperature, $T_d = T_d(V)$, at which the zone emerges, decreases as the velocity increases, ultimately converging with T_c at $V = V_d$.*

6.4.2.2 Bifurcation Line

Solution (6.27) represents the bifurcation condition, while the inequality (6.29) points out the part of the phase diagram occupied by the daughter-phase process zone.

The parameter a_d appearing in (6.27) is the eigenvalue of (6.3). In other words, it is the bifurcation point of the stationary problem. Parameters a/a_d and v are the dimensionless control parameters of the (6.22). The division of the plane, defined by its control parameters, into domains with different types of solutions is referred to as the *parametric portrait* of the bifurcation. In this book, it is sometimes referred to as the *phase diagram*, as both terms play comparable roles.

The parametric portrait of the bifurcation studied above is shown in Fig. 6.3 in the space of the dimensionless parameters a/a_d and v. The expression (6.27) delineates a line in Fig. 6.3 where soft bifurcation occurs. Region III, highlighted by hatching, indicates where the process zone occurs, while the bulk daughter phase occupies the hatched Region II.

Building on (6.9) from (6.27), we can derive the explicit expression for the velocity-dependent bifurcation temperature, $T_d = T_d(V)$:

$$T_d(V) = T_d - (T_d - T_c)\left(\frac{V}{V_d}\right)^2 \tag{6.33}$$

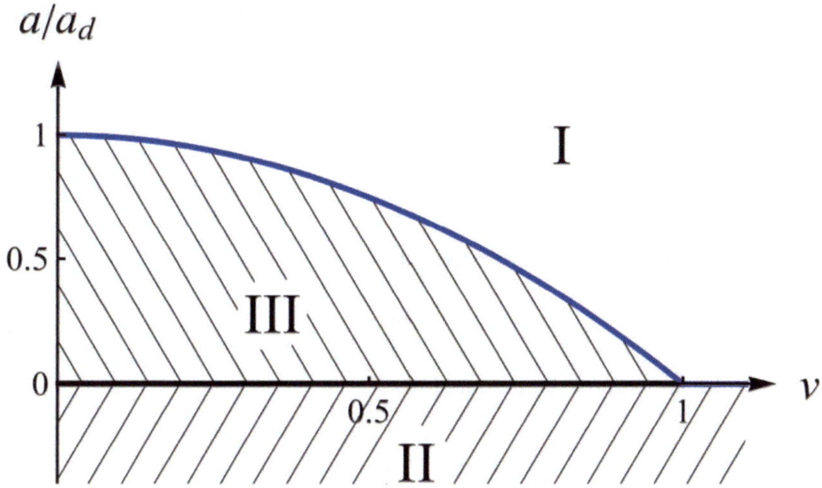

Fig. 6.3 Phase diagram showing the domain occupied by the daughter-phase zone within the mother-phase matrix. The phase diagram is described by expression (6.27). The bulk mother phase with the undressed defect exists within Region I. The bulk daughter phase occupies Region II. The region of existence of the daughter-phase process zone, embedded in the matrix of the bulk mother phase, is indicated as Region III and highlighted by hatching

Here, T_d refers to the bifurcation temperature at a stationary defect (6.10), while $T_d(V)$ is that at the propagating one: $T_d(0) = T_d$. $T_d(V)$ diminishes to T_c as V approaches V_d. Beyond $V = V_d$, the entire crystal undergoes a transition into the daughter phase at $T = T_c$, rendering the notion of a "zone" irrelevant.

Note 6.4.7 The results (6.27) and (6.33) are universally applicable.

***

In conclusion, *the order parameter distribution within the propagating zone remains knot-free. Additionally, the bifurcation point,* $a_d(v)$, *is velocity-dependent for a moving defect. When* v *approaches* v_d, $a_d(v)$ *tends to zero, indicating that beyond the critical velocity, the zone disappears.*

6.5 Propagating Defect: Structure of the Solution

6.5.1 Amplitude

In the dynamic regime, the asymptotically exact solution for the order parameter assumes the conventional form as delineated in Chap. 2:

$$u(x - vt, y) \approx \xi\Psi_{\mathrm{d}}(x - vt, y) + O(\xi^3) \tag{6.34}$$

To calculate the amplitude, ξ_{d}, in the dynamic scenario, we apply the ramification equation (2.7). The operators $\widehat{L}(a)$ and $\hat{N}(u)$ are defined as follows:

$$\widehat{L}(a) = \Delta_1 + 2\sqrt{a_{\mathrm{d}}}v\frac{\partial}{\partial x} - [a + U(\mathbf{r})]\,; \quad \hat{N}(u) = u^3$$

The spectral equation has the following form:

$$\widehat{L}[a_{\mathrm{d}}(v)]\Psi_{\mathrm{d}} = 0$$

Let us build an operator denoted as $\widehat{L}_0$:

$$\widehat{L}_0 = \Delta_1 + 2\sqrt{a_{\mathrm{d}}}v\frac{\partial}{\partial x} - U(\mathbf{r})$$

which is a part of $\widehat{L}(a)$. Using $\widehat{L}_0$, the spectral equation, is expressed as:

$$\widehat{L}_0\Psi_{\mathrm{d}} = a_{\mathrm{d}}(v)\Psi_{\mathrm{d}}$$

From this equation and considering the relationship between $\widehat{L}(a)$ and $\widehat{L}_0$, given by $\widehat{L}(a) = \widehat{L}_0 - a$, it follows that:

$$\widehat{L}(a)\Psi_{\mathrm{d}} = \widehat{L}_0\Psi_{\mathrm{d}} - a\Psi_{\mathrm{d}} \equiv [a_{\mathrm{d}}(v) - a]\,\Psi_{\mathrm{d}}$$

Furthermore, for the case being analyzed, when $\hat{N}(\xi_{\mathrm{d}}\Psi_{\mathrm{d}}) = \xi_{\mathrm{d}}^3\Psi_{\mathrm{d}}^3$. After substituting these findings into the condition (2.7) and integration, one obtains the ramification equation in the following form:

$$\left[a - a_{\mathrm{d}}\left(1 - v^2\right)\right] J_2\xi_{\mathrm{d}} + J_4\xi_{\mathrm{d}}^3 = 0 \tag{6.35}$$

This leads to the solution:

$$\xi_{\mathrm{d}} = \pm\sqrt{\frac{a_{\mathrm{d}}J_2}{J_4}}\left(1 - \frac{a}{a_{\mathrm{d}}} - v^2\right)^{1/2} \times$$
$$\times \begin{cases} 0, & a/a_{\mathrm{d}} > 1 - v^2 \vee v > 1 \\ 1, & a/a_{\mathrm{d}} \le 1 - v^2 \wedge v \le 1 \end{cases} \tag{6.36}$$

where $\vee$ is the logical "or,", $\wedge$ is "and," while J_n are the integrals:

$$J_n = \frac{1}{\Omega_{\mathrm{pz}}}\int \Psi_{\mathrm{d}}^n(\mathbf{r})\,d\Omega \tag{6.37}$$

In principle, the natural number n may take the following values: $n = 2, 3, 4, 5$ and 6. However, in this book, we focus only on J_2, J_4 and J_6.

It's worth distinguishing between j_n (6.18) and J_n (6.37). While j_n are constant values, $J_n = J_n(v)$ are velocity-dependent dimensionless integrals that reduce to j_n when the defect is motionless: $J_n(0) = j_n$. Consequently, the amplitude ξ_d has a more intricate dependency on the velocity v than the explicit term $(1 - a/a_\mathrm{d} - v^2)^{1/2}$, as $(J_2/J_4)^{1/2}$ implicitly depends on the velocity. This dependency, however, is specific to the type of defect under consideration. We provide an example of such specificity in Sect. 10.3.

Note 6.5.1 In contrast, *the structure of the solution* (6.36) *is universal.*

Upon examining (6.36), it becomes evident that the order parameter amplitude ξ_d diminishes with increasing velocity and vanishes when $v = 1$. This reaffirms the conclusions drawn in Sect. 6.4.2.1, albeit from a different angle.

6.5.2 *Asymptote and Zone Asymmetry*

The asymptotic behavior of the solution $\Psi_\mathrm{d}(\mathbf{r})$ for large r one derives combining (6.28) with the asymptote (6.13) of the stationary solution. One finds the following:

$$\Psi_\mathrm{d}(\mathbf{r}) \sim \exp\left(-vx\sqrt{a_\mathrm{d}} - \frac{r}{R_\mathrm{d}}\right) \tag{6.38}$$

Due to the presence of the first term in the exponent, the asymptote becomes anisotropic. Specifically, in the direction of defect propagation, considering (6.12), we can derive the zone dimension R_f ahead of the defect as:

$$R_\mathrm{f} = \frac{1}{\sqrt{a_\mathrm{d}}(1+v)} \tag{6.39}$$

Conversely, behind the defect where $x < 0$, the dimension of the zone R_b is:

$$R_\mathrm{b} = \frac{1}{\sqrt{a_\mathrm{d}}(1-v)} \tag{6.40}$$

Here the subscripts "f" and "b" stand for "front" and "back" respectively. When $v = 1$, the backward dimension of the zone becomes infinite, and the amplitude ξ_d becomes zero, while $R_\mathrm{f} = (2\sqrt{a_\mathrm{d}})^{-1}$ is finite. Figure 6.4 displays the curves (6.39) and (6.40) as the functions of the dimensionless velocity v.

Note 6.5.2 It is important to note that the relations (6.39) and (6.40) are universal.

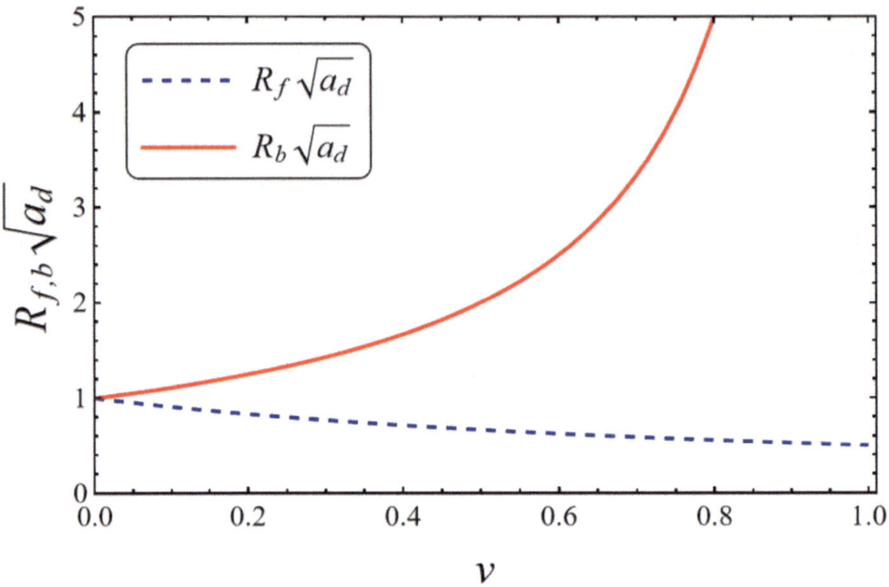

Fig. 6.4 Zone dimensions R_{f} and R_{b} versus the ratio V/V_{d}

6.5.3 What Occurs with the Solution at v ≥ 1?

We have observed that as v approaches 1 from the left ($v \to 1-0$), the distance between the defect and the trailing edge of the zone grows infinitely. Examining the asymptote (6.38), we find that at $v = 1$, the exponent of (6.38) becomes zero, along the direction opposite to that of the velocity v. If v surpasses 1, the exponent becomes positive in this direction. This implies that in this direction, the function $\Psi_{\mathrm{d}}(\mathbf{r})$ diverges. In both scenarios, the boundary condition (6.24) is breached. In simpler terms, as soon as $v \geq 1$, (6.23) and (6.24) lacks a solution.

Note 6.5.3 It is worthwhile to highlight that in Sect. 6.4.2.1, we elucidated why a process zone cannot materialize at $v \geq 1$. This is attributed to the fact that the bifurcation temperature $T_{\mathrm{d}} = T_{\mathrm{d}}(V)$ diminishes to the Curie temperature T_{c}, rendering no room for a process zone in the phase diagram. Additionally, within the present section, we also ascertain that the solution to (6.23) and (6.24), which delineates the characteristics of the process zone, becomes non-existent.

6.5.4 Free Energy and Friction Force of the Propagating Second-Order Zone

6.5.4.1 Free Energy

In this section, we examine the free energy associated with a zone that propagates along with a defect. To begin with, let us reformulate equation (6.23) leveraging (6.27) in a form convenient for the next step:

$$-\Delta_1 \Psi_{\mathrm{d}} + U(\mathbf{r})\Psi_{\mathrm{d}} = -\left(a_{\mathrm{d}} - v^2\right)\Psi_{\mathrm{d}} + 2\sqrt{a_{\mathrm{d}}}\left(\mathbf{V}\cdot\nabla_1\right)\Psi_{\mathrm{d}} \tag{6.41}$$

To find the dimensionless free energy F factor (5.14), we insert the asymptotically-exact solution $u = \xi_{\mathrm{d}}\Psi_{\mathrm{d}}(\mathbf{r})$ (6.34) into the general expression for free energy (6.16). Utilizing the identity (6.41), we get:

$$F = \sqrt{a_{\mathrm{d}}}\xi_{\mathrm{d}}^2 \int \Psi_{\mathrm{d}}\left(\mathbf{V}\cdot\nabla_1\right)\Psi_{\mathrm{d}} d\Omega + \frac{1}{2}\left[a - a_{\mathrm{d}}\left(1 - v^2\right)\right]\xi_{\mathrm{d}}^2 \times$$
$$\times \int \Psi_{\mathrm{d}}^2 d\Omega + \frac{1}{4}\xi_{\mathrm{d}}^4 \int \Psi_{\mathrm{d}}^4 d\Omega$$

Note that in the first term, the expression under the integral represents a total derivative:

$$\Psi_{\mathrm{d}}\left(\mathbf{V}\cdot\nabla_1\right)\Psi_{\mathrm{d}} \equiv \frac{1}{2}\left(\mathbf{V}\cdot\nabla_1\right)\Psi_{\mathrm{d}}^2.$$

According to Gauss's theorem, this integral evaluates to the value of $(\mathbf{V}\cdot\mathbf{n})\,\Psi_{\mathrm{d}}^2/2$ at the infinitely remote surface. In this case, $\mathbf{n}$ is its normal unit vector. Because of the boundary conditions at infinity (6.24), the latter term turns to zero. We are then left with the following expression for the dimensionless free energy factor of the propagating zone:

$$F = \Omega_{\mathrm{pz}}\left[\frac{1}{2}\left[a - a_{\mathrm{d}}\left(1 - v^2\right)\right]\xi_{\mathrm{d}}^2 J_2 + \frac{1}{4}\xi_{\mathrm{d}}^4 J_4\right] \tag{6.42}$$

Inserting here the solution from (6.36), we obtain the universal expression:

$$F = -\frac{a_{\mathrm{d}}^2 \Omega_{\mathrm{pz}}}{4} \frac{J_2^2}{J_4} \left(1 - \frac{a}{a_{\mathrm{d}}} - v^2\right)^2 \times$$

$$\times \begin{cases} 0, \; a/a_{\mathrm{d}} > 1 - v^2 \vee v > 1 \\ 1, \; a/a_{\mathrm{d}} \leq 1 - v^2 \wedge v \leq 1 \end{cases} \tag{6.43}$$

Note 6.5.4 The expressions for free energy (6.42, 6.43) are universally applicable to any defect that gives rise to a process zone.

6.5.4.2 Friction Force

The formation of the propagating process zone proceeds as follows. Initially, the order parameter is zero at a considerable distance ahead of the zone in the direction of propagation. As the zone nears, the order parameter starts to rise from zero, achieving values as dictated by equation (6.36). Once the zone has moved beyond a specific point, the order parameter reverts from its value attained within the zone—back to zero. This evolution of the order parameter is coupled with energy dissipation, which in turn gives rise to a friction force, f exerted on the defect. It is directed opposite to the defect's velocity.

By employing (3.58), we arrive at the following expression for the friction force per unit length in the z direction:

$$f = 2v\sqrt{\frac{a_{\mathrm{d}}}{\gamma}} B^{\frac{2}{2-z}} g^{\frac{2-3z}{4-2z}} \xi_{\mathrm{d}}^2 I_2 \tag{6.44}$$

Here, the dimensionless integral I_2 is defined as:

$$I_2 = \int \left[\frac{\partial \Psi_{\mathrm{d}}(\mathbf{r})}{\partial x}\right]^2 dxdy \tag{6.45}$$

It is important to emphasize that the friction force's dependence on the defect velocity, v, as given by (6.44), is both defect-specific and nonlinear. This is because both ξ_{d} and Ψ_{d} intricately depend on v, as evidenced by (6.25) and (6.36). Regardless of the specific nature of the defect, one can assert that f is zero when $v = 0$. As the velocity increases, the friction force initially grows but later decreases. It vanishes at $v = 1$ because ξ_{d} becomes zero according to (6.36). For $v > 1$, the friction force is again zero as no zone is present.

Note 6.5.5 Here, a clarification is in order. The statement regarding the vanishing of the amplitude for $v \geq 1$ applies strictly to the steady-state motion of the defect-zone complex. If the defect propagates with a variable or periodic velocity, its behavior interacts with the finite decay time of the zone, potentially giving rise to complex scenarios that fall outside the steady-state framework. These situations warrant a dedicated study. One such example the reader finds in Sect. 29.3.

***

In summary, *the order parameter distribution of a moving zone appears stretched in its trailing part and compressed in its leading part. As the defect velocity approaches the critical value, the length of the rear part of the zone extends to infinity, while the amplitude of the order parameter diminishes to zero. Unlike in the motionless case, the free energy of the propagating defect-zone complex cannot provide a correct ramification equation. Instead, it must be derived from the general equation (2.7). In this case, the bifurcation point* $a_d(v)$ *and the coefficients* J_2 *and* J_4 *depend on the velocity.*

Furthermore, energy dissipation within the zone generates a friction force acting on the defect. This force vanishes as the velocity approaches zero, exhibiting behavior analogous to viscous friction at small velocities. The described friction force is a universal and fundamental property of any propagating process zone. It applies to both single-component and multi-component order parameters and is relevant for zones of second- and first-order transitions. Importantly, it displays a nonlinear dependence on velocity and disappears entirely during steady motion when $V \geq V_d$.

References

1. A.A. Bulbich, J. Mater. Sci. **27**, 1070 (1992)
2. A. Boulbitch, A.L. Korzhenevskii, Eur. Phys. J. B **89**, 261 (2016)
3. L.D. Landau, E.M. Lifshitz, *Quantum Mechanics*, 3 edn. (Butterworth-Heinemann, Burlington, MA, USA, 2003). Sect. 97
4. A. Boulbitch, P.E. Pumpyan, Sov. Phys. Crystallogr. **35**(2), 156 (1990)

Chapter 7
Second-Order Mother-Phase Process Zones

Abstract The mother-phase zone embedded in the bulk daughter phase exhibits distinct characteristics from those of the daughter-phase zone, primarily due to the influence of a nonzero order parameter field (Boulbitch and Korzhenevskii, first published here).

7.1 Equations Describing the Mother-Phase Zone

In this section, we derive equations that describe the second-order mother-phase zone embedded in the daughter-phase matrix. This is achieved without specifying the form of the potential, $U(r)$, thereby yielding a universal result. We demonstrate that the presence of the bulk daughter phase introduces a specific term in the equation of state, which acts as a field conjugated to the order parameter. Consequently, the process zone in the daughter phase never vanishes. Finally, we determine the asymptotic solution that describes the process zone at large distances from the defect.

***

7.1.1 The Second-Order Mother-Phase Zone

7.1.1.1 Equation of State

We concentrate on the case where the entire crystal resides in the daughter phase.

In the homogeneous daughter phase, the order parameter distribution represents a plane $u(x, y) = u_0$, where u_0 is the equilibrium order parameter value in the daughter phase. Schematically, the mother-phase zone at the crack tip appears as a depression in the $u(x, y)$ plane, not always reaching the $u = 0$ plane, thereby constituting a mother-phase zone (Fig. 7.1a). However, in certain situations, the depression's depth reaches the $u = 0$ plane, forming a true mother-phase zone, primarily in domains where Landau's potential is 'soft' (i.e., a is small).

© The Author(s), under exclusive license to Springer Nature Switzerland AG 2026
A. Boulbitch and A. Korzhenevskii, *Stress-Induced Local Phase Transformations Near Cracks and Dislocations*, Springer Series in Materials Science 356,
https://doi.org/10.1007/978-3-032-00945-6_7

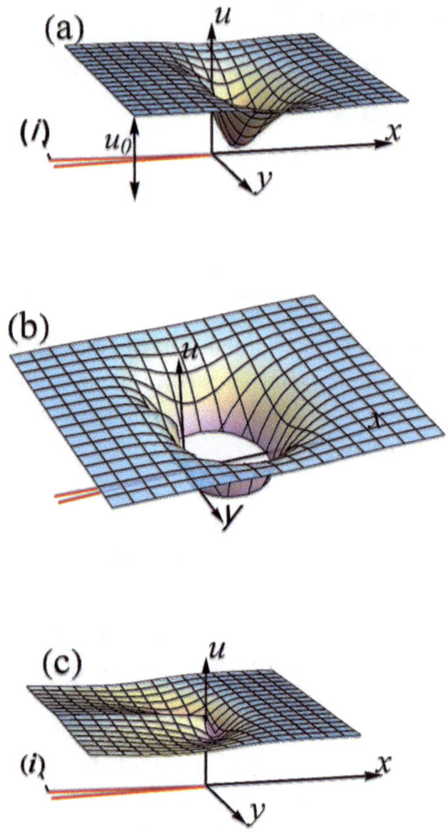

Fig. 7.1 Schematic View of the Mother-Phase Process Zone at the Crack Tip, Embedded in the Daughter-Phase Matrix. The red lines, indicated by (i), mark the crack tip. **a** Depicts the case of a depression in the plane $u(x, y) = u_0$, which is not deep enough to reach the plane $u(x, y) = 0$. **b** Shows the developed mother-phase zone touching the plane $u = 0$. **c** Illustrates the mother-phase zone at a propagating crack

As the defect propagates, the depression extends in the direction opposite to that of motion (Fig. 7.1c).

Figure 7.1 gives an artistic image of the mother-phase zone. The genuine, simulated results for specific defects will be discussed in subsequent chapters.

The bulk daughter phase exists at $a < 0$, a condition we assume throughout this section. The dimensionless order parameter $u(\mathbf{r})$, which may eventually describe the formation of the mother-phase zone, satisfies the following equation:

$$\begin{cases} \Delta u = [a + U(\mathbf{r})]u + u^3 \\ u(\infty) = (-a)^{1/2}; \quad u(\mathbf{r}) < \infty \end{cases} \tag{7.1}$$

varies spatially near the defect due to the effect of the defect field, $U(\mathbf{r})$. Expressing the solution as:

$$u(\mathbf{r}) = \sqrt{-a} + s(\mathbf{r}) \tag{7.2}$$

where $s(\mathbf{r})$ represents the deviation of the order parameter's value from the homogeneous solution $u_0 = \sqrt{-a}$, we formulate the state equation in terms of $s = s(\mathbf{r})$:

$$\begin{cases} \Delta s - [-2a + U(\mathbf{r})]\, s - 3\sqrt{-a}s^2 - s^3 = U(\mathbf{r})\sqrt{-a} \\ s(\infty) = 0; \quad s(\mathbf{r}) < \infty \end{cases} \tag{7.3}$$

highlighting the role of the free term $U(\mathbf{r})(-a)^{1/2}$ analogous to an external field conjugated to the order parameter (cf. Sect. 3.3.4).

This property represents the *non-vanishing nature of the mother-phase zone field* $s(\mathbf{r})$. It has a twofold character. On the one hand, since the potential, $U(\mathbf{r})$, is long-range, the perturbation, $s(\mathbf{r})$, engendered by this potential spans across *the whole crystal*. On the other hand, the mother-phase zone embedded into the daughter-phase matrix is situated at $a < 0$. It manifests itself as a depression in the $u(x, y)$ field (as shown in Fig. 7.1a). *There is no place in the phase diagram where this depression completely vanishes*. In contrast, the daughter-phase zone exhibits no such property.

Due to its complexity, the (7.3) admits no analytical solution akin to that formulated for the daughter-phase zone, necessitating numerical solutions for specific potentials, which we will explore in detail in the corresponding chapters.

***

To summarize, *unlike the daughter-phase zone, the mother-phase zone extends across the entire region of the phase diagram where the bulk daughter phase exists. Near the bulk phase transition line, its influence becomes more pronounced, making it more easily detectable and significantly impacting the system's physical properties.*

This persistent presence of the mother-phase zone has important physical implications. Specifically, it contributes to the free energy associated with defects and exerts a frictional force on these defects. These effects remain nonzero throughout the entire domain of the daughter-phase within the phase diagram.

Chapter 8
Universal Characteristics of First-Order Process Zones

Abstract In this chapter, we focus on the properties of process zones in cases where the mother-to-daughter phase transition is of the first order, commonly referred to as first-order zones. The bifurcation that initiates the formation of these first-order zones is invariably hard. We begin by introducing a concept we call "quasi-second-order process zones." In these zones, bifurcation theory remains applicable even in the face of hard transformation in those regions of the parametric space where the amplitude of the order parameter is small. Without specifying the generating potential, we offer a comprehensive solution that describes these quasi-second-order process zones and exposes their universal properties. Moving on from the small amplitude scenario, we turn our attention to a more general case. Here, we derive estimates for the zone's emergence temperature, its size, and its interface's thickness. The chapter concludes with the proof of a theorem stating that the size of the zone grows infinitely as the temperature approaches the bulk binodal (Boulbitch and Toledano 1999, Boulbitch and Korzhenevskii 2020).

In this chapter we follow the papers [1–3] and [5].

8.1 Rescaling of the First-Order Zone Equations

In this section, we examine the process zones of the first order and identify the condition under which the bifurcation theory approach can be applied to these zones. We then rescale the equation describing such zones, solve it, and use the solution to construct the process zone phase diagram against the backdrop of the bulk phase transformation phase diagram.

***

In Sect. 3.4.1, we derived the equations of motion for first-order phase transitions. After excluding the elastic variables, as discussed in Sect. 4.1, the system's dynamics is entirely described by a single equation (4.10) for the order parameter. For simplicity, we restate this equation here:

© The Author(s), under exclusive license to Springer Nature Switzerland AG 2026

A. Boulbitch and A. Korzhenevskii, *Stress-Induced Local Phase Transformations Near Cracks and Dislocations*, Springer Series in Materials Science 356,
https://doi.org/10.1007/978-3-032-00945-6_8

$$\kappa\frac{\partial\eta}{\partial t}=g\left(\frac{\partial^2\eta}{\partial X^2}+\frac{\partial^2\eta}{\partial Y^2}\right)-[\alpha+U_0(\mathbf{R})]\,\eta+|\beta|\eta^3-\gamma\eta^5=0 \tag{8.1}$$

Here, $U_0(\mathbf{r})$ is defined as in (5.3). We apply the substitution outlined in (5.1), although we choose different values for the coefficient η_0:

$$\mathbb{R}_0=\left(\frac{g}{B}\right)^{\frac{1}{2-z}};\quad \mathbb{T}_0=\frac{\kappa g^{\frac{z}{2-z}}}{2\sqrt{a_\mathrm{d}}B^{\frac{2}{2-z}}};\quad \eta_0=\frac{1}{\gamma^{1/4}}B^{\frac{1}{4-2z}}g^{\frac{z}{4(z-2)}} \tag{8.2}$$

This rescaling preserves the function $f(\varphi)$ unchanged.

Note 8.1.1 It should be noted that the coefficients $\mathbb{R}_0$ and $\mathbb{T}_0$ in the rescaling relations (8.2) are identical to those in (5.5). Therefore, the expressions for the natural scales in the first-order case coincide with those in (5.5) for the second-order one. For the same reason, the parameters a and v in the first-order case are defined by the same expressions as in (5.6) and (5.12). In contrast, the order parameter scale, η_0 in (8.2) differs from that in (5.5) for the reason that in the second-order and first-order cases, the order parameter depends on different sets of parameters: g, α and β in the former and g, α, β, and γ in the latter case.

The application of replacements (5.1), (8.2) yields the rescaled value for the coefficient in front of u and u^3:

$$a=\frac{g^{z/(2-z)}}{B^{2/(2-z)}}\alpha;\qquad b=-\frac{\beta}{\sqrt{\gamma}}\frac{g^{z/2(2-z)}}{B^{1/(2-z)}} \tag{8.3}$$

In an analogous manner, one introduces a dimensionless parameter b_1 related to β_1 as introduced in (3.20):

$$b_1=-\frac{\beta_1}{\sqrt{\gamma}}\frac{g^{z/2(2-z)}}{B^{1/(2-z)}} \tag{8.4}$$

We then arrive at the rescaled equation of motion:

$$2\sqrt{a_\mathrm{d}}\frac{\partial u}{\partial\tau}=\frac{\partial^2 u}{\partial x^2}+\frac{\partial^2 u}{\partial y^2}-[a+U(\mathbf{r},\tau)]\,u+bu^3-u^5 \tag{8.5}$$

The boundary conditions:

$$u(\infty)=0;\quad u<\infty$$

supplement this equation. Here, $U(\mathbf{r},\tau)$ is defined by (5.8).

Shifting to the co-moving system, where $x'=x-v\tau$, $y'=y$ and $\mathbf{r}'=(x',y',0)$ we obtain the automodel equation describing the process zone accompanying the propagating defect:

$$\left(\frac{\partial^2 u}{\partial x'^2}+\frac{\partial^2 u}{\partial y'^2}\right)+2\sqrt{a_\mathrm{d}}v\frac{\partial u}{\partial x'}-\left[a+U(\mathbf{r}')\right]u+bu^3-u^5=0 \tag{8.6}$$

Note 8.1.2 From this point on, we omit the prime signs and use x, y, $\mathbf{r}$ in place of x', y', and $\mathbf{r}'$.

Equation (8.6) is formulated in terms of the dimensionless control parameters a, b, and the rescaled dimensionless velocity v. Setting v to zero yields the equation of state:

$$\frac{\partial^2 u}{\partial x^2} + \frac{\partial^2 u}{\partial y^2} - [a + U(\mathbf{r})]\, u + bu^3 - u^5 = 0 \tag{8.7}$$

Note 8.1.3 Let us emphasize that, unlike the equation of state for the homogeneous crystal (3.25), which depends on the parameter b_1, the (8.6) and (8.7) characterizing the process zone are influenced by the parameter b, as elaborated in (8.3). This difference arises from the elimination of elastic variables in the homogeneous crystal scenario, leading to (3.25), as opposed to the different procedure of their elimination in the inhomogeneous case resulting in the formulations of (8.6) or (8.7).

We further introduce the rescaled Gordon velocities for the plane and curved phase fronts, v_{G} and $v_{\mathrm{G}}^{(\pm)}$, which are related to (3.69) and (3.74) respectively:

$$v_{\mathrm{G}} = \pm\frac{1}{\sqrt{6}}\frac{8a - 3b^2 + 3b\sqrt{b^2 - 4a}}{\left|b^2 - 2a - b\sqrt{b^2 - 4a}\right|^{1/2}} \tag{8.8}$$

$$v_{\mathrm{G}}^{(\pm)} = v_{\mathrm{G}} \pm \frac{1}{2\sqrt{a_{\mathrm{d}}\rho}} \tag{8.9}$$

By incorporating the variables from (5.1), (8.2), (5.6) and (8.3) into the free energy per unit length in the z direction (4.14), we derive the following expression:

$$\mathcal{F} = \frac{1}{\sqrt{\gamma}} B^{\frac{1}{2-z}} g^{\frac{4-3z}{4-2z}} F \tag{8.10}$$

where, as in the second-order case, F is the dimensionless integral:

$$F = \int \left\{\frac{1}{2}\left(\nabla_1 u\right)^2 + \frac{1}{2}\left[a + U(\mathbf{r})\right] u^2 - \frac{b}{4}u^4 + \frac{1}{6}u^6\right\} dxdy \tag{8.11}$$

The transformation of the dissipation function from (3.59) results in:

$$\mathcal{D} = \frac{2a_{\mathrm{d}}v^2}{\sqrt{\gamma}\kappa} B^{\frac{3}{2-z}} g^{\frac{4-5z}{4-2z}} \int \left(\frac{\partial u}{\partial x}\right)^2 dxdy \tag{8.12}$$

Similarly, the expression for the friction force per unit length becomes:

$$f = 2v\sqrt{\frac{a_{\mathrm{d}}}{\gamma}} B^{\frac{2}{2-z}} g^{\frac{2-3z}{4-2z}} \int \left(\frac{\partial u}{\partial x}\right)^2 dxdy \tag{8.13}$$

Note 8.1.4 It is important to note that the integrals in the expressions (8.10), (8.12), and (8.13) are dimensionless, with the dimensionality encapsulated in their pre-factors.

8.1.1 Bifurcation Condition

In our standard methodology, we identify the bifurcation point through an analysis of an auxiliary equation derived from the equation of state (8.7). The form of the auxiliary equation is identical to (6.3) derived for the second-order process zone. Analogously, in the case of the defect propagating with a constant velocity, the auxiliary equation coincides with (6.23). Correspondingly, all the results from the auxiliary equations for the second-order zone persist in the first-order order case. Let us list them below for easy reference.

The first discrete eigenvalue of the auxiliary equation (6.3) for the static case, a_1, denoted by a_d, marks the bifurcation point. Below this bifurcation threshold ($a \le a_\mathrm{d}$), an overcritical solution emerges.

Remind that in the first-order case, we also base our argumentation on the assumption that the eigenvalue, $a_1 = a_\mathrm{d}$, as well as the eigenfunction, $\psi_1(\mathbf{r}_1) = \psi_\mathrm{d}(\mathbf{r}_1)$ are explicitly known.

The ground state eigenfunction in the dynamic case $\Psi_\mathrm{d}(\mathbf{r})$ is related to the static case related eigenfunction $\psi_n(\mathbf{r})$ as follows.

$$\Psi_\mathrm{d}(\mathbf{r}) = \psi_\mathrm{d}(\mathbf{r}) \exp\left(-vx\sqrt{a_\mathrm{d}}\right) \tag{8.14}$$

while its eigenvalue $a_\mathrm{d}(v)$ is expressed in terms of the static eigenvalue and velocity as:[1]

$$a_\mathrm{d}(v) = a_\mathrm{d}\left(1 - v^2\right) \tag{8.15}$$

This implies that an overcritical solution only exists for $a_\mathrm{d}(v) > 0$, valid when:

$$0 \le v \le 1 \tag{8.16}$$

Beyond this range ($v > 1$), the (8.7) admits no nontrivial solutions. Thus, regarding the rescaled parameters, the critical velocity value is $v = 1$.

Besides the expression for V_d (6.31), the relation $v = V/V_\mathrm{d}$ (6.32), the expression for $T_\mathrm{d}(v)$ (6.33) and for R_f and R_b (6.39), (6.40) obtained in the case of the second-order zone, persist in the first-order case.

According to the bifurcation theory (Chap. 2) the asymptotically exact overctitical solution is represented as:

$$u = \xi_\mathrm{d}\Psi_\mathrm{d}(\mathbf{r}_1) \tag{8.17}$$

where ξ_d is the dimensionless amplitude. It will be determined during the next step.

---

[1] The static eigenvalue a_d in (8.14) should not be confused with $a_\mathrm{d}(v)$ (8.15), the eigenvalue of the automodel equation.

Note 8.1.5 One returns to the original notation of Landau's theory by using the rescaling relations for the first-order case (5.1), (8.2). For example, for $\alpha_{\rm d}$ and $\eta(X, Y, t)$ one finds:

$$\alpha_{\rm d} = \frac{B^{\frac{2}{2-z}}}{g^{\frac{z}{2-z}}} a_{\rm d}; \quad \eta(X, Y, t) = \frac{B^{\frac{1}{2(2-z)}}}{\gamma^{1/4} g^{\frac{z}{4(2-z)}}} u(x, y, \tau) \tag{8.18}$$

Let us point out that, in general, these relations differ from those formulated for the second-order zone (see Note 6.4.5).

8.1.2 Emergence of a Quasi-Second-Order Process Zone

Bifurcation theory offers an asymptotically precise framework for characterizing the second-order process zone. The expansion of the nonlinear equation in terms of such a series lies at the core of the method we have employed following [4], as briefly described in Sect. 6.3.1. In the case of second-order zones, there exists a region in the phase diagram where $\xi_{\rm d}$ is zero or infinitesimally small, making the series expansion applicable. In contrast, for first-order zones, $\xi_{\rm d}$ exhibits a finite discontinuity at the bifurcation point, making the series expansion questionable. Given these complexities, computational analysis becomes indispensable for investigating first-order zones, as discussed in Chap. 12.

Within the phase diagram, however, a domain exists where the phase transition is first-order, yet the order parameter's discontinuity at the transition point remains small. Since $\xi_{\rm d}$ is dimensionless, the latter statement has the following form:

$$\xi_{\rm d} \ll 1 \tag{8.19}$$

This domain is located near the local tricritical point,[2] $a = a_{\rm d}$, $b = 0$.

One must combine the inequality (8.19) with those formulated later on, such as (8.16), (8.32) and $a < a_{\rm d} < a_{\rm up}^{(loc)}$ (where $a_{\rm up}^{(loc)}$ is defined by (8.29)). By solving these inequalities together, one obtains an intricate result. We refrain from presenting it here due to its complexity.

However, it is worth noting that the results of our simulations reported in Chaps. 11 and 22 demonstrate that the numerical solution qualitatively coincides with the one obtained in this section within the broad phase diagram domain. The latter suggests that satisfying a softer condition:

$$\xi_{\rm d} < 1 \tag{8.20}$$

enables one to apply the methods of the bifurcation theory. In this case, we label such a process zone as a *quasi-second-order* one. With these considerations in mind, we expound below the theory of the quasi-second-order zone.

---

[2] It should not be confused with the bulk tricritical point $a = b_1 = 0$.

8.1.2.1 Equations for the Quasi-Second-Order Zone

Ramification Equation. Following precisely the same argumentation as those formulated in Sect. 6.5.4, in the quasi-second-order case, one derives the ramification equation:

$$J_2\left[a-a_{\mathrm{d}}\left(1-v^2\right)\right]\xi_{\mathrm{d}}-J_4 b\xi_{\mathrm{d}}^3+J_6\xi_{\mathrm{d}}^5=0 \tag{8.21}$$

with the solution:

$$\xi_{\mathrm{d}}^2=\begin{cases} 0, & a>a_{\mathrm{d}}\left(1-v^2\right)+\frac{3\phi b^2}{16} \\ \frac{J_4}{2J_6}\left\{b+\sqrt{b^2-\frac{4}{\phi}\left[a-a_{\mathrm{d}}\left(1-v^2\right)\right]}\right\}, & a\le a_{\mathrm{d}}\left(1-v^2\right)+\frac{3\phi b^2}{16} \end{cases} \tag{8.22}$$

8.1.2.2 Free Energy

Let us now derive the expression for the free energy of a propagating zone.

Upon rescaling the free energy in accordance with (5.1), (5.6), (8.2), (8.3), we obtain:

$$\mathcal{F}=\frac{1}{\sqrt{\gamma}}B^{\frac{1}{2-z}}g^{\frac{4-3z}{4-2z}}\int\left\{\frac{1}{2}\left(\nabla_1 u\right)^2+\frac{1}{2}\left[a+U(\mathbf{r})\right]u^2-\frac{1}{4}bu^4+\frac{1}{6}u^6\right\}dxdy \tag{8.23}$$

The integral in this formulation:

$$F=\int\left\{\frac{1}{2}\left(\nabla_1 u\right)^2+\frac{1}{2}\left[a+U(\mathbf{r}_1)\right]u^2-\frac{1}{4}bu^4+\frac{1}{6}u^6\right\}dxdy \tag{8.24}$$

we refer to as the *dimensionless free energy factor*. By substituting the solution (8.17) into the expression (8.24) and following the logic formulated in Sect. 6.5.4, we derive an expression for the dimensionless free energy factor, F in terms of the amplitude, ξ_{d}:

$$F=\Omega_{\mathrm{pz}}\left[\frac{1}{2}J_2\left[a-a_{\mathrm{d}}\left(1-v^2\right)\right]\xi_{\mathrm{d}}^2-\frac{1}{4}J_4 b\xi_{\mathrm{d}}^4+\frac{1}{6}J_6\xi_{\mathrm{d}}^6\right] \tag{8.25}$$

Here, Ω_{pz} is the dimensionless volume of the process zone, as the one defined in Example 6.3.

8.1.3 Local and Bulk Phase Diagrams

As we already mentioned[3] various types of rescaling bring the potential (3.22) to the same form (8.25). One of these transformations is given by (5.1)–(5.6), (8.3).

The latter transformation yields the free energy (8.25) associated with the bifurcation. At the same time, (8.21) also bears the structural similarity to (3.25). Both the free energies (3.22) and (8.25) as well as the (3.25) and (8.21) transform into one another by the replacements:

$$a \to a - a_{\mathrm{d}}\left(1 - v^2\right); \quad b_1 \to b\sqrt{\phi} \tag{8.26}$$

where $\phi = \phi(v)$ is the combination of the integrals J_n:

$$\phi = \frac{J_4^2}{J_2 J_6} \tag{8.27}$$

Therefore, one can obtain the expressions for the phase diagram lines related to the bifurcation using equations describing the lines (3.30), (3.29) and (3.32) of the bulk phase diagram already derived earlier. One makes it without effort by replacing the parameters a and b_1 according to (8.26) in (3.30), (3.29), (3.32).

Figure 8.1 delineates the phase diagram for local phase transitions, superimposed onto the overarching bulk phase diagram. In this diagram, markers '*i*' and '*ii*' signify the lower and upper bulk spinodals, respectively, while '*iii*' represents the bulk binodal. These lines coincide with those shown on the phase diagram of the homogeneous solid shown in Fig. 3.3.

The diagram also features lines '*j*' and '*jj*' denoting the lower and upper spinodals of the local phase transition, and '*jjj*' which marks its local binodal.

At $b_1 < 0$, the second-order bulk phase transition takes place at $a = 0$.

At $b_1 > 0$, the line $a = 0$ transforms into the lower spinodal. It is pointed to by *i* in Fig. 8.1.

The upper spinodal *ii* of the bulk transition and its binodal *iii* take place at $b_1 \geq 0$. They are described by the relations $a = b_1^2/4$ and $a = 3b_1^2/16$ correspondingly as we have already derived in Sect. 3.3.2.

With the replacement (8.26) applied to the expression of the lower bulk spinodal, $a = 0$, we obtain the horizontal line (*j*) in Fig. 8.1 described by the relation:

$$a = a_{\mathrm{d}}\left(1 - v^2\right), \tag{8.28}$$

which represents the emergence line of the second-order zone at $b < 0$. We have already obtained this relation in Sect. 6.4. The same line (8.28) becomes the lower spinodal of the local transformation at $b > 0$.

---

[3] See the Note 3.3.2.

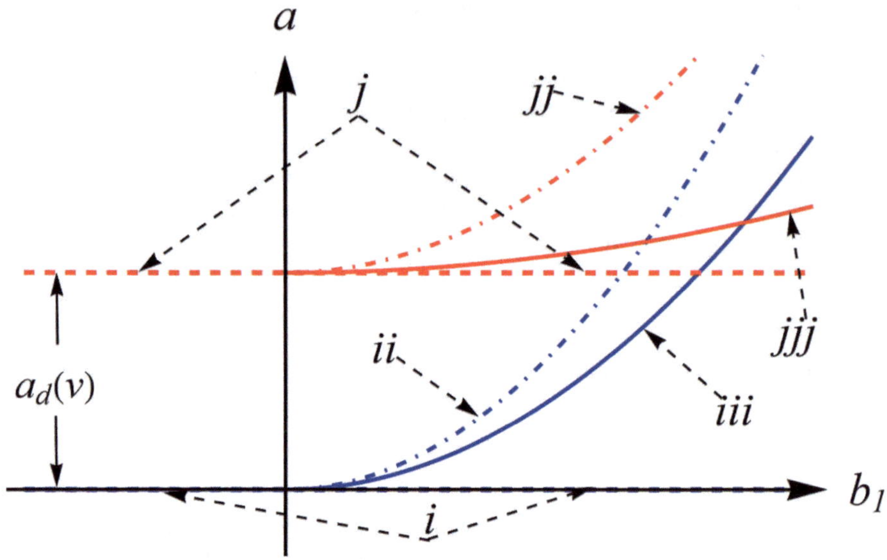

Fig. 8.1 Phase Diagram of the Local Phase Transformation Superimposed on the Bulk Phase Diagram in the (b_1, a)-Plane. *(i)* and *(ii)* indicate the lower and the upper bulk spinodals, (3.30) and (3.29), while *(iii)* points to the bulk binodal (3.32) copying the analogous lines displayed in Fig. 3.3. The lines *(j)* and *(jj)* are the lower and the upper spinodals, (8.28) and (8.29) of the local phase transformation, while *(jjj)* is it's binodal, (8.30)

Further, applying the replacements (8.26) to (3.29), (3.32) one finds the upper spinodal:

$$a_{\mathrm{up}}^{(loc)} = a_{\mathrm{d}}\left(1 - v^2\right) + \frac{b^2\phi}{4} \tag{8.29}$$

and binodal of the process zone formation:

$$a_{\mathrm{b}}^{(loc)} = a_{\mathrm{d}}\left(1 - v^2\right) + \frac{3\phi b^2}{16} \tag{8.30}$$

referred to as the *local upper spinodal* and *local binodal* correspondingly. Let us stress that while the (3.30), (3.29), and (3.32) depend on b_1, the equations of the bifurcation lines (8.29) and (8.30) depend upon b.

The solution (8.22) of the ramification (8.21) is discontinuous at the local binodal (8.30). The non-trivial solution (8.22) for ξ_{d} describing the daughter-phase zone is only real if the inequality holds:

$$a \le a_{\mathrm{up}}^{(loc)}$$

Analogously, the inequality:

$$a \le a_{\mathrm{b}}^{(loc)}$$

controls the stability of the above-mentioned solution.

At $a_{\mathrm{up}}^{(loc)} < a \le a_{\mathrm{b}}^{(loc)}$, the daughter-phase zone is metastable.

The discontinuity $\Delta\xi_{\mathrm{d}}$ at the points of the local binodal is:

$$\Delta\xi_{\mathrm{d}}^2 = \frac{3J_4 b}{4J_6} \tag{8.31}$$

8.1.3.1 Constraints on ϕ and Cauchy-Bunyakovsky-Schwarz Inequality

The value of ϕ is confined by

$$0 < \phi \le 1 \tag{8.32}$$

where $\phi > 0$ is a direct corollary of the statement formulated in the Note 6.1.4, asserting that $\Psi_{\mathrm{d}}(\mathbf{r})$ is everywhere positive.

The upper bound $\phi \le 1$ we derive from Cauchy-Bunyakovsky-Schwarz inequality,[4] which for positive real functions f and g is formulated as:

$$\left(\int f \times g d\Omega\right)^2 \le \int f^2 d\Omega \times \int g^2 d\Omega$$

Taking $f = \Psi_{\mathrm{d}}(\mathbf{r})$ and $g = \Psi_{\mathrm{d}}^3(\mathbf{r})$ one comes to the following result:

$$\left(\int \Psi_{\mathrm{d}}^4(\mathbf{r}) d\Omega\right)^2 \le \int \Psi_{\mathrm{d}}^2(\mathbf{r}) d\Omega \times \int \Psi_{\mathrm{d}}^6(\mathbf{r}) d\Omega$$

equivalent to the following inequality:

$$J_4^2 \le J_2 J_6$$

The latter proves the statement (8.32).

8.1.3.2 Intersection Point of the Binodals

The inequality (8.32) proved in the previous section shows that the local binodal (8.30) is more shallow than the bulk one (3.32). Thus, these two lines intersect. To find the intersection point, one equates the expression for the bulk binodal (3.32) with the local one[5] (8.30). By solving this equation one obtains the coordinates $(b_1^{(*)}, a^{(*)})$ of the intersection point:

---

[4] A. Boulbitch 1992 [1].

[5] Here, it is necessary to remember that the former is expressed in terms of b_1.

$$b_1^{(*)} = \frac{\sqrt{a_{\text{d}}(1-v^2)(1-\phi)+9\phi\Delta b^2} - 3\phi\Delta b}{3(1-\phi)}; \tag{8.33}$$

$$a^{(*)} = \frac{\left[\sqrt{a_{\text{d}}(1-v^2)(1-\phi)+9\phi\Delta b^2} - 3\phi\Delta b\right]^2}{48(1-\phi)^2} \tag{8.34}$$

Δb is defined in Fig. 8.2. One can see that both $b_1^{(*)}$ and $a^{(*)}$ decrease with increasing velocity, and at $v = 1$ (or at $V = V_{\text{d}}$) achieve their minimum values:

$$b_1^{(*)} = \frac{\Delta b\sqrt{\phi}}{1+\sqrt{\phi}}; \quad a^{(*)} = \frac{3\phi\Delta b^2}{16(1+\sqrt{\phi})^2}$$

Figure 8.2 displays the phase diagram of the daughter-phase zone on the background of the phase diagram of the bulk transition. Only the binodals are shown to improve its visibility. The bulk and local binodals are shown by the blue solid line and the dashed red one correspondingly. The green asterisk points out their intersection. The domain where the process zone occurs is shaded in a light blue color.

***

In summary, *the dynamics of the first-order defect-zone complex is governed by a nonlinear differential equation of degree five or higher. This book focuses on*

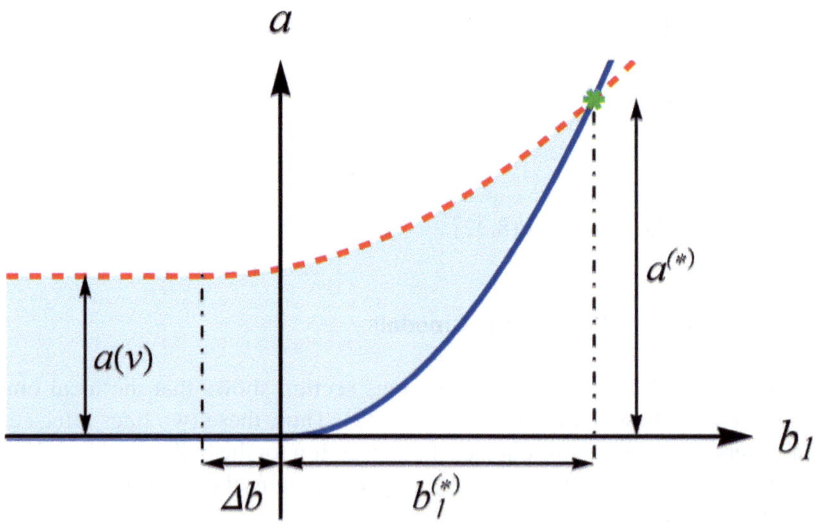

Fig. 8.2 Phase Diagram Illustrating the Domain of Existence of the Process Zone. For better visibility only the binodals are shown. The bulk binodal, (3.32), is shown by the solid blue line, while the local binodal, (8.30), - by the dashed red one. Let us mention that the bulk binodal starts at $b_1 = 0$, while the local one - at $b_1 = -\Delta b$. The binodals intersect at the point $(b_1^{(*)}, a^{(*)})$ marked by a green asterisk

the fifth-degree equation, which requires different rescaling compared to the second-order zone. For the first-order zone to occur, the parameter b must be positive. When this condition is met, key relationships, such as $a_d(v)$ *and* V_d*, are qualitatively and quantitatively similar to those of the second-order case. The free energy and friction force remain qualitatively identical in both cases, though their dimensional pre-factors are different.*

Due to the Cauchy-Bunyakovsky-Schwarz inequality the local binodal has a gentler slope compared to the bulk binodal, limiting the existence of the first-order process zone on the phase diagram to the region where $b_1 \leq b_1^{(*)}$*, as shown in Fig. 8.2. Additionally, the size of the zone decreases as the defect velocity increases. These characteristics of the zone's behavior on the phase diagram are universal.*

8.2 Temperature Range for the Zone Existence

This section is dedicated to estimating the important zone properties, and these estimates are universally applicable, regardless of the nature of defects.

***

Studying the first-order process zone is a challenging endeavor, often requiring computer simulations, which we address in subsequent sections of this book. However, simulations come with limitations, particularly regarding their generality. They often necessitate fixing the numerical values of all independent parameters. The Landau theory of phase transition introduces numerous parameters for the equation of motion, including α_0, T, T_c, β, γ, g, κ, and A. Additionally, mechanical material constants like E and ν, parameters characterizing defects (b or K_I), and various geometric parameters must be considered. This results in a multidimensional parametric space, making detailed numerical examination challenging.

Nonetheless, some properties, such as the temperature at which the zone emerges, the size of the zone, and the interface thickness, can still be analytically estimated, at least by the order of magnitude. For experimentalists, these estimates are often more practical than cumbersome exact expressions. Their simplicity enables their application to the measurement results. Furthermore, as readers will see in subsequent chapters and annexes, these estimates, which require minimal effort, closely align with corresponding numerical results and exact ones when available.

8.2.1 *Size of a Mature Zone*

At the point T_d of its emergence, the process zone exhibits its minimal size, denoted as r_d, representing its sole characteristic scale.[6] Notably, at this point, the thickness

[6] Recall that r_d is a dimensionless parameter corresponding to the physical parameter $\mathbb{R}_d$.

of the zone's interface also matches r_d. Consequently, at this juncture, no discernible distinction exists between the zone's bulk and its interface.

However, once the temperature dips noticeably below T_d, the zone assumes two distinct characteristic dimensions: the zone size, r_d, and the thickness of its interface, d, such that $d \ll r_\mathrm{d}$. This section addresses the estimation of r_d.

Let us begin by assessing the term:

$$\Delta u - \left[a + \overline{U}(\mathbf{r})\right] u \equiv -a_\mathrm{eff}(\mathbf{r}) u,$$

The latter is a key component in the equation of state (8.7) necessary for this estimate. On the scale encompassing the entire zone, we estimate the contribution of Δu as $\Delta u \sim u/r_\mathrm{d}^2$. This leads us to evaluate $a_\mathrm{eff} = a_\mathrm{eff}(r_\mathrm{d})$ as

$$a_\mathrm{eff}(r_\mathrm{d}) \sim a + \overline{U}(r_\mathrm{d}) - \frac{1}{r_\mathrm{d}^2}$$

Here, $a_\mathrm{eff}(r_\mathrm{d})$ represents the coefficient associated with u in (4.12). Leveraging the insights gained in Sect. 3.3.2, in the second-order case, we can establish the condition for the zone size by equating $a_\mathrm{eff}(r_\mathrm{d})$ to zero:

$$\frac{1}{r_\mathrm{d}^2} \sim a_0(T - T_\mathrm{c}) + \overline{U}(r_\mathrm{d}) \tag{8.35}$$

In the case of the first order, one should equate $a_\mathrm{eff}(r_\mathrm{d})$ to the binodal expression (3.32):

$$a + \overline{U}(r_\mathrm{d}) - \frac{1}{r_\mathrm{d}^2} \sim \frac{3b^2}{16} \tag{8.36}$$

Utilizing in this latter estimate the relation (3.33), we can express it in a more concise form in terms of the binodal temperature, T_0:

$$a_0(T - T_0) = a_0(T - T_\mathrm{c}) - \frac{3b^2}{16}$$

Finally, we obtain:

$$\frac{1}{r_\mathrm{d}^2} \sim \begin{cases} a_0(T - T_0) + \overline{U}(r_\mathrm{d})), & \text{First-order zone} \\ a_0(T - T_\mathrm{c}) + \overline{U}(r_\mathrm{d})), & \text{Second-order zone} \end{cases} \tag{8.37}$$

For any specific $\overline{U}(r_\mathrm{d})$, (8.37) represents an equation for the zone size r_d.

8.2.2 Size of a Motionless Zone Near the Binodal

In solids, one frequently encounters various *spatially localized defects*. These defects exist within a confined region of space, which is significantly smaller in at least one dimension compared to the overall size of the solid. Examples of 3D localized defects include inclusions and dislocation loops. In contrast, 2D localized defects include single dislocations, their limited groups, crack tips, and stacking faults. Dislocation networks and forests represent instances of 3D-extended defects.

When one ventures far enough away from most of these localized defects, digressing from their angular dependence, their averaged elastic field, denoted as $\overline{U}(r)$, can be approximated by the following asymptotic expression:

$$\overline{U}(r) \approx -\frac{\langle f \rangle}{r^z} \tag{8.38}$$

Here, the real positive exponent z characterizes how rapidly the elastic field decreases with distance, and $\langle f \rangle$ is a factor obtained by the averaging of the angular dependence of the original potential $f(\varphi)$ over the angle. Substituting this expression into (8.37), we derive the following relationship:

$$\frac{1}{r_{\mathrm{d}}^2} + \frac{\langle f \rangle}{r_{\mathrm{d}}^z} \sim a_0(T - T_0)$$

For values of z within the range $0 < z < 2$, as r_{d} becomes large, the first term on the left side of this equation becomes significantly smaller than the second term. Conversely, when $z > 2$, the first term surpasses the second one. When z equals 2, they are of similar magnitude. This leads to the following asymptotic solution:

$$r_{\mathrm{d}} \sim \begin{cases} \left[\dfrac{\langle f \rangle}{a_0(T - T_0)}\right]^{1/z}, & 0 < z < 2 \\ \left[\dfrac{1 + \langle f \rangle}{a_0(T - T_0)}\right]^{1/2} & z = 2 \\ \left[\dfrac{1}{a_0(T - T_0)}\right]^{1/2} & z > 2 \end{cases} \tag{8.39}$$

As the temperature approaches T_0, the zone size described by (8.39) diverges.

This result, while perhaps anticipated, holds profound significance. It implies that in close proximity to the binodal, the zone substantially impacts the behavior of the generating defect. Consequently, *at the binodal, the zone strongly influences any crystal's macroscopic mechanical and physical properties, determined by the behavior of these defects*.

It's noteworthy that for values of z less than two and as T approaches T_0, the zone size increases following $(T - T_0)^{-1/z}$ law. This means that smaller values of z

lead to faster growth in the zone size. As a result, the zone at the crack tip ($z = 1/2$) exhibits more rapid growth at the binodal than at the dislocation ($z = 1$).

***

To summarize, *the size of the process zone formed at a spatially localized defect inevitably diverges as the temperature approaches the binodal.*

References

1. A.A. Bulbich, J. Mater. Sci. **27**, 1070 (1992)
2. A. Boulbitch, Math. J. **20**, 1–21 (2018)
3. A. Boulbitch, A.L. Korzhenevskii, Phys. Rev. E **101**, 033003 (2020); idem Phys. Rev. E **101**, 033004 (2020)
4. M.M. Vainberg, V.A. Trenogin, *Theory of branching of solutions of non-linear equations* (Noordhoff, Leyden, 1974)
5. A. Boulbitch, P. Tolédano, Phys. Rev. Lett. **81**(4), 838 (1998)

Part II
Zone at the Tip of Mode I Crack. Its Geometrical and Physical Properties

In the preceding part, we studied the concept of a process zone generated by an unspecified defect. If, in some instances, we provided specific examples, then solely for illustrative purposes. This exploration revealed many physical properties associated with this universal scenario. In this part, we focus on examining specific defects, the mode I crack and the single-edge dislocation.

Chapter 9
Crack Tip Zone: Equation of Motion and Properties

Abstract In the above sections, we derived estimates for the emergence temperature and zone's size, which apply to any defect type. These results encompass the influence of various types of elastic defects. In this chapter, we narrow our focus to the process zone at the tip of the mode I crack (Boulbitch and Korzhenevskii, first published here).

9.1 Crack Tip Zone: Emergence Temperature and Size

In the case of the mode I crack, one derives the potential $U_0(\mathbf{r})$ from (1.6). This yields:

$$U_0(\mathbf{r}) = B_c \frac{\cos(\theta/2)}{\sqrt{r}}, \tag{9.1}$$

where

$$B_c = \sqrt{\frac{2}{\pi}} \frac{(1+\nu)(1-2\nu)AK_I}{E}. \tag{9.2}$$

As we already discussed in Sect. 6.1.2, the auxiliary equation describing the emergence of the daughter-phase zone embedded into the mother-phase matrix has the discrete spectrum, only provided at least within a limited domain $U_0(\mathbf{r}) < 0$. In the expression (9.2), K_I, E, and $(1+\nu)(1-2\nu)$ are always positive, while the striction constant, A, can be either positive or negative. If the binodal slope is negative[1] $(A < 0)$, the daughter-phase zone emerges over the temperature T_0. This case is illustrated in Fig. 9.1a. Here, one can see a fragment of the phase diagram with the binodal (indicated by the solid black line) and the emergence line marked by the dashed line. Insets above the dashed line and below the solid line depict a crack with no zone, while the region between the two lines, labeled Region III, shows a dressed crack with a zone containing Phase II surrounded by the matrix of Phase I.

[1] As discussed in Sect. 3.3.3, we assume that the sign of A coincides with the sign of the binodal slope.

© The Author(s), under exclusive license to Springer Nature Switzerland AG 2026
A. Boulbitch and A. Korzhenevskii, *Stress-Induced Local Phase Transformations Near Cracks and Dislocations*, Springer Series in Materials Science 356,
https://doi.org/10.1007/978-3-032-00945-6_9

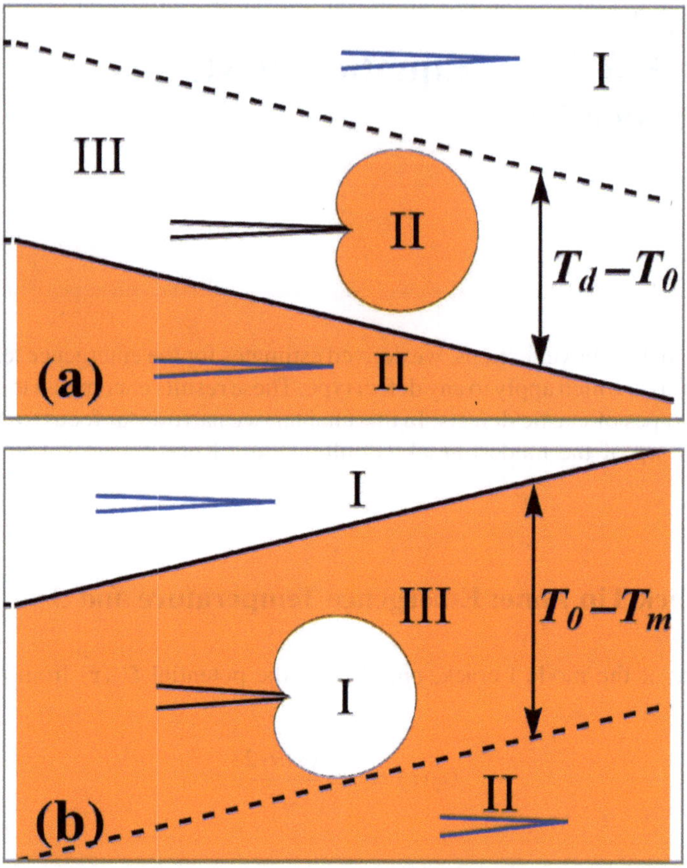

Fig. 9.1 Schematic representation of the process zone's location at the crack tip in relation to the slope of the transition line. **a** The case with a negative slope ($k_0 < 0$) of the binodal, $T_0(p)$. The mother phase is located in the Region I (white). Here the crack tip is undressed. The daughter phase occupies Region II (orange). In the Region III, the crack exhibits the zone with the daughter-phase II (orange) embedded within the matrix of the mother-phase I (white). It is positioned above the binodal (the black solid line). This region is limited by the line of zone emergence (dashed) above which the zone does not exist. **b** The binodal, $T_0(p)$, with a positive slope ($k_0 > 0$) depicted by a solid black line, separates the bulk mother phase I (white) from the daughter phase II (orange). In the region III, the developed mother-phase process zone I is embedded within the matrix of the bulk daughter-phase II touching the bottom level $u = 0$. Below the local binodal (shown by the dashed line) $u(x, y)$ does not reach the bottom level

In the case of the positive binodal slope ($A > 0$), the mother-phase process zone emerges below the binodal. This case is depicted in Fig. 9.1b. As previously discussed in Chap. 7, the mother-phase process zone within the daughter phase matrix manifests itself as a deviation in the order parameter distribution $u = u(x, y)$ from its constant value $u(x, y) = u_0 > 0$. This variation (mainly representing a depression) differs in depth and can diminish but never completely disappears.

The boundary on the phase diagram that separates the domain containing the mother-phase zone from the rest of the diagram is, therefore, conventional. A developed mother-phase zone is one in which the order parameter u reaches the bottom level, $u = 0$, within some lateral region. Consequently, the boundary where the developed mother-phase zone ceases to exist lies along the line where the order parameter touches the bottom at a single point. Beyond this boundary, however, the depression persists and continues to influence the crack properties. A detailed discussion of this boundary can be found in Sect. 16.2.

***

In summary, *when the binodal slope is negative, the process zone at the crack tip can only appear above the binodal, while the crack tip remains bare below it. Conversely, with a positive binodal slope, only the mother-phase zone can be observed below the binodal, embedded within the matrix of the bulk daughter phase, while the crack tip in the bulk mother phase shows no zone.*

Chapter 10
The Second-Order Process Zone at the Crack Tip

Abstract This chapter focuses on the formation of the second-order process zone at the tip of a mode I crack. We begin with a rigorous analytical investigation, deriving the exact eigenfunction solution for the auxiliary equation. This analysis offers precise insights into key characteristics of the zone, including its spatial distribution, size, form factor, and propagation dynamics. Complementing the analytical framework, we use simulations to explore the bifurcation behavior of the process zone across a range of parameters. We construct a phase diagram that demonstrates strong agreement with analytical predictions. Together, these approaches provide a comprehensive understanding of the second-order process zone's formation and behavior at the mode I crack tip.

10.1 Rescaling in the Crack Case

In this section, we apply the general rescaling formulae to the case of the crack by setting $z = 1/2$ in the relations (5.3)–(5.6). The resulting expressions are summarized and provided as a reference for further applications.

***

As a first step, we apply the rescaling relations derived in Sect. 5.1 to the case of the crack and the second-order process zone. This is accomplished by assigning $z = 1/2$ and $f = \cos(\theta/2)$ in the relations (5.3)–(5.6). The resulting expressions are summarized below.

The natural spatial and temporal scales are expressed as follows

$$\mathbb{R}_0 = \left(\frac{g}{B_c}\right)^{2/3}; \quad \mathbb{T}_0 = \frac{B_c^{4/3}}{\kappa g^{1/3}}. \tag{10.1}$$

By substituting (9.2) we find:

$$\mathbb{R}_0 = \left(\frac{\pi}{2}\right)^{1/3} \left[\frac{gE}{|A|K_I(1+\nu)(1-2\nu)}\right]^{2/3} \tag{10.2}$$

© The Author(s), under exclusive license to Springer Nature Switzerland AG 2026
A. Boulbitch and A. Korzhenevskii, *Stress-Induced Local Phase Transformations Near Cracks and Dislocations*, Springer Series in Materials Science 356,
https://doi.org/10.1007/978-3-032-00945-6_10

It plays an important role in this rescaling:

$$(X, Y) \to \mathbb{R}_0(x, y); \quad \eta(X, Y) \to \frac{B_c^{2/3}}{g^{1/6}\beta^{1/2}} u(x, y). \tag{10.3}$$

where x_1 and y_1 are dimensionless independent variables in the co-moving coordinate system, and $u(x_1, y_1)$ is dimensionless order parameter. The rescaled equation takes the following form

$$\Delta u + 2^{1/3} v \frac{\partial u}{\partial x'} - \left[a - \frac{\cos(\theta/2)}{\sqrt{r}} \right] u - u^3 = 0 \tag{10.4}$$

where Δ_1 represents the Laplacian operator in the dimensionless coordinates: $\Delta = \partial^2/\partial x_1'^2 + \partial^2/\partial y_1^2$, $r_1 = (x_1'^2 + y_1^2)^{1/2}$, and $x_1' = x_1 - v\tau$.

Equation (10.4) depends on two dimensionless control parameters, a and $v = V/V_d$. We express a as

$$a = \left(\frac{g}{B_c^4} \right)^{1/3} \alpha \tag{10.5}$$

and V_d we determine later on by (10.22).

In the form (10.4), the equation is convenient to solve analytically or simulate.

10.2 The Exact Solution of the Auxiliary Equation

We present the exact solution of the auxiliary equation applicable to the case of a crack. By using general properties of the process zone summarized in Chap. 6, we use this solution to provide the bifurcation point, as well as the detailed structure of the process zone in the immediate vicinity of the bifurcation.[1]

***

In Chap. 6, we solved the problem of the emergence of the daughter-phase process zone at any defect. The only parameter we still have to define in the specific case of the crack is the bifurcation value a_d of the dimensionless parameter a. We follow the approach of the paper [1].

In this section, we find the exact solution for a_d. We focus on the formation of the second-order process zone at the crack tip using the universal approach detailed in Sects. 6.1 and 6.4. In the second-order case, we derived equation of state (5.7). For the easy reference we restate it here:

$$\left(\frac{\partial^2 u}{\partial x} + \frac{\partial^2 u}{\partial y} \right) - [a - U(\mathbf{r})]\, u - u^3 = 0 \tag{10.6}$$

[1] Boulbitch (Bulbich) 1992.

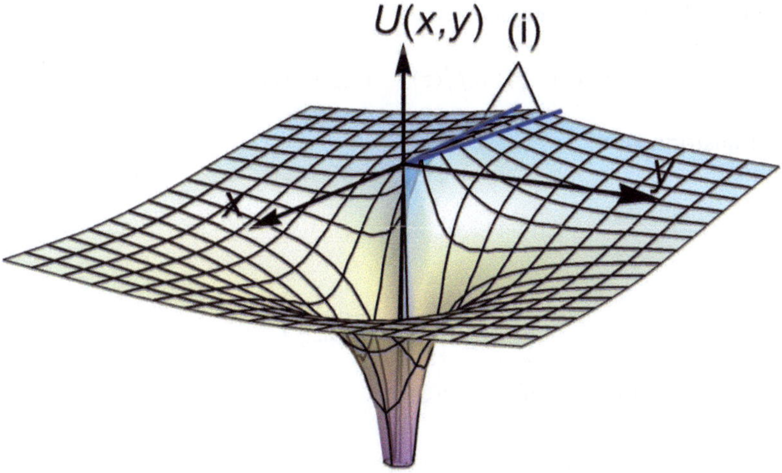

Fig. 10.1 Potential (10.7) of the Schrödinger equation (10.9). (i) indicates the position of the crack surfaces

Our approach relies on solving the auxiliary equation (6.3), (6.4). Specifically, for the mode I crack, with the potential $U(\mathbf{r})$:

$$U(\mathbf{r}) = \frac{\cos(\theta/2)}{\sqrt{r}} \tag{10.7}$$

The latter derives from (9.1) by rescaling. $U(\mathbf{r})$ is depicted in Fig. 10.1.

The auxiliary equation takes the following form:

$$\begin{cases} \Delta\psi_n - \left[a_n - \frac{\cos(\theta/2)}{\sqrt{r_1}}\right]\psi_n = 0 \\ \psi_n(\infty) = 0;\ |\psi_n| < \infty \end{cases} \tag{10.8}$$

It is convenient to look for the solution of (10.8) in polar coordinates:

$$\frac{\partial^2\psi_n}{\partial r^2} + \frac{1}{r}\frac{\partial\psi_n}{\partial r} + \frac{1}{r^2}\frac{\partial^2\psi_n}{\partial\theta^2} - \left(\varepsilon_n^2 - \frac{\cos(\theta/2)}{\sqrt{r}}\right)\psi_n = 0 \tag{10.9}$$

We defined a spectral variable $\varepsilon_n = \sqrt{a_n}$ ($n = 1, 2...$). To find a solution of equation (10.9), we represent $\psi(r, \theta)$ in the form:

$$\psi_n(r, \theta) = f_n(z)\exp(-\varepsilon_n z)$$

Here,

$$z = r^{1/2}\cos(\theta/2)$$

This leads to the following equation:

$$f_n''(z) - 4\varepsilon_n z f_n'(z) - 4(\varepsilon_n - z) f_n(z) = 0$$

Further transformation:

$$f_n = \varphi_n(\zeta)\exp(z/\varepsilon_n)$$

with

$$\zeta = \sqrt{\frac{\varepsilon_n}{2}}\left(z - \frac{1}{2\varepsilon_n^2}\right)$$

yields an equation in terms of $\varphi = \varphi(\zeta)$:

$$\varphi_n''(\zeta) - 2\zeta\varphi_n'(\zeta) + m\varphi_n(\zeta) = 0;\quad m = 2\left(1 - \frac{1}{4\varepsilon_n^3}\right) \tag{10.10}$$

The latter is the Hermitian equation. It exhibits eigenvalues m that are non-negative, integer, and even values: $m = 0, 2, 4, \ldots$ From (10.10), we determine that only the ground state solution,[2] $m = 0$, is compatible with the condition $\varepsilon_1 > 0$, which gives:

$$\varepsilon_1 = 2^{-2/3}$$

and $\varphi(\zeta) = const.$

Returning to the original variables, we find:

$$a_1 = \varepsilon_1^2 = \frac{1}{2\sqrt[3]{2}};\qquad \psi_1(\mathbf{r}) = \exp\left[-\frac{r}{\sqrt[3]{4}} + \sqrt[3]{4}\sqrt{r}\cos(\theta/2)\right] \tag{10.11}$$

Equation (10.11) represents the exact solution of the auxiliary equation (10.9) for the ground state.

Thus,

$$a_{\mathrm{d}} \approx 0.397.$$

[2] According to our convention, the enumeration of the eigenvalues and eigenfunctions starts from the index 1, rather than 0. Therefore, $m = 0$ corresponds to the eigenvalue ε_1.

10.2.1 Estimate of the Ground State Eigenvalue for Mode I Crack by the "Old Quantum Theory" Method

For the mode I crack, the rescaled potential takes the form:

$$U(\mathbf{r}) = -\frac{\cos(\theta/2)}{\sqrt{r}} \tag{10.12}$$

provided $A < 0$. In this case, the potential $U(\mathbf{r})$ is negative for the mode I crack in the interval $-\pi < \theta < \pi$. Upon averaging over this interval, we obtain the following potential $\overline{U}(r)$:

$$\overline{U}(r) = -\frac{2}{\pi}\frac{1}{\sqrt{r}} \tag{10.13}$$

With this, the (6.15) yields:

$$a_\mathrm{d} \approx 0.548 \tag{10.14}$$

10.2.2 Comparison with the Accurate Results

For the mode I crack, the exact solution yields $a_\mathrm{d} = 2^{-4/3} \approx 0.4$. This value closely aligns with the approximation indicated by (10.14).

By making these comparisons, we gain insight into the accuracy of our approximations in relation to more precise computational or exact solutions.

***

In summary, *the result* (10.11) *represents the exact solution of the auxiliary equation for the crack. The "old quantum mechanics" method yields an accurate numerical estimate in this case.*

10.3 Bifurcation Point and Overcritical Solution

In this section, we work with the exact solution of the auxiliary equation alongside the universal results presented in Sects. 6.1.3, 6.3.1, *and Chap.* 6.4. *By simply substituting these results into the universal solutions, we immediately obtain the zone emergence temperature T_d, the overcritical solution $\eta(R)$ for the nonlinear equation of state in the case of a motionless crack, and $\eta(X - Vt, Y)$ for the equation of motion in the case of a propagating crack, without the need for further derivation.*

***

10.3.1 Case of the Motionless Crack

In the case of mode I crack, $B = B_c$ (9.2), and $z = 1/2$. Applying (5.1), (5.5), and (10.11) to the solutions obtained in Chap. 6 we obtain the specific solution for the process zone at the mode I crack tip in terms of the original variables.

One finds the eigenvalue, α_d:

$$\alpha_d = \frac{1}{g^{1/3}}\left(\frac{2}{\pi}\right)^{2/3}\left[\frac{|A|(1-2\nu)(1+\nu)}{E}\right]^{4/3} K_I^{4/3} \tag{10.15}$$

This yields the temperature, T_d, of the bifurcation:

$$T_d - T_c = \frac{1}{\alpha_0 g^{1/3}}\left(\frac{2}{\pi}\right)^{2/3}\left[\frac{|A|(1-2\nu)(1+\nu)}{E}\right]^{4/3} K_I^{4/3} \tag{10.16}$$

The corresponding eigenfunction takes the following form:

$$\psi_d(\mathbf{R}) = \exp\left[-\frac{R}{\sqrt[3]{4}\mathbb{R}_0} + \sqrt[3]{4}\sqrt{\frac{R}{\mathbb{R}_0}}\cos(\theta/2)\right] \tag{10.17}$$

Leveraging (6.20), one finds the overcritical solution taking the form:

$$\eta(\mathbf{R}) = \frac{1}{g^{1/3}}\sqrt{\frac{2}{\beta\pi}}\left[\frac{|A|(1-2\nu)(1+\nu)K_I}{E}\right]^{4/3}\psi_d(\mathbf{R}) \times$$

$$\times \begin{cases} 0, & \alpha > \alpha_d \\ \pm (j_2/j_4)^{1/2}\left(1 - \frac{\alpha}{\alpha_d}\right)^{1/2}, & \alpha \le \alpha_d \end{cases} \tag{10.18}$$

10.3.2 The Solution at the Propagating Crack

Equation of motion (5.10) in the case of the process zone following the tip of the propagating crack has the following form:

$$\Delta_1 u + 2^{1/3} v \frac{\partial u}{\partial x} - [a + U(\mathbf{r})]\, u - u^3 = 0 \tag{10.19}$$

and its auxiliary equation (6.23) in this case, reads:

$$\Delta_1 \Psi_n(\mathbf{r}) + 2^{1/3} v \frac{\partial \Psi_n(\mathbf{r})}{\partial x} = [a_n(v) + U(\mathbf{r})]\Psi_n(\mathbf{r}) \tag{10.20}$$

One effortlessly writes its ground state eigenfunction leveraging (6.25). It takes the following form:

$$\Psi_{\mathrm{d}}(\mathbf{R}) = \exp\left[-\frac{V}{\sqrt[3]{4}V_{\mathrm{d}}}\frac{R}{\mathbb{R}_0}\cos(\theta) - \frac{R}{\sqrt[3]{4}\mathbb{R}_0} + \sqrt[3]{4}\sqrt{\frac{R}{\mathbb{R}_0}}\cos(\theta/2)\right] \tag{10.21}$$

Using (10.11), (5.1), and (5.5) without further calculations, we derive the expression for the critical velocity, V_{d}, for the daughter-phase zone $V_{\mathrm{d}} = \left(2B_{\mathrm{c}}^2 g\right)^{1/3}/\kappa$. After the substitution of the expression (9.2) it yields:

$$V_{\mathrm{d}} = \frac{2^{4/3}g^{1/3}}{\pi^{1/3}\kappa}\left[\frac{|A|\,(1-2\nu)(1+\nu)}{E}\right]^{2/3} K_I^{2/3} \tag{10.22}$$

The overcritical solution describing the propagating process zone takes the following form:

$$\eta(\mathbf{R}) = \pm\frac{1}{g^{1/3}}\left[\frac{|A|(1-2\nu)(1+\nu)K_I}{E}\right]^{4/3}\left[\frac{2J_2}{\beta\pi J_4}\left(1-\frac{\alpha}{\alpha_{\mathrm{d}}}-\frac{V^2}{V_{\mathrm{d}}^2}\right)\right]^{1/2}\times$$

$$\times\,\Psi_{\mathrm{d}}(\mathbf{R})\begin{cases}0, & \alpha/\alpha_{\mathrm{d}} > 1-(V/V_{\mathrm{d}})^2 \vee V/V_{\mathrm{d}} > 1\\ 1, & \alpha/\alpha_{\mathrm{d}} \le 1-(V/V_{\mathrm{d}})^2 \wedge V/V_{\mathrm{d}} \le 1\end{cases} \tag{10.23}$$

Equation (8.15) for the bifurcation line in the (V, α)-plane takes the following form:

$$\alpha_{\mathrm{d}}(V) = \frac{2^{2/3}}{\pi^{2/3}g^{1/3}}\left[\frac{|A|(1-2\nu)(1+\nu)}{E}\right]^{4/3} K_I^{4/3}\left(1-\frac{V^2}{V_{\mathrm{d}}^2}\right) \tag{10.24}$$

Determining the dimensionless size of the zone, R_{d}, in the forward direction ($\theta = 0$) from the solution for ψ_{d} (10.11) and the condition $4^{1/3}\sqrt{R_{\mathrm{d}}} - 4^{-1/3}R_{\mathrm{d}} = 1$, resulting in $R_{\mathrm{d}} = 2^{2/3}$. This leads to the zone's size, $\mathbb{R}_{\mathrm{d}}$, in the physical coordinates:

$$\mathbb{R}_{\mathrm{d}} = \left(\frac{2g}{|B_{\mathrm{c}}|}\right)^{2/3} \equiv (2\pi)^{1/3}\left[\frac{gE}{AK_I(1+\nu)(1-2\nu)}\right]^{2/3} \tag{10.25}$$

It is only $2^{2/3}$ times larger than $\mathbb{R}_0$. One concludes that $\mathbb{R}_{\mathrm{d}} \sim \mathbb{R}_0$. The latter means that the solution (10.17) is characterized by a single spatial scale. Therefore, the solution (10.17) describes the regime of the newly-born overcritical state.[3] It depends on the single characteristic size $\mathbb{R}_0$. The latter yields the dimension of its bulk as well as the thickness of its transition layer.

---

[3] As we have already discussed in Sect. 6.1.3.

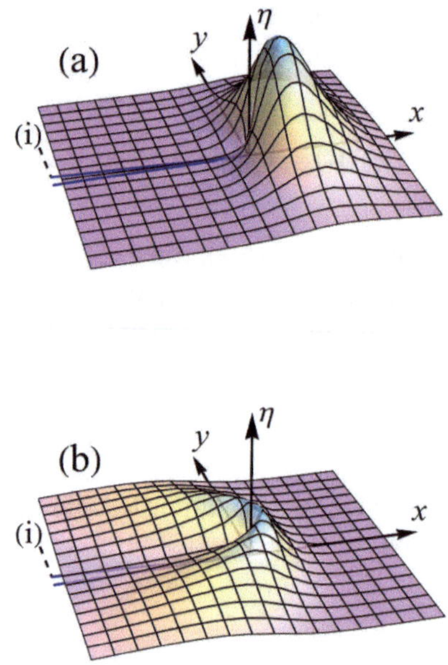

Fig. 10.2 Lateral Distribution of the Eigenfunction $\psi_d(x, y)$. The distribution describes the daughter-phase process zone at the crack tip, surrounded by the bulk mother phase. **a** The process zone near a motionless crack. **b** The process zone near a propagating crack. The two blue lines (i) schematically represent the positions of the crack faces in the vicinity of the tip

Let us also write out the characteristic dimensions in the forward ("f") and backward ("b") directions derived from (6.39), (6.40) in physical coordinates:

$$\mathbb{R}_f = \frac{\mathbb{R}_0}{\sqrt{a_d}(1 + V/V_d)}; \qquad \mathbb{R}_b = \frac{\mathbb{R}_0}{\sqrt{a_d}(1 - V/V_d)} \tag{10.26}$$

Figure 10.2a illustrates the process zone at the tip of the motionless crack, while panel (b) shows the zone propagating with the crack. The eigenfunction $\Psi_d(\mathbf{r})$ configuration aligns with the image presented.

It's worth noting that the distribution shown in Fig. 10.2b is compressed in front of the crack tip and stretched backward compared to that in (a). According to the universal result (6.40), the distribution size in the backward direction diverges as $V \to V_d$. These characteristics are manifestations of the universal properties of the process zone discussed in Sects. 6.1 and 6.4.

***

In conclusion, *the results describing the crack tip zone, derived from the exact solution (*10.11*), align with the general predictions presented in Chap.* 6 *and exhibit all the qualitatively described features. The application of the specific potential also facilitates the numerical estimation of these results.*

Reference

1. A.A. Bulbich, J. Mater. Sci. **27**, 1070 (1992)

Chapter 11
Numerical Study of the Second-Order Process Zone at the Crack Tip

Abstract In this study, we investigate the formation of the second-order daughter phase process zone within the matrix of the mother phase using numerical simulations based on the finite element method (Boulbitch and Korzhenevskii 2016). Subsequently, we compare our simulation results with the analytical solution. We demonstrate that the simulation results are consistent with the analytical findings, both qualitatively and quantitatively. This alignment between the analytical and numerical outcomes provides strong support for the field-theoretical approach.

11.1 Simulation of the Equation of Motion

In this section, we address the numerical solution of the rescaled equation of motion (10.19) *and report the results of the simulation process.*

***

11.1.1 Emergence of a Non-trivial Solution

In this section, we follow results reported in the paper [1]. We only provide the results of the simulation process. Technical details of the simulation the reader finds in the Annex (Appendix B).

At large positive values of a, (10.6) exhibits a trivial solution, $u = 0$. A non-trivial solution describing the daughter-phase zone, emerges below a critical value a_d. Figure 11.1 shows the distributions of the rescaled order parameter u at the tip ($x = y = 0$) of a motionless crack for various values of $a_d - a > 0$, just below the bifurcation point ($a = a_d$).

Figure 11.1 displays the motionless process zone at the crack tip ($v = 0$) obtained by the simulations. One can see that the distributions which are shown in Fig. 11.1 qualitatively look like the one obtained within the framework of the bifurcation theory (see Fig. 10.2a). However, they are different in detail since we performed the simulation at some final distance, $a_d - a$, from the bifurcation point. Here, the modes with $n > 1$ contribute to the formation of the zone.

© The Author(s), under exclusive license to Springer Nature Switzerland AG 2026
A. Boulbitch and A. Korzhenevskii, *Stress-Induced Local Phase Transformations Near Cracks and Dislocations*, Springer Series in Materials Science 356,
https://doi.org/10.1007/978-3-032-00945-6_11

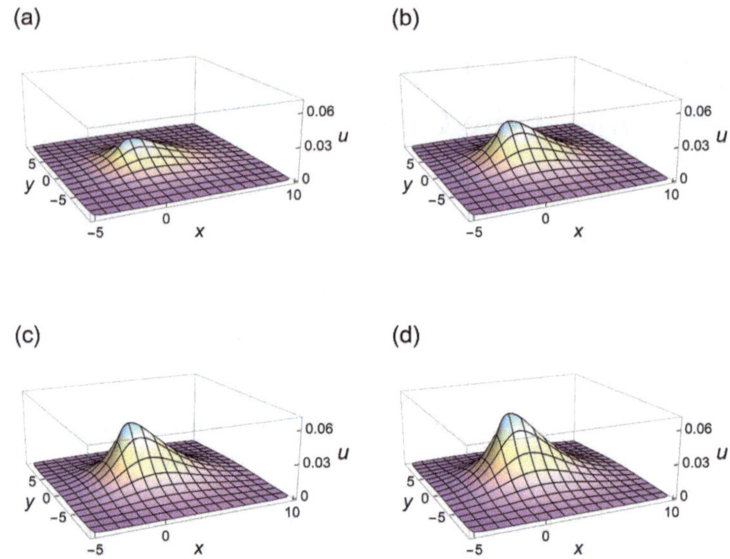

Fig. 11.1 Distributions of the Order Parameter in the Vicinity of the Tip of a Motionless Crack. The images are obtained by simulations at $\nu = 0$ and various values of $a_\text{d} - a$: **a** 1.5×10^{-4}, **b** 6.5×10^{-4}, **c** 12×10^{-4} and **d** 16×10^{-4}

11.1.2 Effect of Crack Motion

We also investigated the effect of crack motion ($\nu \geq 0$) on the distribution of the rescaled order parameter $u(x, y)$. In Fig. 11.2, we observe that the motion of the crack leads to several notable changes in the distribution:

- With increased dimensionless velocity $v = V/V_\text{d}$, the distribution decreases and eventually vanishes when the velocity reaches a critical value.
- The order parameter distribution is compressed in front of and stretched out behind the crack tip compared to the tip of a motionless crack.

11.1.3 Analysis of the Bifurcation

To identify the bifurcation points $a_\text{d} = a_\text{d}(v)$, we conducted simulations while keeping v constant and varying the control parameter a. We observed that at large values of a, the maximum value of the rescaled order parameter u_max vanishes but emerges beyond a certain threshold (Fig. 11.3).

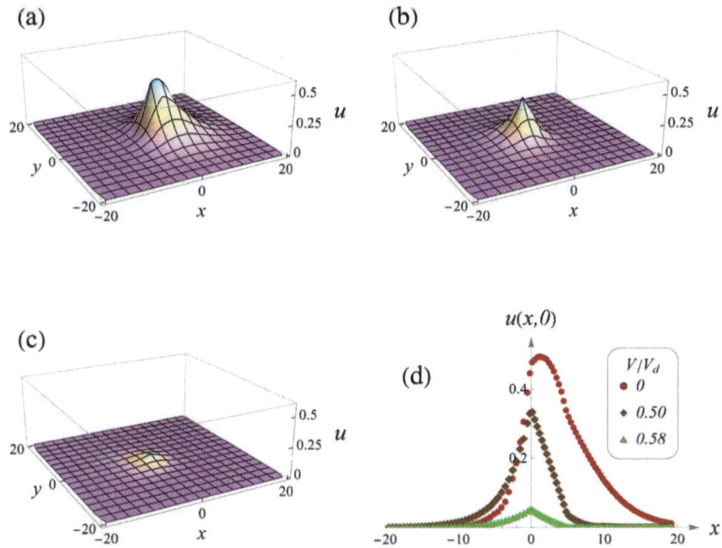

Fig. 11.2 Distribution of the Order parameter in the Vicinity of the Tip of a Propagating Crack. The crack tip propagates in the positive direction of x along the line $y = 0$. Images: **a** Motionless case $v = 0$, **b** $v = 0.5$, **c** $v = 0.58$. Image **d** shows the cross-section of the distributions shown in (**a–c**) by the plane $y = 0$. All images have been obtained by simulation with $a = 0.3$

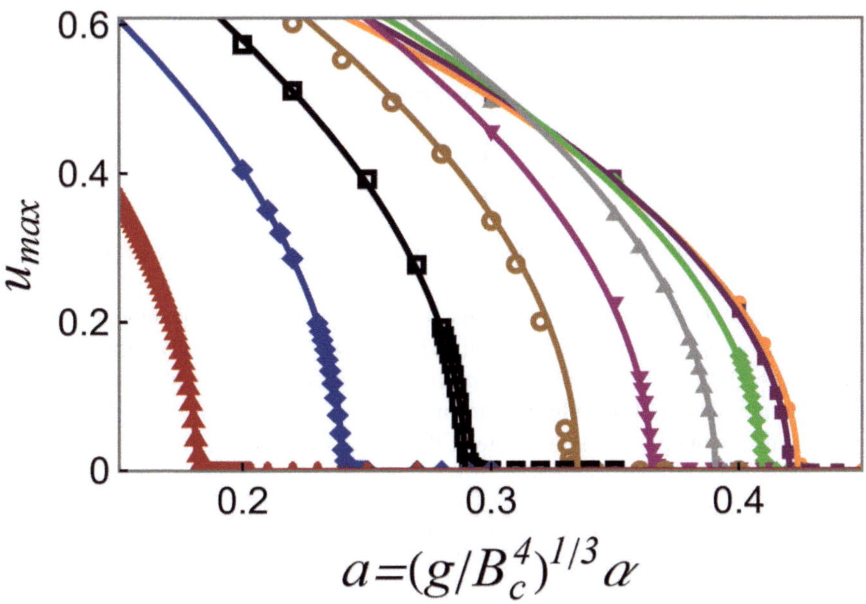

Fig. 11.3 Maximum Value of u Versus a at Various Velocities. Disks represent $\nu = 0$, squares—0.1, filled diamonds—0.2, filled vertex-up triangles—0.3, vertex-down triangles—0.4, open circles—0.5, open squares—0.6, open diamonds—0.7, open vertex-up triangles—0.8. The solid lines show their fitting to the function (11.1)

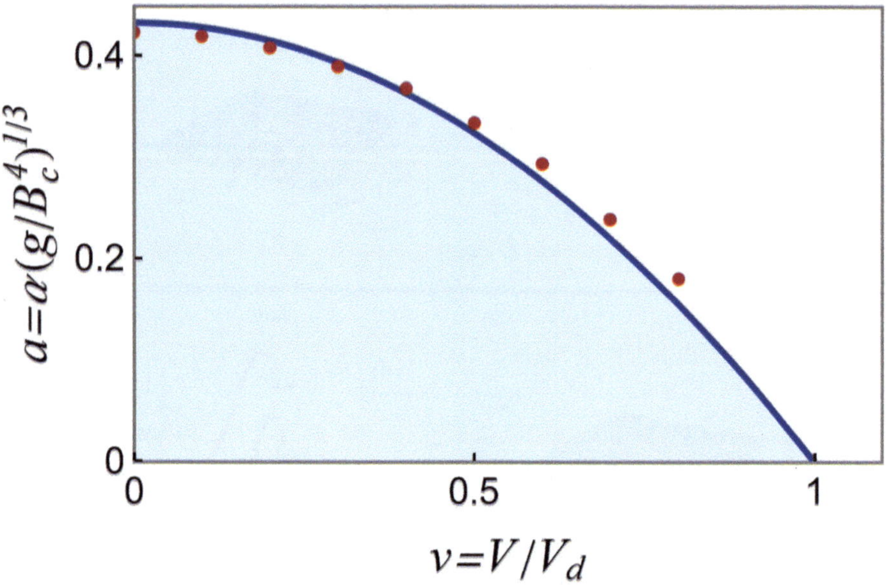

Fig. 11.4 Phase Diagram in the Plane of Dimensionless Parameters $a = \alpha(g/B_c^4)^{1/3}$ and $v = V/V_{cr}$. The nontrivial solution, $u(x_1, y_1) \neq 0$, exists within the shaded region, below the curve representing the bifurcation line $\alpha_d = \alpha_d(V/V_{cr})$. Dots show the simulation results, while the solid line displays the fit (11.2)

We determined these thresholds by fitting the data to the function:

$$u_{max}(a) = \begin{cases} 0, & a > a_d \\ u_0\,(a_d - a)^{1/2}, & a \leq a_d \end{cases} \tag{11.1}$$

Here, u_0 and a_d are fitting parameters, and a is the variable. Figure 11.4 presents the obtained dependence $a_d = a_d(v)$.

Dots indicate the simulation results, while the solid line represents the fit to the function $a = a_d(1 - v^2)$ (8.15). The fitted factor $a_d \approx 0.43$ closely approximates the value $\varepsilon_1^2 \approx 0.397$ (10.11) obtained in bifurcation theory and is also consistent with the estimate value (21.4), which is around 0.55.

In terms of the physical parameters, the fitting results have the following form:

$$\alpha_d \approx 0.43 \left(\frac{B_c^4}{g}\right)^{1/3} \left[1 - \left(\frac{V}{V_d}\right)^2\right] \tag{11.2}$$

***

In conclusion, *the simulation results are consistent with the analytical findings, both qualitatively and quantitatively. This alignment between the analytical and numerical outcomes provides strong support for the field-theoretical approach.*

Reference

1. A. Boulbitch, A.L. Korzhenevskii, Eur. Phys. J. B **89**, 261 (2016)

Chapter 12
First-Order Zone at the Crack Tip

Abstract In this chapter, we focus on the first-order process zones. The framework of bifurcation theory is unsuitable for analyzing such zones; therefore, we study them numerically (Boulbitch and Korzhenevskii 2020).

In the preceding discussion, we explored the characteristics of the process zone at the tip of a brittle crack under the assumption of a second-order process zone. In such cases, the well-established framework of bifurcation theory proves valuable for uncovering key features of the second-order process zone [1].

However, the majority of transformations in nature involve first-order transitions [2]. Hence, we study cracks in brittle solids exhibiting these first-order transformations.

This chapter addresses the emergence of process zones at the crack tip during first-order transitions, which we will refer to as the *first-order process zone* for brevity [3]. Unlike second-order zones, the first-order process zone emerges through a hard bifurcation, leading to a discontinuity in the order parameter at the bifurcation point.

While bifurcation theory remains useful for quasi-second-order zones, as discussed in Sect. 8.1.2, simulations become the method of choice when it is not applicable. We will explore this further in the next chapter.

Within this chapter, we will derive approximate analytical formulae for critical geometric parameters within the first-order process zone. Our approach involves employing reasonable approximations applied to the Ginzburg-Landau equation and utilizing the exact solution describing the propagating phase front.

First, we simplify the problem by directly adopting the previously derived rescaling, reducing the number of independent control parameters to three. Subsequently, we provide a condensed explanation of the theory behind the propagating phase front, followed by the derivation of approximate analytical expressions for the most required geometric zone measures.

This endeavor enhances our comprehension of first-order process zone behavior but also aids in determining transformation toughness.

© The Author(s), under exclusive license to Springer Nature Switzerland AG 2026
A. Boulbitch and A. Korzhenevskii, *Stress-Induced Local Phase Transformations Near Cracks and Dislocations*, Springer Series in Materials Science 356,
https://doi.org/10.1007/978-3-032-00945-6_12

12.1 Adoption of Rescaling for the First-Order Zone at the Crack Tip

In this section, we apply the previously derived rescaling relations for the first-order zone to the case of the crack tip.

***

In Sect. 8.1, we performed the rescaling in the general case, which is given by the relations (8.2), (8.3). To apply these results to the case of the crack, we substitute $z = 1/2$, $f = \cos(\theta/2)$, and $B = B_c$ (9.2)[1] we come to the scaling parameters (5.1) valid for the first-order process zone at the crack tip:

$$\mathbb{R}_0 = \pi^{1/3}\left[\frac{gE}{2AK_I(1+\nu)(1-2\nu)}\right]^{2/3} \tag{12.1}$$

$$\mathbb{T}_0 = \frac{\kappa g^{1/3}}{2^{1/3}}\left[\frac{E}{2AK_I(1+\nu)(1-2\nu)}\right]^{4/3} \tag{12.2}$$

$$\eta_0 = \frac{1}{\pi^{1/6}\gamma^{1/4}g^{1/12}}\left[\frac{2AK_I(1+\nu)(1-2\nu)}{E}\right]^{1/3} \tag{12.3}$$

The relations between the dimensionless control parameters α and a as well as β and b have the following form:

$$\alpha = \frac{a}{\pi^{2/3}g^{1/3}}\left[\frac{2AK_I(1+\nu)(1-2\nu)}{E}\right]^{4/3}, \tag{12.4}$$

$$a_0 = (9\pi)^{2/3}g^{1/3}\left[\frac{E}{6AK_I(1+\nu)(1-2\nu)}\right]^{4/3}\alpha_0, \tag{12.5}$$

$$\beta = -\frac{b\sqrt{\gamma}}{\pi^{1/3}g^{1/6}}\left[\frac{2AK_I(1+\nu)(1-2\nu)}{E}\right]^{2/3}. \tag{12.6}$$

The critical velocity in physical coordinates for the crack case has the form:

$$V_d = \frac{1}{\kappa}\left(\frac{2g}{9\pi}\right)^{1/3}\left[\frac{6AK_I(1+\nu)(1-2\nu)}{E}\right]^{2/3} \tag{12.7}$$

The rescaled Gordon velocity and the velocity for the zone's interface with the dimensionless curvature ρ are given by:

---

[1] Due to technical reasons, we rescaled equation (4.10) slightly differently, resulting in the potential $U(\mathbf{r})$ taking the form $U(\mathbf{r}) = \sqrt{2}\cos(\theta/2)/\sqrt{r}$ rather than $U(\mathbf{r}) = \cos(\theta/2)/\sqrt{r}$.

$$v_{\mathrm{G}} = \frac{1}{2^{1/6}\sqrt{3}} \frac{8a - 1 - \sqrt{1-4a}}{\left|1 - 2a + \sqrt{1-4a}\right|^{1/2}} \tag{12.8}$$

$$v_{\mathrm{G}}^{(\pm)} = v_{\mathrm{G}} \pm \frac{1}{2\sqrt{a_{\mathrm{d}}\rho}} \tag{12.9}$$

The rescaled critical velocity is:

$$v_{\mathrm{d}} = 1 \tag{12.10}$$

Remind that since $\beta < 0$, we obtain $b > 0$. Finally, α_{d}, and a_{d} are as previously given by the relations (10.15), (10.11), and (10.22) correspondingly. The rescaled potential has the form:

$$U(\mathbf{r}) = -\frac{\sqrt{2}\cos(\theta/2)}{\sqrt{r}} \tag{12.11}$$

12.1.1 Equation of Motion in Rescaled Coordinates

The rescaled equation (8.6) takes the following form:

$$\frac{\partial^2 u}{\partial x^2} + \frac{\partial^2 u}{\partial y^2} + 2^{1/3} v \frac{\partial u}{\partial x} - [a + U(\mathbf{r})]\, u + bu^3 - u^5 = 0 \tag{12.12}$$

with the boundary condition, $u\,|_{\partial\Omega} = 0$.

In the case of the motionless crack, setting $v = 0$, we obtain the equation of state:

$$\frac{\partial^2 u}{\partial x^2} + \frac{\partial^2 u}{\partial y^2} - [a + U(x, y)]\, u + bu^3 - u^5 = 0 \tag{12.13}$$

***

In conclusion, *the rescaled equation of motion* (12.12) *depends on three parameters: a, b, and v, whereas the equation of state* (12.13) *depends only on a and b. This formulation makes them well-suited for simulation.*

References

1. A. Boulbitch, A.L. Korzhenevskii, Eur. Phys. J. B **89**, 261 (2016)
2. E.Y. Tonkov, *High Pressure Phase Transformations: A Handbook*, vols. 1–3 (Gordon and Breach, Amsterdam, 1992); E.Y. Tonkov, E.G. Ponyatovsky, *Phase Transformations of Elements Under High Pressure*, vol. 4 (CRC Press, Boca Raton, FL, 2005)
3. A. Boulbitch and A.L. Korzhenevskii, Phys. Rev. E **101**, 033003 (2020); idem Phys. Rev. E **101**, 033004 (2020)

Chapter 13
Morphological Transformation of the First-Order Process Zone

Abstract This chapter explores the characteristics of the first-order process zone at the crack tip within the hysteresis region on the phase diagram. First, our simulations confirmed the existence of the setback in the motionless regime as theoretically predicted previously. Subsequently, they reveal a significant morphological transformation of the propagating process zone with increasing crack velocity (Boulbitch and Korzhenevskii 2018).

This chapter presents the results of the papers [1, 2].

13.1 Zone at the Tip of the Motionless Crack

In this section, we demonstrate the results of our simulations of the first-order process zone at the tip of the motionless crack.

***

In our investigation of the process zone at the tip of a motionless crack, we employed simulations based on equation (12.13). Figure 13.1 offers a visual representation of the process zones at the tips of motionless cracks for three distinct values of parameter a: $a = 0.22$ (i), $a = 0.23$ (ii), and $a = 0.24$ (iii). In Fig. 13.1, the zones are positioned in close proximity to enable one a meaningful comparison of their sizes and configurations, although each simulation involved a single crack-zone complex. The straight solid lines behind the zones (marked in red in the color version) serve as schematic markers for the crack surfaces positions.

Throughout our further presentation, we write x and y instead of x_1 and y_1. Thus, in Fig. 13.1, x and y are expressed in dimensionless units, with one such unit corresponding to the characteristic length $\mathbb{R}_0$. For an in-depth discussion and estimates regarding the characteristic length's potential values, please refer to Sect. 14.2.2.

An intriguing observation emerges as we examine these zones: their size exhibits a pronounced dependence on parameter a, rapidly increasing as we approach the

© The Author(s), under exclusive license to Springer Nature Switzerland AG 2026
A. Boulbitch and A. Korzhenevskii, *Stress-Induced Local Phase Transformations Near Cracks and Dislocations*, Springer Series in Materials Science 356,
https://doi.org/10.1007/978-3-032-00945-6_13

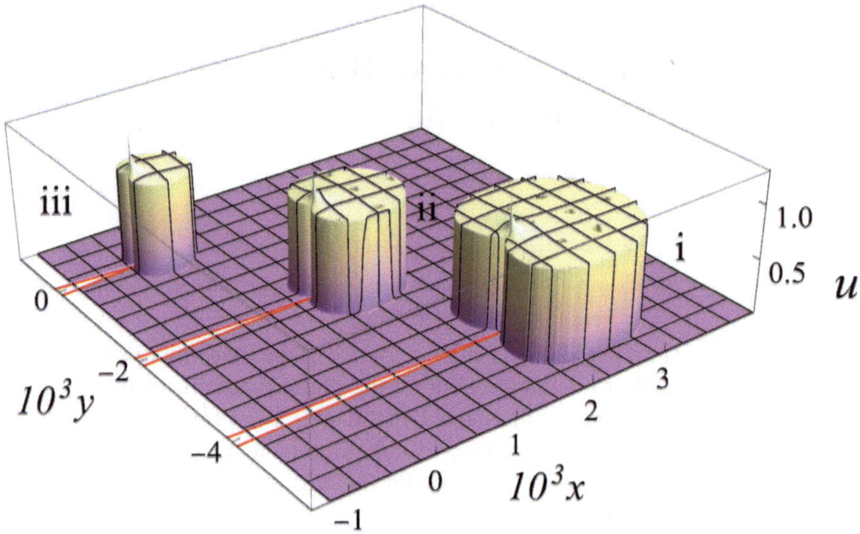

Fig. 13.1 Process Zones at the Tips of the Motionless Cracks Obtained by Simulations. The simulations were performed at various a values: **i** $a = 0.22$, **ii** $a = 0.23$, and **iii** $a = 0.24$

binodal. Of particular note is an invagination in the rear section of the zone, where it interfaces with the crack surfaces.

It is worth noting that the zone boundaries appear abrupt in Fig. 13.1. This is an illusion resulting from the relation $R_d \gg d$, where R_d is the dimensionless zone size and d is its thickness. Indeed, as discussed later (see Fig. 13.11), the boundary width is about 1 to 2 dimensionless units, whereas the zone size shown in Fig. 13.1 ranges from a few hundred to a few thousand such units.

Figure 13.2a illustrates the contour of the process zone at the tip of the motionless crack, specifically for the case where $a = 0.24$. We derived this contour by identifying points on the crest of the distribution $(\nabla u)^2$, which correspond to the middle of the kink. While the zone's overall shape closely resembles a cardioid on a larger scale, subtle distinctions emerge upon closer examination. Crucial among these is the shape of the contour in the rear section of the interface, near the crack surface (Fig. 13.2b). When $a = 0.24$, it intersects the crack surface ($y = 0$) at approximately $x \approx -18.85$, deviating from the expected $x = 0$ position for a true cardioid. This distinctive setback phenomenon also occurs for other values of a. We have comprehensively summarized these setbacks at $v = 0$ in Sect. 14.2.2.

***

In summary, *the process zone displays the following distribution of the order parameter: within the zone, the order parameter forms a "roof" with a gentle slope, sharply dropping to zero at the boundary, which defines the zone's edge. The boundary*

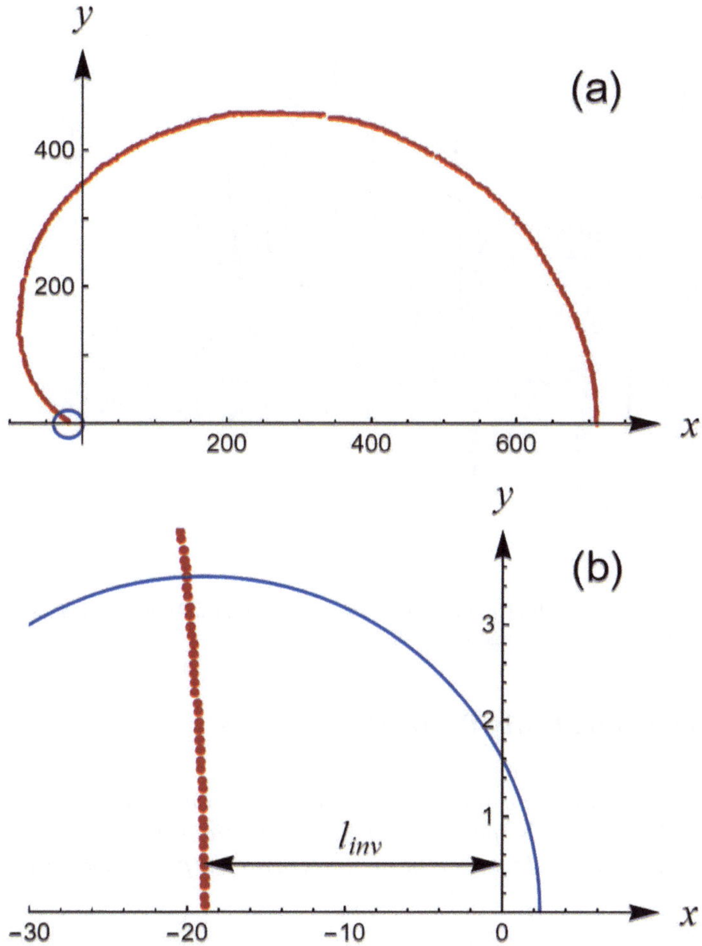

Fig. 13.2 Contour of the Interface of the Motionless Process Zone ($a = 0.24$). **a** The image shows only the part of the zone contour in the upper semi-plane, $y \geq 0$. Red dots indicate the contour points corresponding to the crest of $(\nabla u)^2$. The blue circle indicates the rear part of the contour. **b** An enlarged view of the rear section of the same contour in close proximity to the x-axis

closely resembles a cardioid, differing near the crack tip, where it is shifted slightly backward from the tip, unlike a true cardioid.

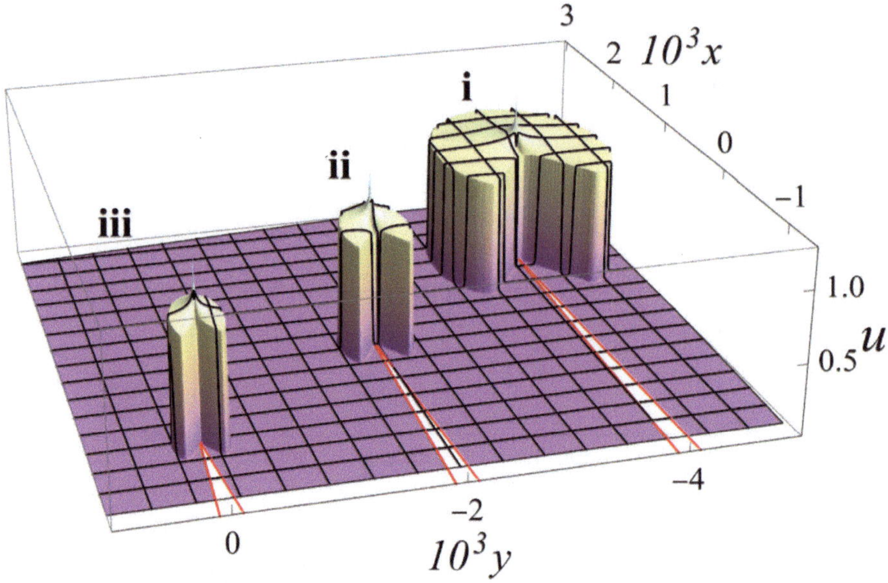

Fig. 13.3 Process Zones at the Tips of Slowly Propagating Cracks. Parameter a takes the following values: **i** $a = 0.22$, **ii** $a = 0.23$, and **iii** $a = 0.24$

13.2 Zone at a Slowly Propagating Crack

In this section, we focus on simulations of the crack-zone complex propagating at velocities below Gordon's speed. Under these conditions, the zone displays a distinctive configuration and set of properties, which are the primary focus of this study.

***

In the following, we refer to the crack propagating with the velocity, $0 < V < V_G$ as a slowly propagating one. Here, V_G is Gordon's velocity (see Sect. 3.5).

Figure 13.3 presents the process zones at the tips of slowly propagating cracks for three different values of parameter a: (i) $a = 0.22$, (ii) $a = 0.23$, and (iii) $a = 0.24$. Since the shapes of these zones exhibit a strong dependence on a, we carefully selected the velocities. In this way, each zone exhibits a shape typical of a small crack tip speed. Specifically, zone (i) propagates at a velocity of $v = 0.06$, while zones (ii) and (iii) move at $v = 0.127$. This perspective allows for a clearer view of the invaginations in the rear part of the zones, which are significantly more pronounced compared to the motionless crack-zone complex.

For a deeper understanding, we provide an example of the evolution of $u(x, y)$ as the velocity, v, increases, depicted in Fig. 13.4 for $a = 0.24$.

Figure 13.4a portrays the zone at the tip of a motionless crack, while panels (b–d) illustrate zones as the crack's velocity gradually increases. The crack extends along the semi-infinite line: $y = 0$ and $-\infty < x - v\tau \leq 0$ with its tip located at the

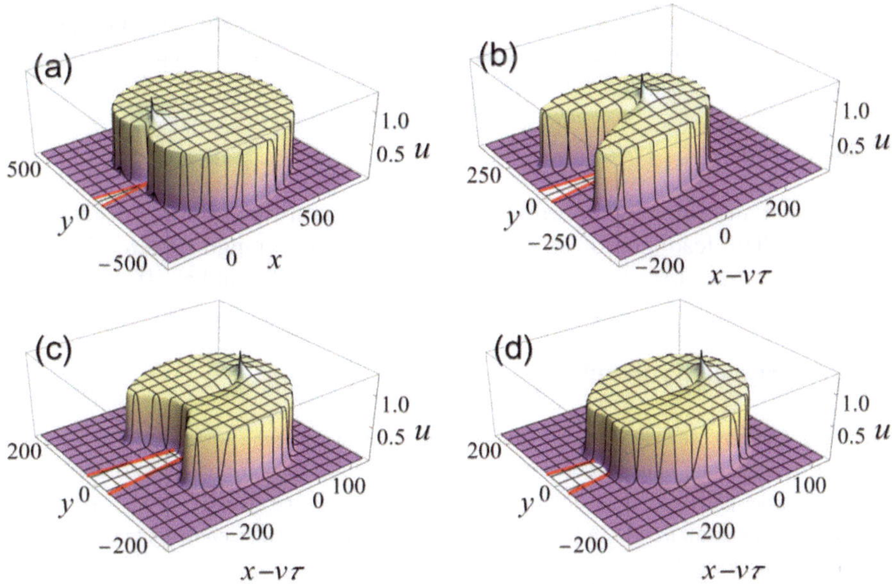

Fig. 13.4 Evolution of the Order Parameter Distribution Near the Crack Tip as Crack Velocity Increases. The sequence of distributions of the order parameter, $u(x - v\tau, y)$, within the process zone with the parameters $a = 0.24$, $b = 1$. We performed simulation at four velocity values: **a** $v = 0$, **b** $v = 0.159$, **c** $v = 0.238$ and **d** $v = 0.254$. The solid lines behind the distribution (shown in red in the color version) schematically indicate the positions of the crack surfaces

origin of the coordinates. The velocities of crack propagation for the order parameter distributions shown in Fig. 13.4 are (a) $v = 0$, (b) $v = 0.159$, (c) $v = 0.238$, and (d) $v = 0.254$. As we explore later on in Sect. 13.3, it is important to note that all these velocities are smaller than the speed corresponding to the morphological transformation, denoted as v_G. The velocity $v = 0.254$ (Fig. 13.4d) is only slightly below v_G.

13.2.1 Away from the Zone

Far from the process zone, the order parameter remains at zero.

13.2.2 "Roofs"

At the boundary of the zone, the order parameter exhibits a steep kink, which gradually transitions into an almost flat top surface, referred to as the 'roof' (Fig. 13.4a–d).

Figure 13.5a illustrates a cross-section of the order parameter distributions presented in Fig. 13.4 along the $y = 0$ plane, revealing two distinct types of roofs.

13.2.2.1 Roof α

One of them, indicated by the arrow α in Fig. 13.5a, corresponds to the leading part of the zone. This leading roof, characterized by a height u_1 ranging from 0.79 to 1.2, possesses a gentle slope,[1] $\partial u_1/\partial x$, varying between $\sim 10^{-3}$ and $\sim 10^{-4}$.

13.2.2.2 Secondary Details

The steepest, narrowest zone yields the former slope estimate, while the latter applies to the widest zone. The zone's width influences the roof's slope; wider zones exhibit shallower α-roofs. In this region, the order parameter gradually increases as we approach the crack tip at $x = y = 0$ and typically remains below $u = 1$. A peak reaching the value of $u = 1.2$ becomes noticeable in close proximity to the crack tip.

13.2.2.3 Roof β

The roof marked by the arrow β is situated behind the crack tip, categorizing it as the trailing part of the zone. This portion forms an almost flat plateau with a height of $u_2 \approx 0.8$ and an almost negligible slope. It's important to note that the values u_1 and u_2 provided above are specific to $a = 0.24$ and $b = 1$. In general, u_1 and u_2 depend on both a and b, as elaborated in Sect. 14.2.

Figure 13.5a also reveals that the speed of the crack does not significantly influence the heights, u_1 and u_2, of these roofs.

13.2.2.4 Topknot

In the immediate vicinity of the crack tip, a distinct feature that we refer to as the *topknot* appears in the order parameter distribution. This feature manifests as a narrow peak with a height of approximately $u_{\text{topknot}} \approx 1.2$, which is about 0.4 units higher than the β-roof, and about 0.2 units higher than the α-roof. The origin of the topknot we discuss in detail in Sect. 23.1.3 within the Chap. 23.1.3 dedicated to dislocations.

---

[1] Remind of the fact that u_1 and x are dimensionless together with their derivative.

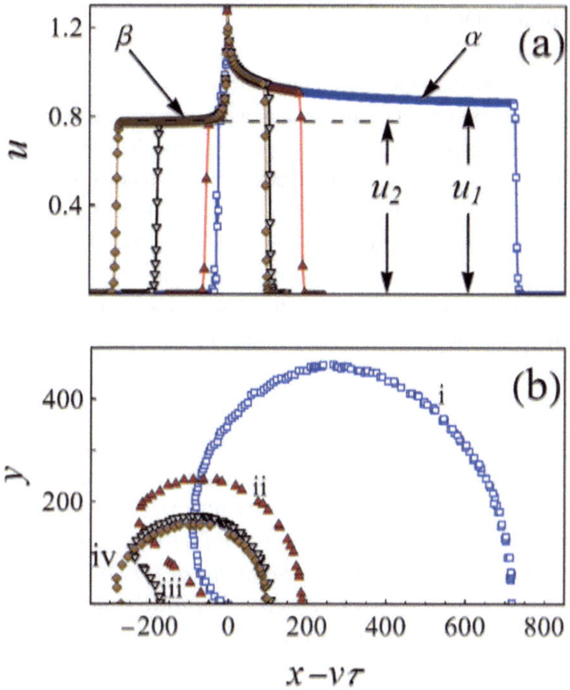

Fig. 13.5 **a** Cross-Section of the Zone Along $y = 0$ at $a = 0.24$, $b = 1$, for Various Velocities. The arrow (α) indicates the "roof" of the leading part of the distribution, with height u_1, while (β) marks the "roof" of the trailing part, with height u_2. The dashed line is the continuation of the level of the "roof" (β). **b** Zone Boundaries in the Upper Half-Plane ($y > 0$) corresponding to the cross-sections shown in (**a**) with matching plot markers and colors in the colore version. In (**a**), open squares and in (**b**), contour (i) represent a velocity of $v = 0$. Red upward-pointing triangles in (a) and curve (ii) in (**b**) correspond to $v = 0.159$. Brown diamonds in (**a**) and curve (iii) in (**b**) represent $v = 0.238$, while downward-pointing triangles in (**a**) and curve (iv) in (**b**) indicate $v = 0.254$

13.2.2.5 Zone Shape and Size as a Function of Velocity

Conversely, the zone's size decreases as the crack tip's velocity increases. This phenomenon is evident in Fig. 13.5b, where we depict the zone boundaries in the upper half-plane $y > 0$.

At lower crack tip speeds, the zone's contour looks like a distorted cardioid, as if the process zone at a motionless crack tip were compressed in size and slightly stretched backward. The point where the zone boundary intersects the crack surface, $y = 0$, also shifts backward to a certain position at $(-l_{\text{inv}}, 0)$, with the distance $l_{\text{inv}} = l_{\text{inv}}(v)$ increasing with velocity. Additionally, the zone exhibits an invagination in its rear portion. Notably, the depth of this invagination varies with crack tip velocity and gradually diminishes as v approaches v_G (refer to Fig. 13.5b, contour iv).

13.2.2.6 Secondary Details

In the following we use smaller letters to highlight details that one might wish to skip during the first reading, to prevent any potential misunderstanding, it's important to clarify that the distribution shown in Fig. 13.5a is taken along the line $y = 0$, passing through the invagination. The endpoint of roof β, as shown in Fig. 13.5a, aligns with the crack surface and possesses coordinates $x - v\tau = -l_{\text{inv}}$ and $y = 0$. This point lies behind the crack tip, situating at $x - v\tau = 0$, $y = 0$. The length $l_{\text{inv}} > 0$ equals the distance from the deepest point of the invagination $(-l_{\text{inv}}, 0)$ to the crack tip $(0, 0)$. In contrast, the size of the distribution's "petals" may be more substantial, and the x-coordinate projection of the back point of the distribution's petal, x_{a}, may exceed l_{inv} in absolute value. This distinction becomes apparent when comparing Fig. 13.5a with Fig. 13.5b.

***

In summary, *in cases of a stationary or slowly advancing crack, the process zone closely resembles a cardioid, characterized by an invagination at its trailing edge. The slowly propagating process zone exhibits several characteristic features. These include the α-roof with a gentle slope, the β-roof with no slope, and an invagination in the trailing part of the distribution. The depth of this invagination shows a non-monotonic dependence on the crack velocity: it initially increases at low velocities, then decreases, and reaches zero at Gordon's velocity.*

13.3 Morphological Transition

A morphological transformation occurs when $V = V_{\text{G}}$. At this point, the invagination at the trailing edge of the zone disappears, and the zone becomes convex. This section focuses on the characteristics of the process zone at Gordon's velocity.

***

As the crack tip velocity reaches a critical value, denoted as v_{G}, the order parameter distribution loses its concave shape. At this point, the rear part of the zone boundary becomes perpendicular to the y-axis. In the cases of $a = 0.22$, 0.23, and 0.24, the distribution $u(x - v\tau, y)$ is presented in Fig. 13.6a–c, precisely at the crack tip velocity that triggers the morphological transformation.

13.3.1 Secondary Details

The velocities of the morphological transformation, v_{G}, depend on the specific location within the phase diagram. As we maintain $b = 1$, this dependence becomes apparent in the variation with a. The values $v_{\text{G}} = v_{\text{G}}(a)$ are summarized in Table 13.1.

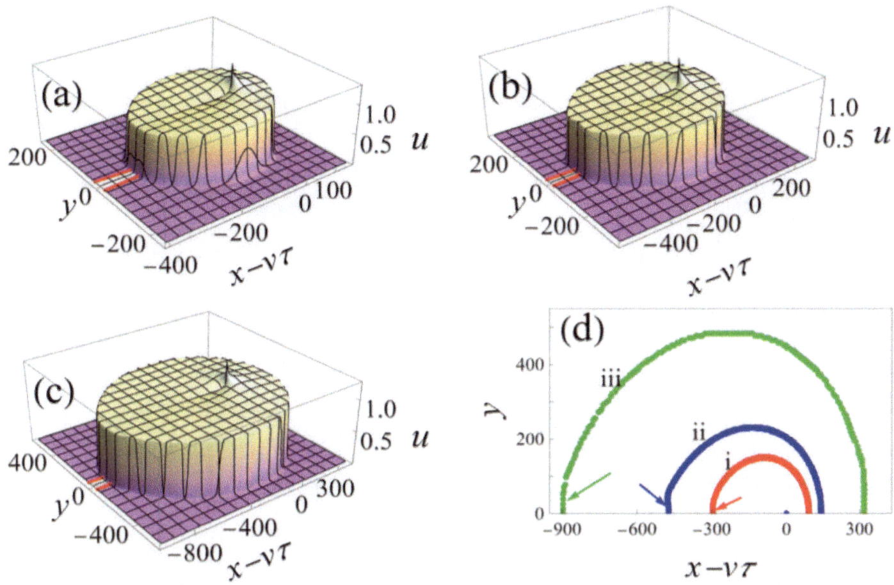

Fig. 13.6 **a–c** Order Parameter Distribution at the Crack Tip Propagating at the Morphological Transformation Velocity. The dimensionless Gordon's velocity, v_G, is equal to 0.264 at $a = 0.24$ (**a**), 0.22 at $a = 0.23$ (**b**) and 0.149 at $a = 0.22$ (**c**). The solid lines behind the distribution (red in the color version) schematically show the position of the crack surfaces. **d** The boundary contours in the upper semi-plane ($y > 0$) of the zones propagating with the velocities of the morphological transformation. (i) corresponds to the image (a), (ii) to (b), and (iii) to (c). The arrows indicate the trailing zone's end, where the zone boundary is perpendicular to the y-axis

Figure 13.6d shows the boundaries of zones propagating at the velocities of the morphological transformation for the cases depicted in Fig. 13.6a–c. The arrows highlight the trailing zone's endpoints, where the zone boundary assumes a flat, perpendicular orientation to the y-axis.

***

In summary, *when* $V = V_G$, *the invagination vanishes, and the process zone adopts a convex configuration.*

Table 13.1 Velocities of the morphological transformation

a	0.22	0.23	0.24
v_G	0.149	0.220	0.264

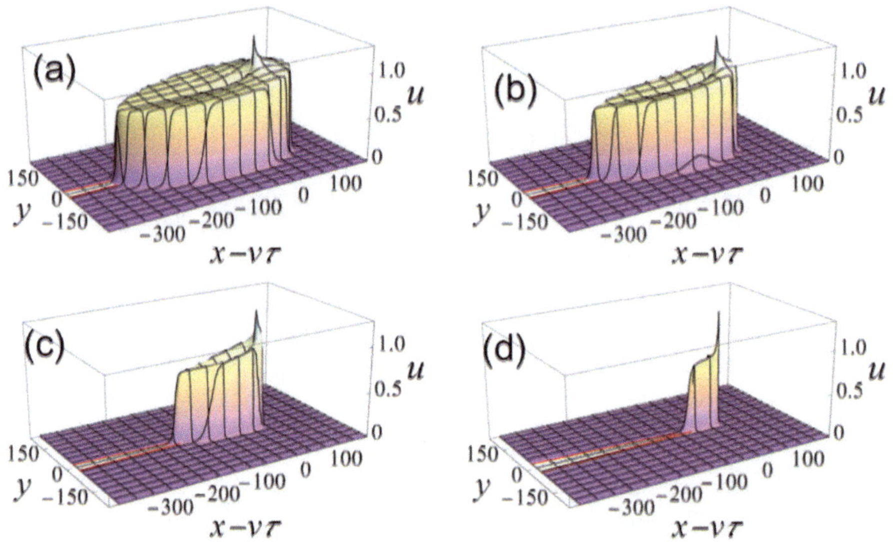

Fig. 13.7 Order Parameter Distribution Around the Tip of a Crack Propagating with the Velocity $v > v_G$. Simulation results with $a = 0.24$ at $b = 1$ are shown. Panel **a** displays the configuration just above the transition $v \approx 0.294$, **b** at $v \approx 0.4$, **c** at $v \approx 0.63$, and **d** depicts the distribution at $v \approx 0.82$, close to the critical velocity, $v_d = 1$, at which the zone vanishes. The solid lines (red in the color version) schematically indicate the position of the crack surfaces

13.4 Process Zone at the Tip of a Rapidly Propagating Crack

In this chapter, we investigate the process zone at the tip of a rapidly propagating crack through simulations, focusing on velocities in the range $V_G < V \leq V_d$. Our analysis explores the peculiarities of the order parameter distribution in this regime, examining the structure of the process zone, and identifying the critical velocity, V_d.

***

13.4.1 Dependence of Zone Configuration on Crack Velocity

When the velocity surpasses v_G, the zone transforms into a convex shape. Figure 13.7 illustrates the process zone configurations at different velocities $v > v_G$. The zone takes on the appearance of a droplet with a pointed end, significantly elongated along the x-axis in the direction opposite to the crack tip velocity.

The relationship between zone geometry and crack velocity becomes apparent in Fig. 13.8a, showcasing the contours of the process zone for the order parame-

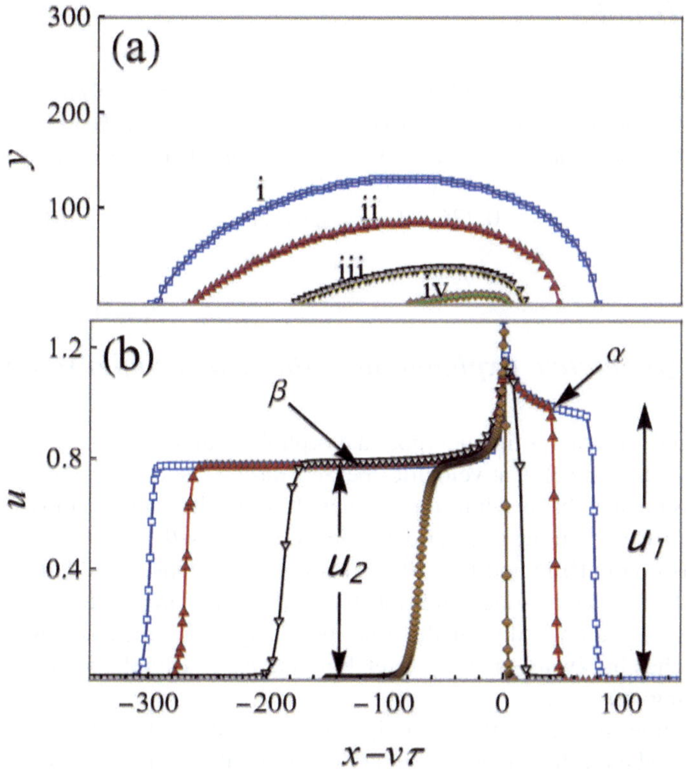

Fig. 13.8 **a** Boundaries of the Process Zones Shown in Fig. 13.7. (a) The line indicated by i corresponds to the distribution shown in Fig. 13.7a. ii—to Fig. 13.7b, iii—to Fig. 13.7c, and iv corresponds to Fig. 13.7d. **b** Cross-Sections of the Order Parameter Distributions, u, Along the Crack, as Shown in Fig. 13.7 (α) represents the 'roof' with the height u_1 in front of the crack tip, while (β) designates the one with the height u_2 behind it

ter distributions presented in Fig. 13.7a–d. Specifically, contour (i) corresponds to Fig. 13.8a, contour (ii) to (b), contour (iii) to (c), and contour (iv) to (d).

13.4.2 Key Features of the Rapidly Moving Zone

- The zone size diminishes as the velocity increases.
- Similar to the previous scenario, these distributions exhibit two 'roofs:' α and β like those described in Sect. 13.2.2.

13.4.3 Details

The one marked by arrow α (Fig. 13.8b) possesses a height of approximately $u_1 \approx 0.99$, located in front of the crack tip, and displays a shallow slope, $\partial u_1/\partial x$, varying between 5×10^{-3} and 2×10^{-2} (see the footnote at the Sect. 13.2.2). This slope only becomes pronounced, forming a small peak near the crack tip ($0 \leq x - v\tau \lesssim 10$). The rear roof (indicated by arrow β) has a height of approximately $u_2 \approx 0.776$ and an insignificant slope $\partial u_2/\partial x \sim 10^{-4}$ up to the kink region.

13.4.4 Qualitative Explanation of the Zone Configuration

The region at the crack's tip undergoes a morphological transformation, as described in Chap. 13. Specifically, at velocities below Gordon's speed v_G, the crack tip zone assumes a concave shape, while at velocities above v_G, it becomes convex. This transformation arises from the angular dependence $U \sim \cos(\theta/2)$ of the crack potential, denoted as $U(\mathbf{r})$ (10.7) and Fig. 10.1. This potential leads to the concave configuration of the stationary zone. Additionally, as the crack propagates, an invagination persists at the trailing edge of the zone boundary. The crack advances while the zone stretches backward, and this stretching increases with the velocity, eventually disappearing at $v = v_G$.

At velocities above v_G, the crack tip zone takes on a droplet-like shape. Its wake can be described as having approximately a triangular shape (see Figs. 14.1c, 13.7, 13.9). This triangular configuration is again a consequence of the angular dependence of the potential $U \sim \cos(\theta/2)$. The potential contribution becomes negligible near $\theta = \pm\pi$. Consequently, near such angles, the wake is unaffected by the potential and behaves similarly to the metastable material at a constant temperature and pressure. The reduction in the volume of this metastable phase occurs through the propagation of a planar phase front at Gordon's velocity, v_G, ultimately resulting in the formation of the triangular configuration, as detailed in Sect. 3.5.

Both the morphological transition and the wake configuration at velocities above v_G are attributable to the specific angular dependence of the potential, $U = U(\theta)$ (10.7). This form arises from the one-component nature of the order parameter under study, which limits consideration in the Landau potential to the single interaction invariant $\eta^2\varepsilon_{\alpha\alpha}$ (3.12), thereby accounting for dilatational spontaneous transformation. In contrast, if the order parameter is multi-component, the interaction term can be more intricate, involving several components $\varepsilon_{\alpha\beta}$. Such a dependence implies various possible configurations of the process zones, including those with multiple petals that exhibit a concave shape. We refer the reader to the paper [3] examples.

13.4.4.1 Details

However, in the elastically-isotropic plane problem, the stress components $\sigma_{\alpha\beta}$ ($\alpha\beta = 11, 22, 33$ and 12) determined by the relations (1.1) always become zero as θ approaches $\pm\pi$, along with $\varepsilon_{\alpha\beta}$ and any linear combinations thereof. Consequently, under the conditions outlined in the preceding sentence, the process zone, even in the case of a multicomponent order parameter, will exhibit an invagination at its rear. This invagination depth decreases with increasing crack speed and disappears at $v = v_G$. Furthermore, at velocities above v_G, the potential near $\theta \approx \pm\pi$ is small, facilitating the formation of the triangular wake.

13.4.5 Variation of Order Parameter Distributions with Phase Diagram Position

It is significant to examine how the configurations of order parameter distributions change depending on the location of the crystal within the phase diagram while keeping velocity constant. Figure 13.9 illustrates such distributions obtained at a velocity $v \approx 0.79$. We performed the simulations for a single crack-zone complex, positioning these distributions closely in the image for easy comparison. Notably,

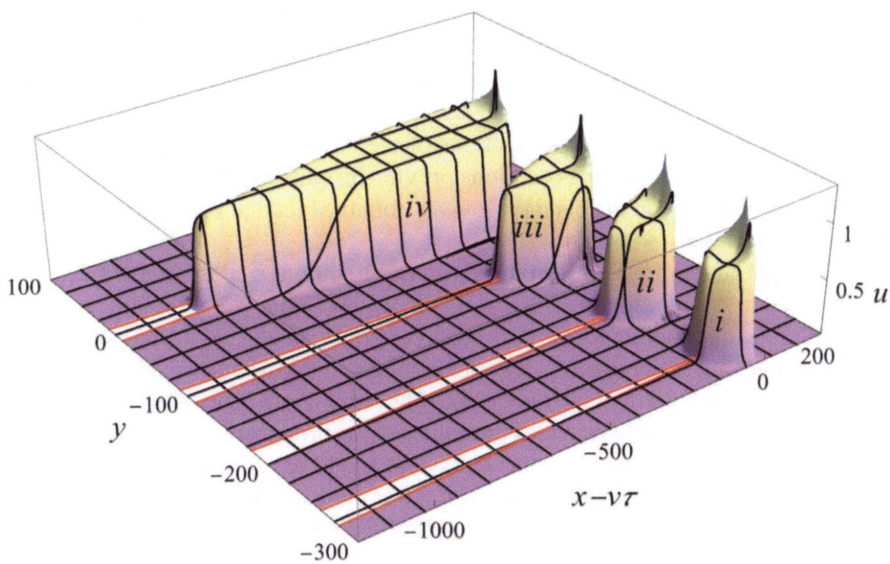

Fig. 13.9 Comparison of Zone-Wake Complex Configurations Propagating at the Same Velocity Exceeding v_G, but at Different Distances from the Binodal **i** $a = 0.24$, **ii** $a = 0.23$, **iii** $a = 0.22$ and **iv** $a = 0.20$. As one approaches the binodal $a_b \approx 0.18$, the length of the wake dramatically increases, and the 'roof' height of the $u = u(x - v\tau, y)$ distribution remains constant. The solid lines (red in the color version) indicate the crack tips' positions

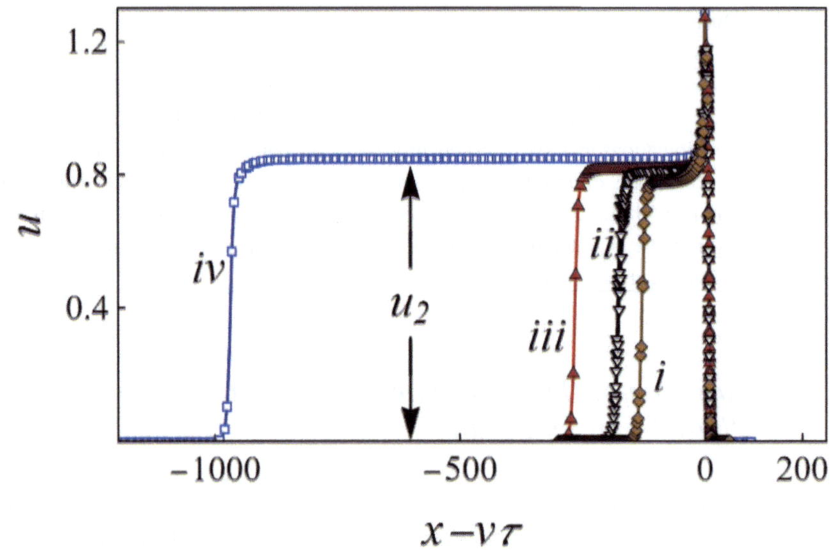

Fig. 13.10 Cross-Section of the Order Parameter Distributions Shown in Fig. 13.9 *Along the Line* $y = 0$. **i** $a = 0.24$, **ii** $a = 0.23$, **iii** $a = 0.22$ and **iv** $a = 0.20$

when approaching the binodal, the distribution's dimension in the x-direction experiences a significant increase. Some limited growth is also observed in the so-called "zone height," traditionally defined as the maximum dimension of the zone in the y-direction.

Let us look closer at the cross-sections of the order parameter distributions presented in Fig. 13.9. These cross-sections provide insights into how the "roof" of the distributions depends on the distance $a - a_b$ from the binodal. Figure 13.10 displays these cross-sections.

The height of the 'roof' β forming the wake is summarized in Table 13.2.

The second row of Table 13.2 presents the values of the "roof" height, u_2, at the rear part of the distribution for various values of the parameter a. As one can observe, when approaching the binodal, the "roof" height, denoted as u_2, experiences a slight increase. It is worth noting that the values of u_2 remain independent of velocity. The third row of the same table provides the values of the order parameter in the daughter phase, u_d, in the stationary homogeneous case (3.28) at $b = 1$ for different values of a. A close agreement between u_2 and u_d is evident, suggesting that the order parameter relaxes to u_d within the inner part of the wake.

Table 13.2 Dependence of the 'Roof' height on a

a	0.2	0.22	0.23	0.24
u_2	0.85	0.82	0.81	0.79
u_d	0.85	0.82	0.80	0.78

Table 13.3 Interface width as a function of velocity

v	0.24	0.40	0.56	0.79	0.95
d	1.83	1.63	1.45	1.22	1.08

13.4.6 Determination of Phase Boundary Width

Another intriguing question pertains to the width of the interface boundaries. We extracted this information from the quantity $(\nabla u)^2$. Throughout almost the whole domain's part, except at the kink, $(\nabla u)^2$ remains nearly zero. At the kink, $(\nabla u)^2$ exhibits a bell-shaped behavior, as depicted in Fig. 13.11a.

Figure 13.11a illustrates the behavior of $(\nabla u)^2$ in the leading part of the utmost zone for $a = 0.24$ at various velocities: $v = 0.24$, 0.40, 0.56, 0.79, and 0.95. We have fitted the simulation data points with a Gaussian curve,

$$(\nabla u)^2 \approx s \exp\left[-(x - x_0)^2 / d^2\right],$$

where s, x_0 and the front width d are fitting parameters.

The relationship $d = d(v)$, as summarized in Table 13.3, is depicted in Fig. 13.11b.

Note 13.4.1 Notably, in this regime, the thickness of the phase boundary remains consistently much smaller than the size of the zone, a trend observed for other values of the parameter a as well.

Note 13.4.2 Furthermore, one can observe that the phase front width decreases as the velocity increases.

13.4.7 Progressive Changes in Zone Configuration Toward the Critical Velocity

Once the velocity surpasses a critical threshold, $v_d = 1$, the process zone collapses abruptly.

Figure 13.12 illustrates a comparison of three zone-wake complexes, each taken at $a = 0.24$, with velocities that gradually increase: (a) $v = 0.82$, (b) 0.88, and (c) 0.94. The latter is the closest to the critical velocity value we have achieved. As we continued to increase the velocity beyond this point, the solution for u decreased,

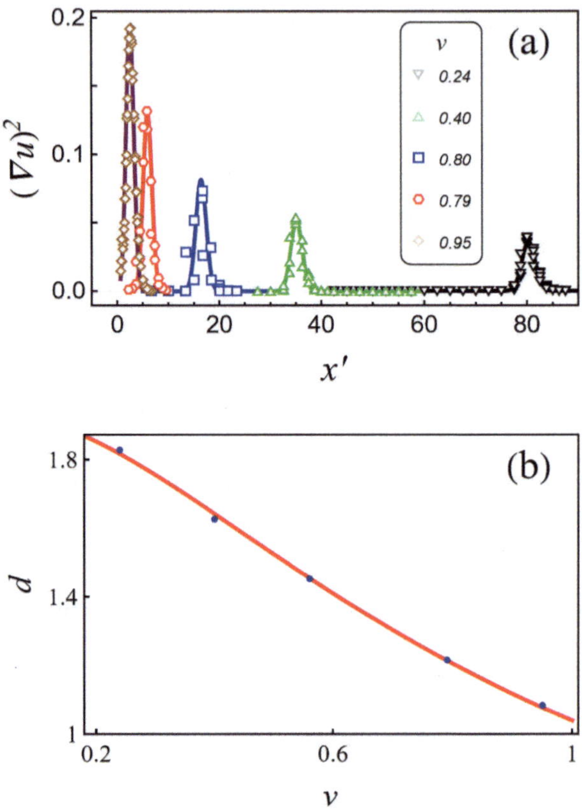

Fig. 13.11 **a** Behavior of $(\nabla u)^2$ as a Function of x. We show the cross-section of the 3D plots along the line $y = 0$ in the leading part of the process zone. The data points represent simulation results, while the solid lines display their fitting using a Gaussian curve. The panel **b** demonstrates the dependence of the phase interface width, d, on the crack velocity, v. Both parameters are dimensionless. The dots represent the front thicknesses obtained by fitting the curves shown in (**a**) and summarized in Table 13.3, while the solid line serves as a visual guide

approaching the limits of numerical precision. Notably, the transition from $v = 0.82$ to $v = 0.94$ dramatically reduced the wake's length. Similarly, the zone's height (h) and width (r_0) also decreased, albeit not as dramatically as the reduction in the wake length. Importantly, the 'roofs' height, represented by $u_{1,2}$ in Fig. 13.8, showed no significant variation up to the critical velocity.

***

In summary, *surpassing Gordon's velocity induces a notable change: the trailing edge elongates backward, transforming the zone into a droplet-like shape of finite size. In this regime, the process zone adopts a configuration consisting of two distinct parts: a leading section with the α-roof and a trailing section with the*

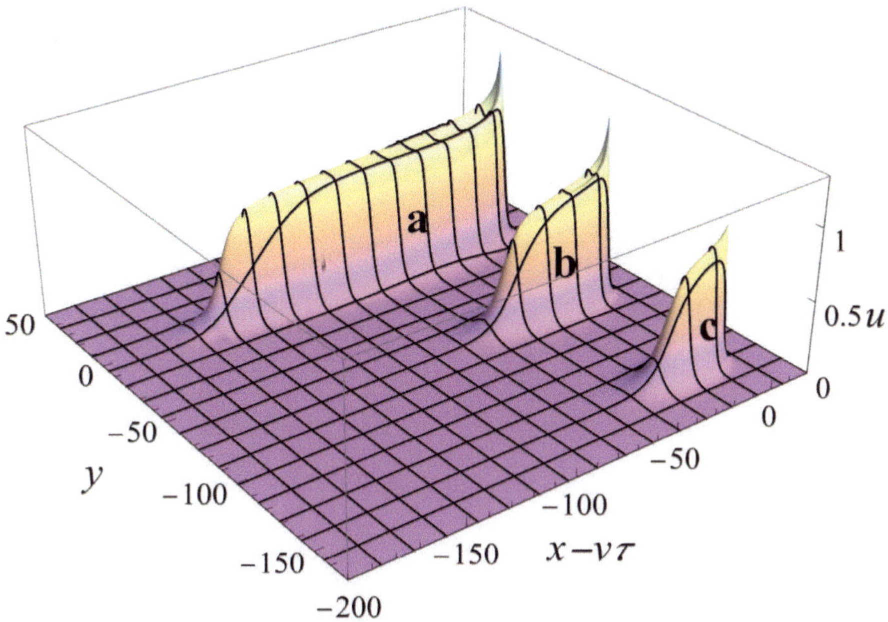

Fig. 13.12 Gradual Evolution of the Process Zone–Wake Complex Near the Critical Velocity at $a = 0.24$. **a** $v \approx 0.82$, **b** $v \approx 0.88$, and **c** $v \approx 0.94$. Further increases in v result in a pronounced vanishing of u, rendering it imperceptible in the plot

β-roof. The α-roof exhibits a gentle slope, while the β-roof maintains a constant height, closely corresponding to the order parameter value in the metastable daughter phase, thereby forming the metastable wake. The size of this wake is finite and depends on the crystal's position in the phase diagram, infinitely increasing as the system approaches the binodal. Both the leading and trailing parts end abruptly, beyond which the order parameter drops nearly to zero. The interface width ranges from 1 to 2 dimensionless units, two to three orders of magnitude smaller than the overall zone size. As the velocity approaches the critical value, the zone shrinks, and at $V \geq V_d$, it ceases to exist.

References

1. A. Boulbitch, Mathematica Journal **20**, 1–21 (2018)
2. A. Boulbitch, A.L. Korzhenevskii, Phys. Rev. **E101**, 033003 (2020); idem Phys. Rev. **E101**, 033004 (2020)
3. A. Boulbitch, Y.M. Gufan, A.L. Korzhenevskii, Phys. Rev. E **96**, 013005 (2017)

Chapter 14
Key Geometrical Parameters of the Crack Tip Zone

Abstract This chapter provides analytical estimates of the geometric parameters of the process zone, highlighting the physical phenomena that govern the peculiarities of its shape. Additionally, we compare these analytical findings with the results from simulations (Boulbitch and Korzhenevskii 2020).

14.1 List of Essential Parameters of the Zone

This section outlines the key geometrical parameters that will be the focus of our analysis in the subsequent discussion.

***

To begin, let us enumerate the geometric parameters of the process zone that interest us. These parameters will be discussed in terms of dimensionless units. Additionally, when we refer to a parameter as "large" or "small," we mean it is significantly larger or smaller than unity. We will convert to "physical" (dimensional) units in Sect. 16.3.

Characterizing the process zone involves several key sizes, all expressed in dimensionless terms, yielding an understanding of the crack tip zone behavior. They include:

1. Zone *width* r_0: The distance from the crack tip to the leading edge of the zone, which signifies its width. See Fig. 14.1a–c.
2. Zone *height* h: The maximum ordinate value of the zone contour, representing its height. See Fig. 14.1a–c.
3. The distance l_{inv}: At lower velocities, the dimension l_{inv} characterizes the *length of the direct contact* between the crack faces and the zone. It is equal to the distance from the tip to the base of the zone invagination, marked by point O_1 in Fig. 14.1a.
4. At large velocities, l_{wake} represents the distance from the crack tip to the trailing end of the wake, indicated by point O_3 in Fig. 14.1c.
5. Curvature radius ρ_1: The curvature radius at the bottom of the invagination, which is a key characteristic of the zone, is depicted in Fig. 14.3a, not in Fig. 14.1.

© The Author(s), under exclusive license to Springer Nature Switzerland AG 2026

A. Boulbitch and A. Korzhenevskii, *Stress-Induced Local Phase Transformations Near Cracks and Dislocations*, Springer Series in Materials Science 356,
https://doi.org/10.1007/978-3-032-00945-6_14

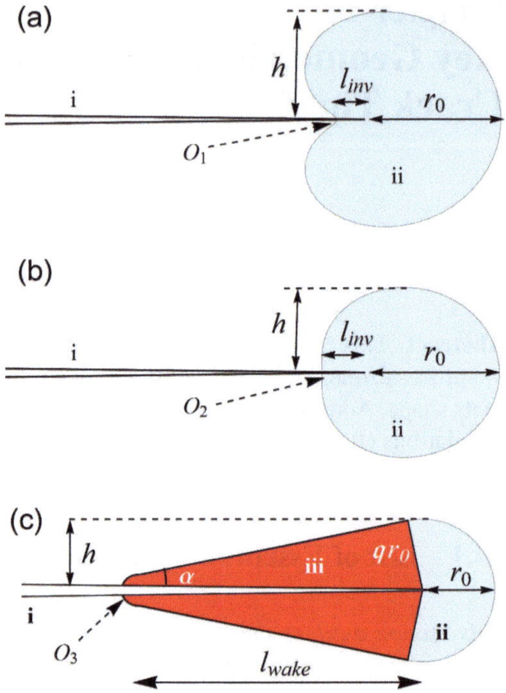

Fig. 14.1 Illustrative View of the Process Zone and Wake Formed at a Propagating Crack Tip. In all panels: (i) shows the crack position schematically, (ii) highlights the zone, and (iii) marks the metastable wake. Panel **a** depicts the zone configuration for small propagation velocities ($v < v_G$). Panel **b** shows the configuration at $v = v_G$. Panel **c** illustrates the zone-wake configuration at high propagation velocities ($v_G < v < v_{cr}$). The parameters l_{inv} and l_{wake} represent the distance from the crack tip to the deepest invagination point in (**a**), the rear zone end in (**b**), and the wake length in (**c**)

All of these parameters are indicated in Fig. 14.1 for visual reference.

Figure 14.1b illustrates a situation where the invagination disappears. In this case, we still use the same parameters mentioned above, as shown in (a). The curvature radius at point O_2 is $\rho_2 = \infty$.

Additionally, the parameters that describe the wake include width r_0, height h, the semi-vertex angle α, the distance l_{wake} between the tip and the trailing end of the wake (point O_3), also referred to as the *setback* or the *contact length*, and the rounding radius ρ_3 (not indicated) of the trailing edge at point O_3 (Fig. 14.1c).

***

In summary, *the geometric parameters outlined in this section govern the properties of the process zone. Their analytical descriptions will be provided in the following sections.*

14.2 Geometry of the Zone at the Tip of a Motionless Crack

This section presents an analytical study of the geometry of the process zone at the tip of a stationary crack, with a focus on (i) the size of the zone and (ii) the contact length - the distance from the crack tip to the point where the zone's boundary separates from the crack surface.

***

14.2.1 Boundary of the Mature Zone at the Motionless Crack Tip

Examining the motionless regime is of particular significance, as it plays a vital role in shaping what is known as *transformation toughness*.[1]

In the context of a motionless zone, let us estimate the zone size and shape, assuming that the zone is large. In this case, in (12.12), (12.13), the term with the Laplacian, Δu, is small compared to $U(\mathbf{r})u$ and can be neglected. The estimate is obtained by equating the term $a + U(\mathbf{r})$ from (12.12)—where a is taken at the actual temperature T: $a = a_0(T - T_c)$ and $U(\mathbf{r})$ is defined by relation (12.11)—to the value of a at the binodal point: $a_b = a_0(T_0 - T_c)$ (3.32):

$$a_0(T - T_c) - \frac{\sqrt{2}\cos(\theta/2)}{\sqrt{r}} = a_0(T_0 - T_c) \tag{14.1}$$

This equation yields $r = r(\theta)$:

$$r(\theta) = r_d \cos^2(\theta/2) \tag{14.2}$$

where the width of the motionless zone, r_d, is approximately:

$$r_d \approx \frac{512}{(16a - 3b^2)^2} \equiv \frac{2}{a_0^2\,(T - T_0)^2} \tag{14.3}$$

One can see the cardioid and related notations in Fig. 14.3. It's important to note that (14.2) holds well when $r_d \gg 1$, as surface tension was disregarded in this approximation. The cardioid's curvature radius in its leading section, ρ_{lead}, is given by:

$$\rho_{lead} = \frac{2}{3} r_d; \;\; \theta = 0 \tag{14.4}$$

while its height, h, occurs at $\theta = \pi/3$ and is represented as:

$$\theta = \pi/3; \;\; h = \frac{3\sqrt{3}}{8} r_d \tag{14.5}$$

---

[1] The reader should refer to Chap. 17 below for its more comprehensive discussion.

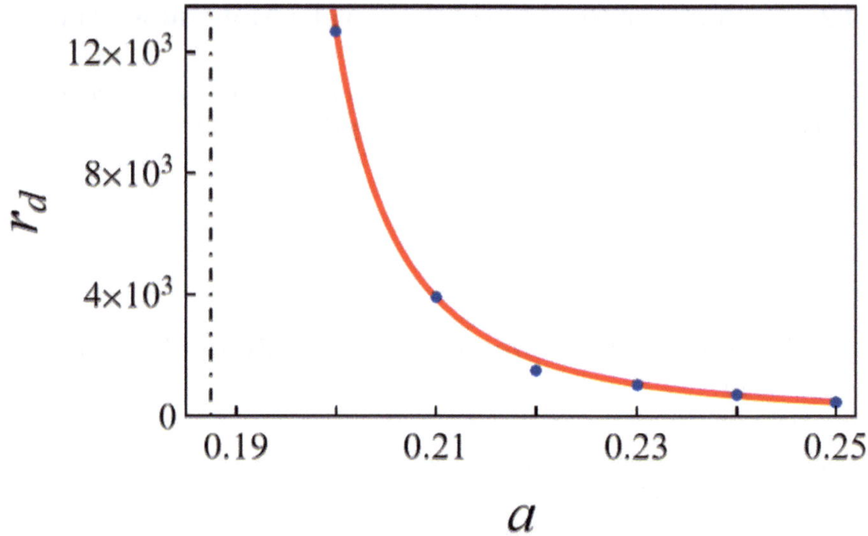

Fig. 14.2 Dependence of Cardioid Size $r_{\rm d}$ on Parameter a at Fixed $b = 1$. Dots depict the simulation results, while the solid line represents the curve from (14.3). The dot-dashed vertical line indicates the binodal position where the curve diverges

Equations (14.2), (14.4), (14.5) provide a reliable approximation for the zone's contour at the tip of a motionless crack. In the next section, we present simulations that confirm the validity of the assumption $r_{\rm d} \gg 1$, which, in turn, justifies the use of relations (14.2), (14.3).

Figure 14.2 illustrates the behavior of the function $r_{\rm d} = r_{\rm d}(a)$ at $v = 0$ (14.3) (solid line) alongside the simulation results (dots). Equation (14.3) serves as an excellent approximation for characterizing the size of the motionless zone. The extensive zone size is particularly noteworthy, with $r_{\rm d}$ spanning from approximately 10^3 to 10^4 dimensionless units.

It is important to note that this result for the zone contour is derived from (12.12) under the assumption of $\Delta u = \nabla u = 0$, a scenario referred to as the *gradientless approximation*. When compared with the rescaled free energy (8.10), (8.11), we observe that the gradientless approximation overlooks the surface tension term $\sim(\nabla u)^2$. This leads to the conclusion that the approximation presented in (14.2), (14.3) remains valid until surface tension begins to dominate.

For references, we give the expression for the zone's size $\mathbb{R}_{\rm d}$ corresponding to $r_{\rm d}$ (14.3) in terms of the original dimensional variables:

$$\mathbb{R}_{\rm d} = \frac{2}{9\pi}\left[\frac{6AK_I(1+\nu)(1-2\nu)}{\alpha_0 E(T-T_0)}\right]^2 \tag{14.6}$$

14.2.2 Contact Length

In our investigation of the crack tip zone, especially in the vicinity of point O_1 Fig. 14.1, it's necessary to address the concept of contact length, which we also refer to as the "*setback*." At this specific location, the cardioid shape described in (14.2) exhibits a cusp, as denoted by (ii) in Fig. 14.3a. However, this cusp is energetically unfavorable due to its infinitesimally small curvature radius when considering the influence of surface tension. Consequently, surface tension intervenes and transforms this cusp into a smooth boundary segment with a finite curvature radius ρ_1, indicated as (iii) in Fig. 14.3a. This transformation process results in forming a setback and a contact length, termed as l_{inv}. It's this mechanism that underlies the observed setback in our simulations.

Below, we derive an analytical approximation for the contact length l_{inv}, as also illustrated in Fig. 14.3a. We achieve this by approximating the boundary invagination as a segment of a circle with a radius of ρ_1. The circular arc must merge smoothly into the cardioid contour (Fig. 14.3a iii), so that the transition from the cardioid to the circle is continuous, preventing an increase in surface energy.

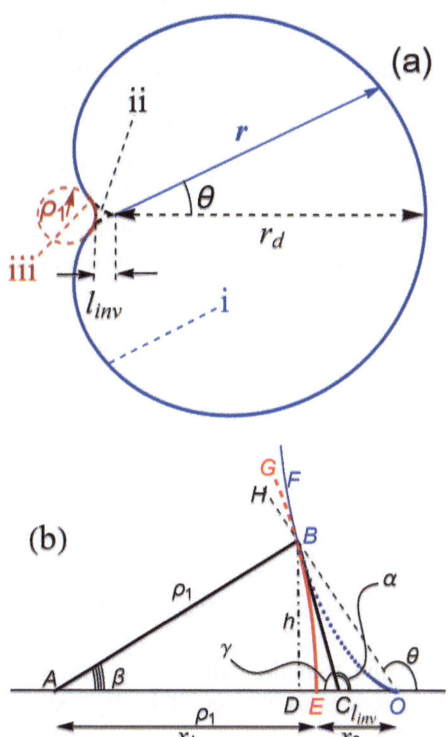

Fig. 14.3 Configuration of the Process Zone Boundary. **a** The cardioid contour (i), its cusp (ii), and the continuous, smooth continuation of the cardioid contour by the circular arc (iii) (red in the color version). We have indicated the zone width, r_0, the contour radius-vector, r, its polar angle, θ, the curvature radius of the invagination, ρ_1, and the setback l_{inv}. **b** A magnified view of the rear part of the image (**a**) with the geometric notations used in the derivation below. Further details are provided in the text

Note 14.2.1 The boundary smoothness condition is essential, as a kink in the boundary curve significantly increases the surface energy cost.

This smoothness condition determines the curvature radius ρ_1, as we demonstrate below. The contact length signifies the interval size where the zone directly interfaces with the crack surface.

14.2.2.1 Derivation of the Contact Length Relation to ρ_1 and r_d^2

Now, let us derive this contact length.

Figure 14.3b provides an enlarged view of the region surrounding[2] the cusp, offering a detailed perspective of the circular arc EBG inscribed into the cardioid OBF while maintaining the zone contour's continuity and smoothness. The dimensions are indicated in terms of the dimensionless units introduced earlier. The zone interface comprises the circular arc EB (highlighted as (iii) in Fig. 14.3a), followed by the section of the cardioid BF (pointed out as (i) in Fig. 14.3a). Solid lines represent both portions of the interface. It's important to clarify that the dashed line BG extends from the circle. In contrast, the dotted line OB continues from the cardioid at point B to the coordinate origin O, corresponding to the portion (ii) in Fig. 14.3a. However, these lines do not conform to the zone boundary.

The arc GBE contacts the contour OBF of the cardioid at point B, where they share a common tangent BC. This tangent exhibits a slope angle α. The contact length, l_inv, is the distance between the cardioid's origin O and point E, where the circular arc intersects the x-axis. The angle subtended at point A by point B is labeled as β, and the angle at point O, subtended by point B, is marked as θ. Additionally, h represents triangle ABC's height BD (dot-dashed line). The lengths AB and AE are equal to the curvature radius of the arc, ρ_1. We denote AD as x_1 and DO as x_2.

One parametrically describes the cardioid as:

$$\mathbf{r}(\theta) = r_\mathrm{d}\cos^2\left(\frac{\theta}{2}\right)(\cos\theta, \sin\theta, 0) \tag{14.7}$$

Here r_d is the dimensionless zone width. Utilizing (14.7), we can calculate the slope, $\tan(\alpha)$, of the tangent BC:

$$\tan(\alpha) = -\cot\left(\frac{3\theta}{2}\right)$$

Let us introduce the angle $\varphi = \pi - \alpha$, from which we can derive from the relation: $\beta = (\pi - 3\varphi)/2$.

In the triangle ABC $h = \rho_1 \sin(\beta) = \rho_1 \cos(3\varphi/2)$. Simultaneously, h equals the Y-coordinate at point B. By expressing h in terms of the angle φ, we find

[2] This section may be skipped during a first reading.

$h = r_d \sin^2(\varphi/2)\sin(\varphi)$. Equating these expressions for h leads to the following expression:

$$\frac{\rho_1}{r_d} = \frac{\sin^2(\varphi/2)\sin(\varphi)}{\cos(3\varphi/2)} \tag{14.8}$$

Next, we determine the values of x_1, x_2, and l_{inv}. It's apparent that $x_1 = \rho_1\cos(\beta) = \rho_1\sin(3\varphi/2)$, and x_2 represents the projection of point B onto the X-axis, resulting in $x_2 = r_d\sin^2(\varphi/2)\cos(\varphi)$. Ultimately, the dimensionless contact length, l_{inv}, is calculated as $l_{inv} = x_1 + x_2 - \rho_1$. Consequently, we find that $l_{inv}/r_d = (\rho_1/r_d)[\sin(3\varphi/2) - 1] + \cos(\varphi)\sin^2(\varphi/2)$. By substituting the expression for ρ_1/r_d from (14.8) into this equation, we obtain:

$$\frac{l_{inv}}{r_d} = \frac{\sin^2(\varphi/2)}{1 + 2\sin(\varphi/2)} \tag{14.9}$$

Equations (14.8) and (14.9) work in concert to establish the relationship between l_{inv}/r_d and ρ_1/r_d in a parametric form. This relationship provides valuable insights into the behavior of contact length concerning the curvature radius.

During our extensive simulations, we consistently observed that the inequality $\rho_1/r_d < 0.1$ remains valid. As a result, we chose to focus exclusively on this specific case.

In Fig. 14.4, we illustrate the relationship between l_{inv}/r_d and ρ_1/r_d, as described by (14.8) and (14.9). The data points, represented by dots, align with this relationship. Remarkably, at $\rho_1/r_d < 0.1$ this data closely adhere to the fitting formula (14.10).

The following approximate formula accurately fits the data shown in Fig. 14.4:

$$l_{inv} \approx 0.2(\rho_1 \times r_d)^{1/2} \tag{14.10}$$

Note 14.2.2 It becomes evident that a non-zero contact length, $l_{inv} \neq 0$, is intimately tied to a non-zero curvature radius, $\rho_1 \neq 0$. In simpler terms, the emergence of the

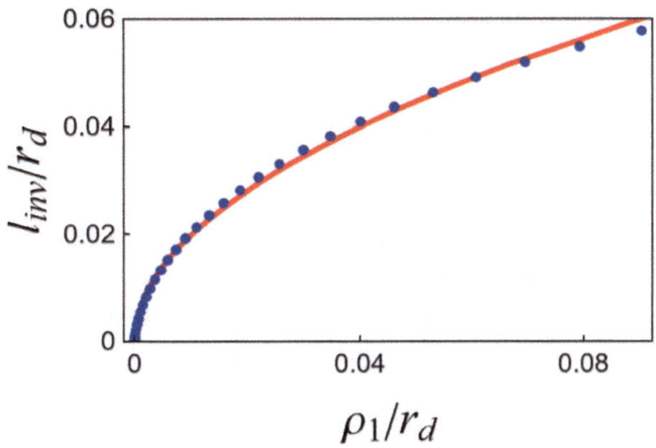

Fig. 14.4 Relationship Between l_{inv}/r_d and ρ_1/r_d. The dots are obtained with (14.9) while the solid line shows it fitting with (14.10)

contact length is an inevitable consequence of the contribution of surface tension of the process zone interface.

14.2.2.2 The Invagination Curvature of the Motionless Zone

Here, we determine the curvature radius, ρ_1, at the bottom of the invagination (Fig. 14.3a), located at the rear of the stationary zone. This point, marked as O_1 in Fig. 14.1. Although we discuss the motionless zone, this derivation approaches it from a dynamic perspective. Specifically, consider point O_1 as the location where surface tension reduces the forward velocity of the curved boundary to zero ($v_G^{(-)} = 0$). This perspective allows us to apply (3.74) yielding:

$$\rho_1 = \frac{1}{2^{2/3} v_G}$$

Furthermore, employing (3.69), we can express the curvature radius, ρ_1, as follows:

$$\rho_1 = \frac{\sqrt{6}\left|b^2 - 2a + b\sqrt{b^2 - 4a}\right|^{1/2}}{8a - b^2 - b\sqrt{b^2 - 4a}} \tag{14.11}$$

It is worth noting that at the binodal $a = 3b^2/16$, the radius ρ_1 (14.11) diverges.

14.2.2.3 The Contact Length in Terms of *a* and *b*

In Sect. 14.2.2 we demonstrate a lengthily derivation of the approximate formula for the dimensionless contact length of the invagination expressed in terms of ρ_1 and r_d. The expressions (14.11) for ρ_1 at rest and (14.3) for r_d now enable us to derive the expression for the contact length l_{inv} within the motionless zone in terms of a and b. This contact length represents the interval where the zone directly interfaces with the crack surface:

$$l_{inv} \approx \frac{l_0}{16a - 3b^2} \frac{\left[b\left(b^2 - 4a\right)^{1/2} - 2a + b^2\right]^{1/4}}{\left[8a - b^2 - b\left(b^2 - 4a\right)^{1/2}\right]^{1/2}} \tag{14.12}$$

Here, $l_0 \approx 6.69$. It's important to note that the invagination contact length, l_{inv}, diverges as it approaches the binodal $a \to a_b = 3b^2/16$, following $l_{inv} \sim (a - a_b)^{-1}$. At the upper spinodal $a = a_{up} = b^2/4$, it takes a finite value $l_{inv}^{(up)} \approx 8.76 b^{-5/4}$. This latter value diverges at the tricritical point $b = 0$ of the phase diagram.

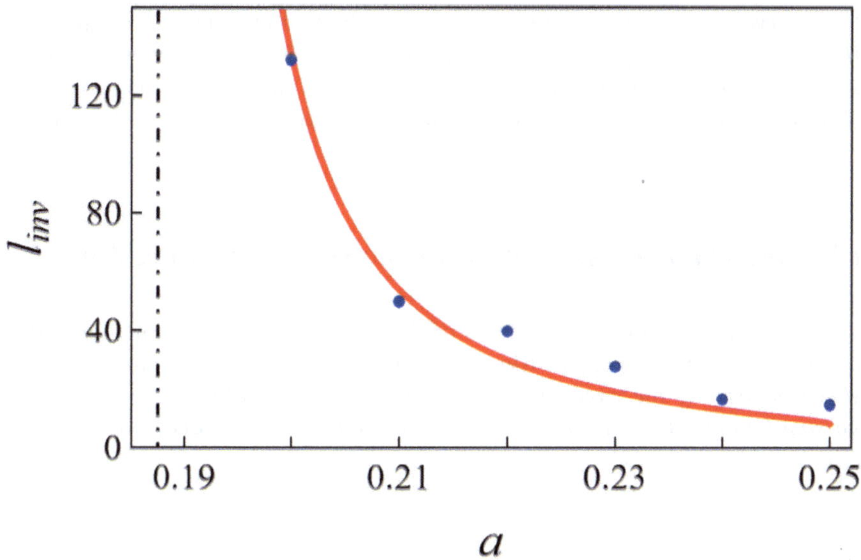

Fig. 14.5 Dependence of l_{inv} on a. Simulation results for different values of a at $b = 1$ are indicated by data points. The solid line represents the approximate relation (14.12) with $l_0 \approx 10.42$. The dot-dashed vertical line indicates the binodal position where the curve diverges

Figure 14.5 provides validation for the approximation (14.12) with a value of the fitting parameter[3] $l_0 \approx 10.42$. The curve's behavior is depicted by the solid line in Fig. 14.5.

The setback l_{inv} forms the transformation toughness, as we demonstrate in Chap. 17 below.

***

In summary, *the large size of the process zone allows us to neglect the Laplacian term in the equation of state, except for the immediate vicinity of the crack tip.*

- *Away from this region, the process zone interface is well described by the cardioid* (14.2), (14.3), *whose size diverges as* $(T - T_0)^{-2}$, *as the temperature approaches the binodal.*

---

[3] It's worth mentioning that due to surface tension, the concave meniscus (Fig. 14.1a, point O_1) causes deformation at the back of the zone, resulting in a configurational force at the boundary. This force deforms the zone in the direction opposite to the x-axis. In the above derivation, we have neglected this effect. Consequently, its result slightly underestimates the correct value of the contact length. Indeed, the curve $l_{inv} = l_{inv}(a)$ (14.12) with $l_0 \approx 6.69$ consistently falls slightly below the contact length values obtained from simulations depicted as dots in Fig. 14.5. To improve this, we conducted a fitting procedure with l_0 as the fitting parameter. It resulted in an adjusted value of $l_0 \approx 10.42$. The curve displaying this adjustment is represented by the solid line in Fig. 14.5.

- *The curvature radius,* ρ_1, *of the zone's invagination* (14.11) *and, consequently, the contact length,* l_{inv} (14.12), *also diverge at the binodal. The latter plays an key role in determining the transformation toughness, as discussed later.*
- *The derived expressions match simulation results, as shown in Figs.* 14.2 *and* 14.5.

14.3 Geometric Properties of the Slowly Propagating Zone

This section addresses two key aspects of the slowly propagating crack tip process zone ($V \leq V_G$)*: (i) deriving Gordon's velocity by analyzing the shape of the zone and (ii) measuring the curvature radius of the zone's trailing edge invagination from simulations. Both parameters are compared with theoretical predictions for validation.*

***

14.3.1 Gordon's Velocity Derived from the Simulations

We conducted direct observations of the order parameter distribution within the zone. Specifically, we noted that at a particular velocity the invagination of the zone disappeared, as evident in Figs. 13.4d and 13.6. At this critical velocity, the rear portion of the zone boundary near the crack edges became flat and oriented perpendicular to the crack surface. This configuration is illustrated by point O_2 in Fig. 14.1b. We documented these velocity values[4] in Table 13.1 and argued that they represent Gordon's velocity v_G. Figure 14.6 displays the dependence of $v_G = v_G(a)$.

The data points in Fig. 14.6 represent the velocities obtained by simulations, by selecting the crack tip speed where the morphological transformation occurred for $a = 0.22$, 0.23, and 0.24. Meanwhile, the solid line in the same figure represents Gordon's velocity (12.8), $v_G = v_G(a)$, at $b = 1$:

$$v_G = \frac{1}{2^{1/6}\sqrt{3}} \frac{8a - 1 - \sqrt{1-4a}}{\left|1 - 2a + \sqrt{1-4a}\right|^{1/2}} \tag{14.13}$$

as derived from (12.8).

Note 14.3.1 Remarkably, these numerical results align closely with the analytical prediction, confirming that the transformation in the zone shape indeed occurs when the crack speed reaches Gordon's velocity, $v = v_G$.

[4] It's worth noting that the boundary's perpendicularity was determined visually, introducing some degree of error into our v_G observations.

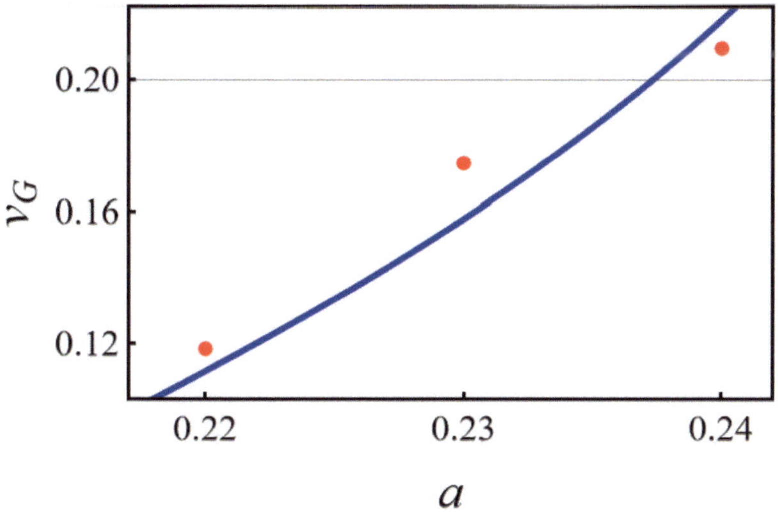

Fig. 14.6 Dependence of Transformation Velocity on a. The data points represent the velocities at which we observed zone shape transformations during simulations at $a = 0.22$, 0.23, and 0.24. The solid line corresponds to Gordon's velocity, $v_{\mathrm{G}} = v_{\mathrm{G}}(a)$, at $b = 1$ according to (14.13)

14.3.2 Slow Regime of the Crack Propagation and Invagination Curvature

Let us examine the behavior of the phase interface behind the zone in the regime of slow crack propagation. At $\theta = \pm\pi$, this interface is exactly stress-free. When θ is close to $\pm\pi$, the stress at the phase front is minimal, owing to both the contribution of $\cos^2(\theta/2)$ and the decrease in the potential U as $r^{-1/2}$. Consequently, we can conclude that an angular aperture where the phase boundary is approximately or exactly stress-free is close to $\pm\pi$.

However, the stress-free phase interface can propagate only at Gordon's velocity (see Sect. 3.5.2). Since we are dealing with a curved interface in this section, it propagates at the velocity $v_{\mathrm{G}}^{(-)}$ (3.74).

The reasoning outlined above leads us to understand the origin of the zone invagination. The rear part of the zone interface at $\theta = \pm\pi$ can only propagate at the same speed, v, as that of the crack tip. However, since $v_{\mathrm{G}} > v$, this can only occur if surface tension decreases the velocity of the interface to $v = v_{\mathrm{G}}^{(-)}$. The curvature of the invagination is responsible for generating surface tension, which, in turn, slows down the interface as described in (3.74). Consequently, the zone adopts a concave configuration (Fig. 14.1a, indicated by O_1). This condition leads to the curvature radius:

$$\rho_1 = \frac{1}{2^{1/3}(v_{\mathrm{G}} - v)}; \quad v < v_{\mathrm{G}} \tag{14.14}$$

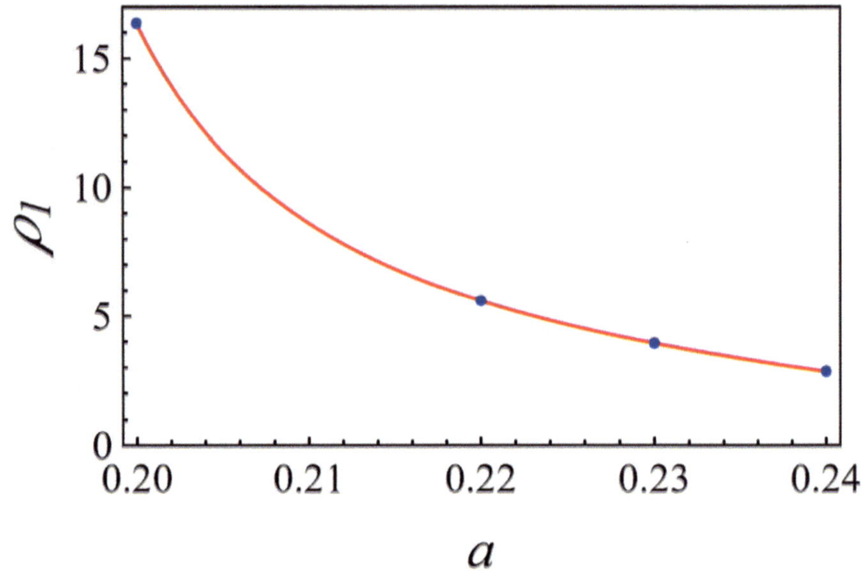

Fig. 14.7 Curvature Radius of Invagination as a Function of a at $b = 1$. Simulation results are depicted as data points, while the solid line corresponds to the curve generated by the formula (14.14)

where dimensionless Gordon's velocity $v_G = v_G(a, b)$ is defined by the (12.8).[5] Thus, ρ_1 also depends on a and b.

The curvature radius values of the invagination, obtained through simulations, are represented by data points in Fig. 14.7.

The solid line in the figure corresponds to the mathematical function $\rho_1(a, b)$ (14.14) at $b = 1$.

As v approaches v_G, the curvature radius diverges, corresponding to a flat portion of the rear edge of the boundary that is normal to the crack surface. This is indicated by O_2 in Fig. 14.1b.

***

In conclusion, *the close match between the velocity of the morphological transformation and the theoretical expression provides evidence that the transformation, indeed, occurs at Gordon's velocity. Furthermore, the comparison between the invagination radius from simulations and theoretical predictions sheds light on the physical origin of the zone's shape peculiarities. Specifically, the surface adopts a shape that generates surface tension, reducing the surface velocity to align with the crack tip speed.*

[5] In the present case, $b = 1$ and the expression (14.13) for the dimensionless Gordon's velocity $v_G(a, 1)$ holds.

14.4 The Regime of Fast Crack Propagation $v_G < v < v_{cr}$

We associate fast crack propagation with a speed V that exceeds Gordon's velocity V_G . In this regime, the rear portion of the phase boundary between the daughter and mother phases lags behind the propagating zone, creating a wake. This wake consists of a metastable daughter phase trailing behind the crack tip zone (Fig. 14.1c). In this section, we also derive analytical expressions for several geometric features of the zone.

***

14.4.1 The Rounding of the Wake's Point

The point of the wake travels at the same speed, v, as the entire system. To keep pace with the crack tip at $v > v_G$, the rear portion of the wake must adopt a convex shape with a rounded tip (Fig. 14.1c, indicated by O_3). The latter increases its speed due to the surface tension according to (12.9). The rounding radius, ρ_2, is expressed as follows:

$$\rho_2 = \frac{1}{2^{1/3}(v - v_G)} \tag{14.15}$$

14.4.2 The Semi-vertex Angle of the Wake's Pointed End

In our simulations, we observed that within the range of $v_G < v < v_d$, the zone's contour assumes a droplet-like shape, as illustrated in Fig. 14.1a. The rear portion of the wake closely resembles a triangular configuration, with its shape becoming increasingly triangular as the wake is stretched further.

The curvature of the rear end of this triangular configuration is inevitable to keep pace with the zone, in accordance with (14.15). This rear portion of the wake propagates at the crack velocity v. Crucially, the stress levels in this region, particularly at angles θ near $\pm\pi$ measured from the crack tip, are small. The boundaries of the zone limiting domain (iii), as shown in Fig. 14.1c, exhibit a nearly planar configuration due to the low stress at the crack tip in this region. In this scenario, the only driving force for the phase interface propagation is related to its overcooling. Under these conditions, the interface moves at Gordon's velocity v_G, as discussed in Sect. 3.5. This leads to the determination of the angle α, following a straightforward relationship:

$$\sin(\alpha) = \frac{v_G}{v} \tag{14.16}$$

Figure 14.8 portrays the behavior of the semi-vertex angle α at various crack velocities. The dots in Fig. 14.8 depict the dependence $\alpha = \alpha(v)$ derived from our simu-

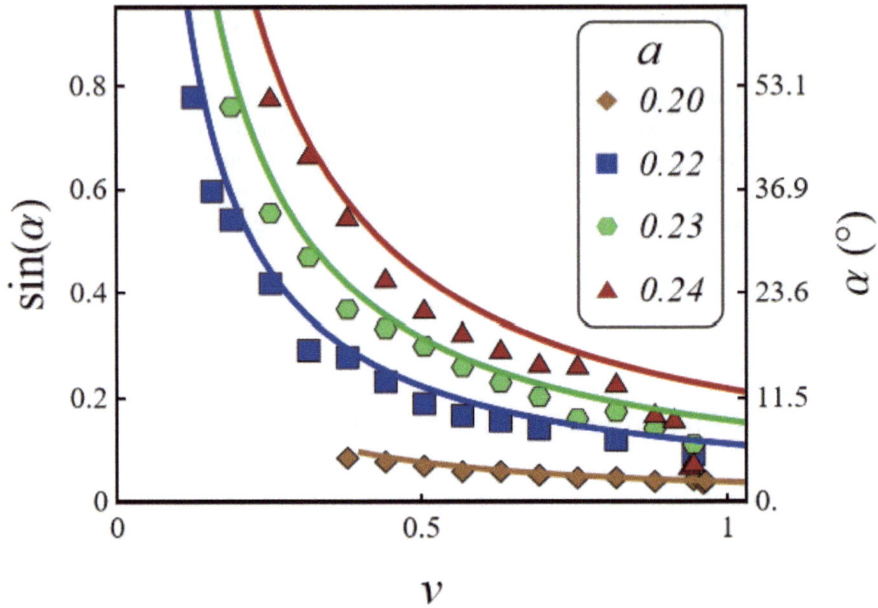

Fig. 14.8 Semi-Vertex Angle α versus Crack Velocity. The dots denote simulation results, while the solid lines represent the predictions from (14.16)

lations. The solid lines represent the expected $\alpha(v)$ behavior according to (14.16), showing good agreement with the simulation data.

The semi-vertex angle becomes increasingly acute as the crack tip velocity rises. At velocities close to $v = 1$, the semi-vertex angle ranges from approximately 11° at $a = 0.24$ down to 2° at $a = 0.20$, corresponding to a stretched-out wake with a sharp triangular shape.

14.4.3 Zone Width and Wake Contact Length

14.4.3.1 Zone Width

This section may be skipped during a first reading. To estimate the width of the zone, propagating with velocity v, we focus on its leading part. This estimate is based on a simple idea that the leading edge of the propagating zone moves under the action of the thermodynamic force, since no other force acts on the boundary. Thus, on the one hand, its speed is equal to a local Gordon's velocity of a curved phase front, on the other hand it is equal to v. Utilizing the expression for the curvature radius (14.4), and equating Gordon's speed (12.9) to the velocity v we arrive at the following equation:

$$v_G - \frac{3}{2^{5/3} r_d} = v \tag{14.17}$$

Now, let us introduce $a_{eff}(r, \theta)$, representing the local value of the parameter a. It is defined as the combination $a_{eff}(\mathbf{r}) = a + U(\mathbf{r})$ preceding u in (12.12) and is given by:

$$a_{eff}(r, \theta) = a - \sqrt{\frac{2}{r}} \cos(\theta/2) \tag{14.18}$$

To account for the inhomogeneity introduced by the potential U (refer to (12.12)), we replace a with $a_{eff}(r, \theta)$. At the leading edge, where $\theta = 0$ and $r = r_d$. We obtain:

$$a_{eff} = a - \sqrt{\frac{2}{r_d}}$$

Substituting $a \to a_{eff}$ into (8.8) and further into (14.17), and primarily focusing on the limit of a large zone ($r_d \gg 1$), we neglect the surface tension term $\sim r_d^{-1}$ in (14.17). This simplification yields:

$$v = \frac{1}{2^{7/6} 3^{1/2}} \frac{b^2 + b\sqrt{b^2 - 4a_{eff}} - 8a_{eff}}{\left|b^2 - 2a_{eff} + b\sqrt{b^2 - 4a_{eff}}\right|^{1/2}} \tag{14.19}$$

One can readily solve this equation for a_{eff} and then use the obtained result to solve for r_d. Following this procedure one obtains the result:

$$r_d \approx \frac{512}{\left(16a - 3b^2 + 2\sqrt{3}bv + 3v^3\right)^2} \tag{14.20}$$

At $v = 0$, the expression (14.20) simplifies to the one describing the motionless zone (14.3). When $v = v_G$, the zone width becomes:

$$r_d \approx \frac{32}{\left(b^2 - 2b\sqrt{b^2 - 4a}\right)^2} \tag{14.21}$$

Figure 14.9 illustrates the relationship between the zone width, r_d, and the crack tip velocity, v.

The data points correspond to the outcomes obtained through simulations, while the solid lines represent the behavior of (14.20) with $b = 1$. The different colors of the curves indicate various values of parameter a. Remarkably, the curves described by (14.20) align closely with the numerical results.

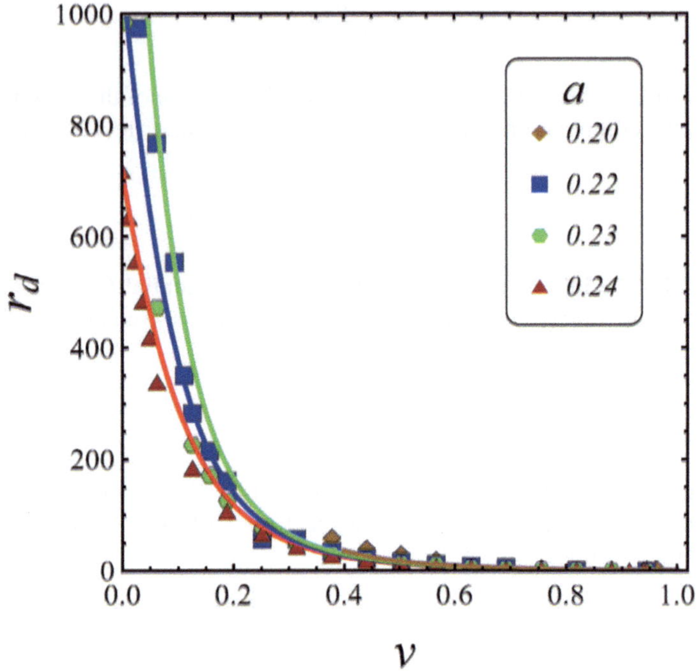

Fig. 14.9 r_d as a Function of Crack Velocity, v. Dots show the simulation results, while the solid lines depict $r_\mathrm{d}(v)$ according to (14.20) with $b = 1$, where the colors correspond to different values of parameter a

At $v = 0$, we observe that the zone width, r_d, is on the order of $\sim 10^3$ dimensionless units. As the crack velocity approaches 1, the zone width decreases significantly, converging to $r_d \sim 1$.

14.4.3.2 Wake Contact Length

This section may be skipped during a first reading. In the regime of fast crack propagation, the wake assumes a shape that closely resembles two triangles on both sides of the crack. This wake configuration is schematically depicted in Fig. 14.1c and is highlighted by the pink color. The hypotenuses of both triangles share a common length, denoted as l_{wake} (Fig. 14.1c). These hypotenuses are in direct contact with the crack faces; we refer to this length as the wake contact length, l_{wake}.

Unfortunately, we could not derive a reasonable analytical expression for the length of the triangle's cathetus. Nevertheless, it is of the same order of magnitude as the zone width, r_d, although slightly exceeds it. Let us denote the cathetus as qr_d

(Fig. 14.1c), where $\mathfrak{q} \sim 1$ is a dimensionless geometric factor and r_d is described by (14.20). The idea of this calculation is trivial. The hypotenuse, l_{wake}, one derives as the length of the cathetus, $\mathfrak{q} r_d$ (14.20), divided by $\sin(\alpha)$ (14.16):

$$l_{wake} = \frac{\mathfrak{q} r_d v}{v_G} \tag{14.22}$$

This expression yields the following contact length of the wake:

$$l_{wake} \approx \frac{512 \times 2^{5/6}\sqrt{3}\mathfrak{q}\left(b^2 - 2a - b\sqrt{b^2 - 4a}\right)^{1/2} v}{\left(b^2 + b\sqrt{b^2 - 4a} - 8a\right)\left(16a - 3b^2 + 2\sqrt{3}bv + 3v^3\right)^2} \tag{14.23}$$

This approximation holds when $v_G < v < v_{cr}$. We introduced the factor $\mathfrak{q}$ intentionally. Even with $\mathfrak{q} = 1$, (14.23) already provides an estimate of the wake contact length. However, setting $\mathfrak{q} = 1.9$ yields a closer agreement with the simulation data, as we demonstrate in Fig. 14.10.

Recasting in Terms of Physical Variables

At low crack velocities ($v \leq v_G$), the process zone assumes a concave shape, characterized by the contact length, l_{inv}, between the crack tip ($x = y = 0$) and the bottom of the invagination (point O_1 in Fig. 14.1a). It is on this segment of the crack surface that spontaneous stress exerts compressive forces, contributing significantly to transformation toughening.

Once the crack tip velocity exceeds v_G, the wake forms, and the contact length, l_{wake} (Fig. 14.1c), represents the distance between the crack tip and the trailing end of the wake. We unify both contact lengths as follows:

$$l_{cont} = \begin{cases} l_{inv}, & v \leq v_G \\ l_{wake}, & v_G < v \leq v_d \end{cases} \tag{14.24}$$

where the subscript "cont" stays for "contact," the analytical expression for l_{wake} one finds in (14.23), while we did not succeed to derive any analytical expression for $l_{inv} = l_{inv}(v)$.

Figure 14.10 illustrates the dependence of l_{cont} on the crack tip velocity, v.

Within the range $0 \leq v \leq v_G$, the contact length l_{inv} increases with velocity, reaching its maximum at $v = v_G$. Its maximum value in terms of the physical variables, $L_{cont}^{(max)}$ has the following form:

$$L_{cont}^{(max)} = \frac{128}{\pi}\left[\frac{AK_I(1 - 2\nu)(1 + \nu)}{7\alpha_0(T - T_0)E}\right]^2 \tag{14.25}$$

In the velocity interval $v_G < v \leq v_d$, the wake takes on a droplet-like configuration. In this regime, we refer to its contact length as l_{wake}. Figure 14.10 reveals that l_{wake} monotonously decreases with velocity.

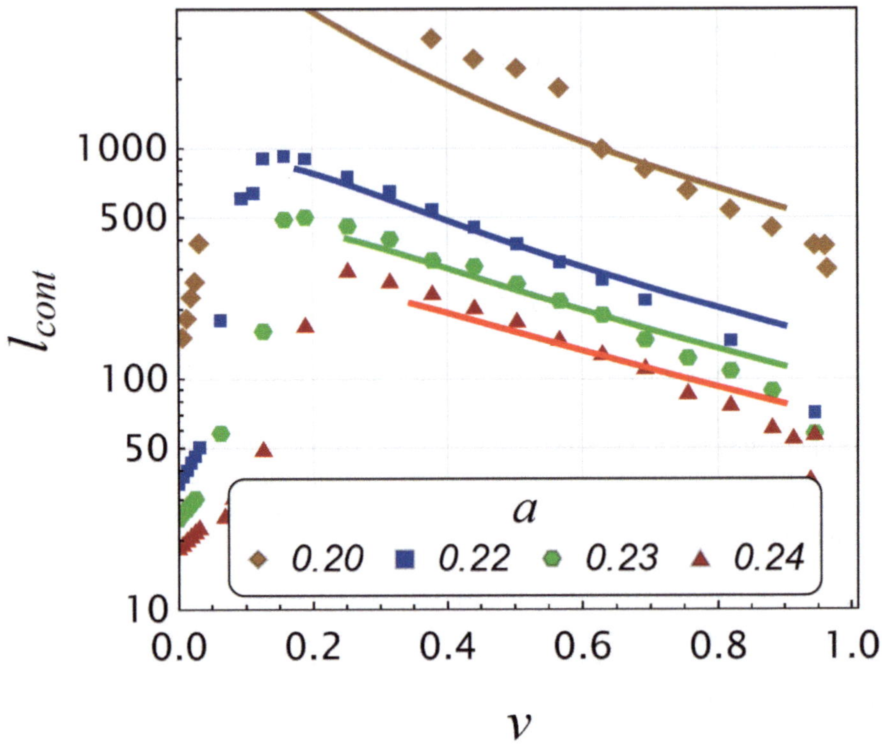

Fig. 14.10 Variation of Contact Length with Crack Velocity. Each curve has two branches. At $0 \le v \le v_G$, the contact length increases, achieves its maximum at $v = v_G$ and then decreases. This dependence of $l_{cont} = l_{cont}(v)$ is shown here in a semi-logarithmic plot at various values of $a = 0.2$, 0.22, 0.23, and 0.24 and $b = 1$. The solid lines represent the behavior of the function $l_{wake}(v)$ (14.23) for these values of a and $\mathfrak{q} = 1.9$

The solid lines in the figure describe l_{wake} (14.23), color-coded to correspond with $a = 0.20$, 0.22, 0.23, and 0.24, with $b = 1$, closely match the numerical data.

***

In summary, *the same physical principle governs both the rounding radius of the wake tip, ρ_2, and the zone width, r_d. This principle is based on the equality between the crack speed and Gordon's velocity for the curved phase front. Additionally, the wake's interface, being free from crack-generated stress, propagates at Gordon's velocity for the planar phase front, which defines the angle α of the wake tip. Since $r_d \sim h$, this relationship determines the wake length, l_{wake}. Our analysis shows that the contact length* (14.24) *reaches its maximum at Gordon's velocity. Understanding the contact length and its dependence on the crack tip velocity allows for the derivation of the resistance to crack motion generated by the wake.*

14.5 The Critical Velocity

Within the framework of bifurcation theory, we proved that the process zone vanishes once the velocity reaches its critical value V_d*, for both soft and quasi-soft process zones (see Sects.* 6.4.2.1 *and* 8.1.1)*. In this section, we investigate this phenomenon through simulations to confirm its persistence in the fully developed first-order zone.*

***

The process zone disappears when the crack velocity reaches a specific value, denoted as v_d. This vanishing of the zone can be explained as follows: let us consider point A, located in front of the process zone, at some distance in the direction of its motion. While the crack tip remains far from this point, the order parameter at point A remains at zero, indicating that the material at this location stays in the mother phase.

The order parameter at this point begins to respond to the approaching crack, when the distance between the leading edge of the zone and point A becomes equal to the thickness of the phase interface, $\mathbb{R}_0$. At this moment, the order parameter starts to increase in response to the crack's approach. However, this growth occurs according to its intrinsic dynamics, taking a characteristic time denoted as $\mathbb{T}_0$ (refer to the relations (12.1) and (12.2)). The ratio $\mathbb{R}_0/\mathbb{T}_0$ provides an order of magnitude estimate for the critical velocity:

$$v_d \sim 1 \tag{14.26}$$

When the crack velocity exceeds v_{cr}, the crack tip becomes undressed, indicating a significant change in the material's behavior.

To effectively analyze the zone vanishing in simulations, we study the behavior of its Hilbert norm, denoted as

$$||u|| = \left[\int u^2(x, y)dxdy\right]^{1/2}$$

Figure 14.11a depicts $||u||$ as a function of v, whereas the detailed behavior in the immediate vicinity of the critical velocity is shown in Fig. 14.11b. It is observable that $||u||$ decreases precipitously as v increases and displays a discontinuity where $||u||$ drops to zero. The velocity at this discontinuity is identified as the *critical velocity*, v_d. When v exceeds v_d, aside from the numerical noise, the norm $||u||$ is zero. v_d slightly decreases as the parameter a increases, as shown in Fig. 14.11c. Consistent with (14.26), v_d remains nearly unity, deviating by only a few percent.

***

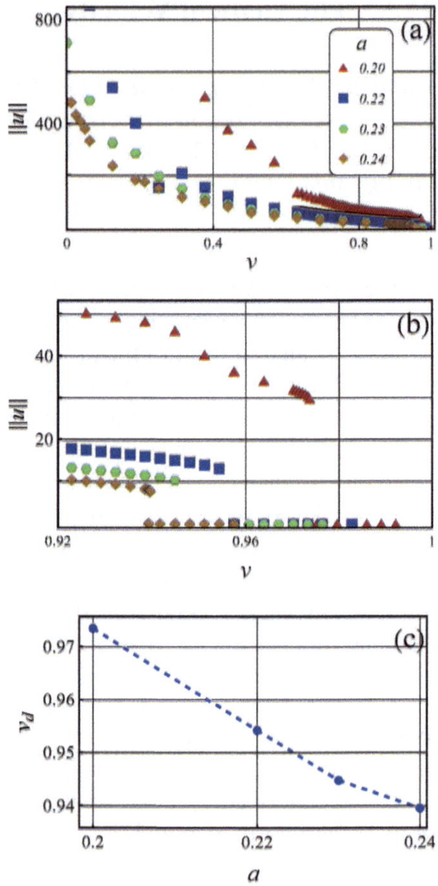

Fig. 14.11 Hilbert Norm of the Solution versus Velocity. **a** The behavior of the Hilbert norm, $||u||$, in relation to the crack tip velocity, v. **b** The behavior near the critical velocity, highlighting the point at which the nontrivial solution ceases to exist. **c** The dependency of the critical velocity on the control parameter a. A dashed line provided to aid the eye

In conclusion, *as predicted by bifurcation theory, we observed the critical velocity in the case of the fully developed first-order process zone. Beyond this point* ($V > V_\text{d}$), *the process zone consistently vanished.*

Chapter 15
Observations of the Wake and Zone at the Crack Tip

Abstract In this chapter, we briefly summarize the wake observation in various materials.

15.1 Wake Decay Time

The relationship between physical time, $\mathbb{T}$, and the dimensionless time, τ, is described in (5.1), (12.2). Considering the typical size of the process zone observed in ferroelectrics, $\mathbb{R}_{\mathrm{d}} \sim 1$ to 10 μm [1–3], and Gordon's velocity $V_{\mathrm{G}} \sim 10^{-3} cm/s$ estimated earlier, we find that the time for wake decay in these materials, $\mathbb{T}_{\mathrm{d}} \sim \mathbb{R}_{\mathrm{d}}/V_{\mathrm{G}}$, falls within the range of approximately $\sim$0.1–1 s. Such a short-lived wake implies that it can only be observed in situ, making 'post-mortem' wake detection in these materials unfeasible.[1]

Detecting the triangular wake is challenging because it is only present in the dynamics of a crack propagating with a velocity $V > V_{\mathrm{G}}$. The lack of a theoretical prediction has hindered the recognition of the metastable wake in experiments. Nevertheless, some observations have provided indications of the metastable wake.

15.2 Early Wake Observations

At the crack tip, the emergence of a daughter phase exhibiting dilatation becomes energetically favorable under the stretching stresses. Consequently, this phase is thermodynamically stable at the crack's tip, where these localized stresses are most pronounced. As the crack grows, the stress behind the tip diminishes, leaving the material in this zone in a metastable state, which we refer to as the *wake*.

[1] The latter statement seemingly contradicts multiple 'post-mortem' observations of the wake in the early experimental papers. We discuss the origin of such a long-lasting wake in Sect. 17.4.1.5.

© The Author(s), under exclusive license to Springer Nature Switzerland AG 2026
A. Boulbitch and A. Korzhenevskii, *Stress-Induced Local Phase Transformations Near Cracks and Dislocations*, Springer Series in Materials Science 356,
https://doi.org/10.1007/978-3-032-00945-6_15

Early experiments, such as those on ZrO_2 [4, 5], detected the first-order transformational process zone through so-called 'post-mortem' studies. These studies revealed the presence of a wake, a long-lived layer on the crack surface post-fracture. This layer, typically micrometers thick, contained a monoclinic phase, while the bulk of the solid remained in the tetragonal phase. Additional insights into the observation of wakes can be found in reviews by Birkby and Stevens, Todd and Saran, and Kelly and Rose [4, 6, 7].

Wakes have been observed in various materials, including PZT ceramics [1, 8], polycrystalline ferroelectric PZT [9], sapphire [10], stishovite [11], silica, and borosilicate glass [12], during thermal fracture of diamonds [13], in polycrystalline silicon [14], high-density polyethylene [15], nitinol [16], single crystals CuAlNi and Ni–Mn–Ga with the shape memory effect [17, 18], TRIP steels [19, 20], antiferroelectric ceramics [2], high-T_c superconductors YBCO and BSCCO [21], metallic glass Zr-Ti-Cu-Nb-Be [22] and in the 2D crystal $MoWSe_2$ [23].

It is widely accepted that the metastable wake gives rise to the *transformation toughness*. On the one hand, the energy of this metastable state adds to the fracture energy, as initially discussed by Antolovich [24] to explain the transformation toughening effect in TRIP steels [4, 19, 25, 26]. On the other hand, the less dense wake material generates compressive stress on the crack surfaces, which must be subtracted from the contribution of the external load to the stress intensity factor [27].

The debate persists on whether these two mechanisms represent the same mechanism [4]. Regardless, for both mechanisms to contribute to fracture toughness, the wake must persist at the crack face for a significant duration. If this holds true, the wake's contribution to K_{IC} can be substantial. For instance, in stishovite [11] and ZrO_2 [28], the increase in K_{IC} reaches an order of magnitude.

However, the dynamics of the order parameter introduce decay in the wake, a phenomenon often overlooked in the cited studies. In contrast, our focus here is on the propagation of the first-order zone and the subsequent behavior of the wake, considering its internal dynamics.

15.3 Direct Observation of Zone and Wake

V.I. Isotov (Institute of Iron and Steel Industry, USSR) captured an electron micrograph of a crack in a Fe-Ni-Ti-Al alloy foil around 1986–1987. The crack formed unexpectedly due to sample fixation conditions and propagated precisely during the measurement. This image unequivocally revealed the presence of the process zone and triangular wake. With gracious permission from V.I. Isotov, one of us published this image in Fig. 1 in [29]. Here, we present another illustration obtained from V.I. Isotov, vividly exhibiting the triangular wake.

Figure 15.1 depicts an electron micrograph of the same Fe-Ni-Ti-Al alloy austenitic-martensitic foil. In the center of the image, one observes a crack indicated by (a).

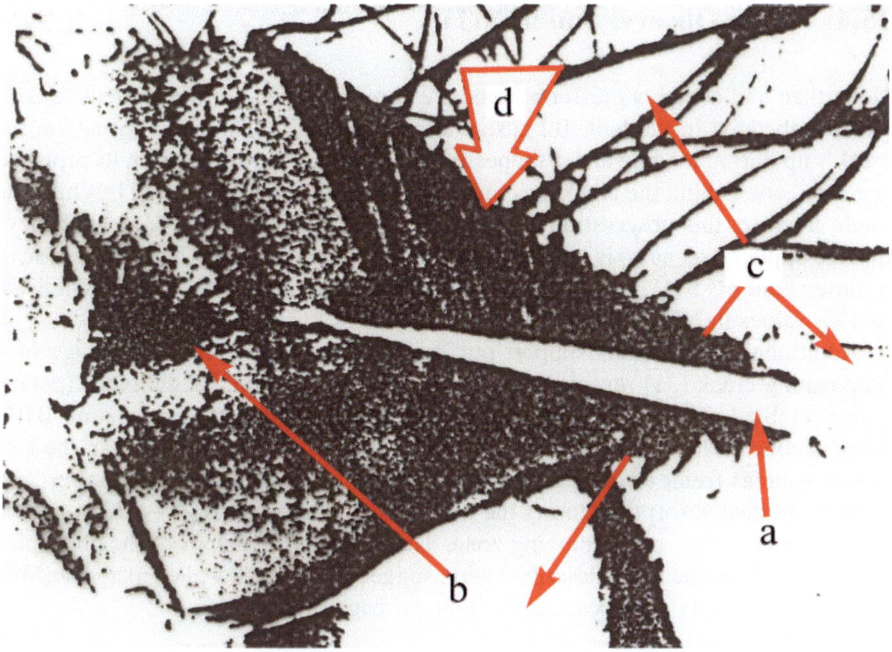

Fig. 15.1 Electron Micrograph of a Fe-Ni-Ti-Al Alloy Showing Martensite and Austenite Phases. The image displays the martensite daughter-phase process zone at the crack tip embedded into the austenite mother phase. **a** indicates the crack, **b** points out the martensitic phase, appearing darker, while the austenite mother-phase **c** appears whitish. The arrow **d** indicates the triangular wake. It is evident that the wake is divided into twins. Courtesy of V.I. Isotov (Institute of Iron and Steel Industry, USSR, 1987)

In this instance, the mother phase is austenite, indicated by (c), displaying an almost white appearance. The dark lines within this phase are interference lines resulting from slight bending or corrugations of the foil.

The arrow (b) identifies the process zone occupied by the martensitic phase, exhibiting a darker hue, its configuration conforming qualitatively to our predictions. Unfortunately, we lack sufficient information to derive any quantitative results from this image concerning the size and shape of the zone.

In its trailing part (d), the process zone transitions into a triangular configuration representing the wake. Besides the report of V.I. Isotov, this triangular wake serves as clear evidence that the image was captured either during the crack propagation process or immediately after its arrest, preventing the wake from "melting down." The wake is visibly divided into a stack of planar entities representing twins. This division into twins signifies a secondary process concerning wake formation, aimed at minimizing elastic energy.

By directly measuring the zone height and the wake semi-vertex angle in the image, it was possible to estimate $V \approx 2.85 V_G$ in this instance.

15.4 Wake Observation in NiTi

Gollerthan and coauthors [30] observed a 2 mm-sized zone of the martensite B19′ phase embedded in the bulk B2 austenite phase in NiTi. This zone formed at the crack's tip during loading and disappeared upon unloading [30]. During its propagation, the zone behind the crack was reported to vanish immediately [31]. While one might interpret this observation as a case of a concave zone ($V < V_G$), we believe that the opposite scenario is occurring. Indeed, if V is sufficiently high in the case of a convex zone ($V > V_G$), the wake becomes challenging to observe. Reports indicate that V_G values in NiTi can reach as high as $370\,\mathrm{m}/s$ [32].

Additional observations support our hypothesis. A thermographic image of a propagating crack [31] revealed two approximately ≈ 1 mm-long regions (petals) that resembled droplets on both faces of the crack behind the propagating B19′' process zone. In this scenario, the material was considerably colder than the surrounding areas (refer to Figs. 3 and 4 in [31]). This temperature variation resulted from latent heat absorption during the backward endothermal B19′ →B2 transition at the rear boundary of the moving zone. The droplet-like shapes of the cold spots closely resemble the wake described here, suggesting that the wake emerges behind the zone but rapidly disappears as the zone propagates.

15.5 Zone and Wake Observation in Zirconia

Kim and coauthors [33] utilized the mechano-luminescence effect to visualize stress-induced transformations in yttria-zirconia ceramics in situ during crack propagation. Intense luminescence irradiation indicated that a region of the solid transformed into the monoclinic phase, representing a 590 μm spot at the tip of the stationary crack. As the crack propagated at velocities ranging from 20 to 140 m/s, this spot elongated approximately tenfold, reaching lengths of up to 6 mm. In light of our findings, this observed evolution can be interpreted as a zone at the stationary crack tip stretching out into a zone-wake complex. Subsequently, this complex follows the motion of the crack tip.

Transmission electron microscopy (TEM) enables observation of the process zone at the nanoscale. A study monitoring stable crack growth in Yttria-Stabilised Zirconia ceramics has been reported in [34].

Another instance of process zone and wake observation in the same material was described in [35], employing a wedge-driven double cantilever beam within a TEM setup. The reported process zone dimensions were $\mathbb{H} \approx 500$ nm and $\mathbb{R}_d \approx 300$ nm with the monoclinic phase forming a herringbone structure.

The work of Meschke and co-authors [3] presented an atomic force microscopy study of magnesia-doped partially stabilised zirconia. It reported the observation of a wedge-shaped process zone, exhibiting the maximum width (referred to here as height $\mathbb{H}$) of approximately 20 μm and a length in the crack direction about three times greater. Notably, the zone dimensions were found to vary with the applied load, and the reported sizes correspond to the specific loading conditions used during the measurements.

15.6 Zone with No Wake in NiAl

An immediate reverse transformation of martensite into austenite behind a slowly propagating crack-zone complex has been observed in NiAl [36], suggesting a zone with no wake ($V < V_G$).

***

In summary, *early studies reported crack tip process zones observed postmortem, where the zone presence was identified by a thin layer of a new phase along the fracture surface. More recently, advanced techniques have enabled researchers to observe a finite-length wake traveling with the crack tip. Two distinct types of wake have, thus, been observed: one that decays during the experiment and another that remains stable, persisting for days or even months. This book focuses on the decaying wake, while we do not study the persistent one. We briefly discuss its potential physical origin in Sect.* 17.4.

References

1. S.O. Kramarov, N.Y. Egorov, L.M. Katsnel'son, Sov. Phys.—Solid State **28**, 1602 (1986)
2. Z. Tan, Y. Peng, J. An, Q. Zhang, J. Zhu, Inorg. Chem. **60**, 7961 (2021)
3. F. Meschke et al., J. Am. Ceram. Soc. **83**, 353 (2000)
4. P.M. Kelly, L.R.F. Rose, Prog. Mater. Sci. **47**, 463 (2002)
5. R.C. Garvi, Hannink, R.T. Pascoe, Nature **258**, 703 (1975); L.K. Lenz, A.H. Heuer, J. Am. Ceram. Soc. **65**, 192 (1982); R.P. Indel, R.W. Rice, D. Lewis, J. Amer. Ceram. Soc. **65**, 108 (1982); F.F. Lange, J. Mater. Sci. **17**, 240 (1982)
6. I. Birkby, R. Stevens, Key Eng. Mater. **122–124**, 527 (1996)
7. R.I. Todd, M.P.S. Saran, Transformation toughening, in *Materials Processing Handbook*, vol. 20, ed. by J.R. Groza (CRC Press, Boca Raton, FL, 2007), pp.1–20
8. Y.-H. Seo, J. Koruza, A. Benčan, B. Malič , J. Rödel, K.G. Webber, C. Landis, J. Amer. Ceram. Soc. **97**, 1582 (2014); C.S. Lynch, R.M. McMeeking, Z. Suo, in *Second International Conference on Material Science, 1994*, ed. by C.A. Rogers, G.G. Wallace (Technomic Publishing Co., Lancaster, PA, 1994), p. 856
9. J.E. Huber, F. Hofmann, S. Barhli, T.J. Marrow, C. Hildersley, Scr. Mater. **140**, 23 (2017)
10. T. Sasaki, N. Shibata, K. Matsunaga, T. Yamamoto, Y. Ikuhara, J. Ceram. Soc. Jpn. **120**, 473 (2012)
11. N. Nishiyama et al., Sci. Rep. **4**, 1 (2014)
12. C. Wuenshe, E. Raedlein, G.H. Frischat, Glastech. Ber. Glass Sci. Techol. **72**, 4954 (1999)
13. V.G. Garin, Sverhtverdye materialy **2**, 17 (1982) (Superhard Materials, in Russian)
14. M. Takeda et al., J. Elect. Microscopy **48**, 609 (1999); K. Youssef, P. Kulshreshtha, G. Rozgonyi, Photovoltaics for the 21st Century **25**, 49 (2010); P.K. Kulshreshtha, K.M. Youssef, G. Rozgonyi, Solar Energy Mater. Solar Cells **96**, 166 (2012)

15. I. Brough, R.N. Haward, G. Healey, A. Wood, Polymer **45**, 3115 (2004)
16. K. Tozawa et al., J. Elect. Microscopy **48**, 613 (1999)
17. G.M. Loughran, T.W. Shield, P.H. Leo, Int. J. Solids Struct. **40**, 271 (2003)
18. F. Xiong, Y. Liu, Acta Mater. **55**, 5621 (2007)
19. E. Hornbogen, Acta Metall. **26**, 147 (1978)
20. D. Hennessy, G. Steckel, C. Altstetter, Met. Trans. **A7**, 415 (1976)
21. H.-J. Sue, J.D. Earls, R.E. Hefner Jr., J. Mater. Sci. **32**, 4039 (1997)
22. C.J. Gilbert, V. Schroeder, R.O. Ritchie, in *Bulk Metallic Glasses*, ed. by W.L. Johnson, A. Inoue, C.T. Liu, vol. 554 (Materials Research Society, Warrendale PA, 1999), p. 34
23. A. Apte et al., ACS Nano **12**, 3468 (2018)
24. S.D. Antolovich, Trans. Met. Soc. AIME **242**, 2371 (1968)
25. H. Oettel, U. Martin, Int. J. Mater. Res. **97**, 1642 (2006)
26. S.D. Antolovich, D. Fahr, Eng. Fract. Mech. **4**, 133 (1972)
27. R.M. McMeeking, A.G. Evans, J. Amer. Ceram. Soc. **65**, 242 (1982)
28. N. Simha, L. Truskinovsky, Acta Metall. Mater. **42**, 3827 (1994)
29. A.A. Bulbich, J. Mater. Sci. **27**, 1070 (1992)
30. S. Gollerthan, M.L. Young, K. Neuking, U. Ramamurty, G. Eggeler, Acta Mater. **57**, 5892 (2009)
31. S. Gollerthan, M.L. Young, A. Baruj, J. Frenzel, W. Schmahl, G. Eggeler, Acta Mater. **57**, 1015 (2009)
32. J. Niemczura, K. Ravi-Chandar, J. Mech. Phys. Solids **54**, 2136 (2006)
33. J.S. Kim et al., Met. Mater. Int. **14**, 165 (2008)
34. O. Gavalda-Diaz, M. Emmanuel, A. Berenov, K. Marquardt, E. Saiz, F. Giuliani, Proc. Natl. Acad. Sci. **121**(43), e2408292121 (2024)
35. M.J. Lai, J.M. Wheeler, J. Michler, P.G. Laity, I.M. Reaney, S. Korte, Appl. Phys. Sci. **121**(43), e2408292121 (2024)
36. U.D. Hangen, G. Sauthoff, Intermetallics **7**, 501 (1999)

Chapter 16
The Complete Phase Diagram of the Process Zone

Abstract In this chapter, we explore the mother-phase process zone located at the crack tip. Our results, derived from simulations, make it possible to build the phase diagram of the process zone on the background of the phase diagram of the bulk phase transition (Boulbitch and Korzhenevskii, first published here).

16.1 The Mother-Phase Process Zone at the Crack Tip

In this section, we formulate the equation of state and boundary conditions that describe the mother-phase process zone. Building on this formulation, we present the qualitative simulation results for the mother-phase zone at the crack tip.[1]

***

16.1.1 Equation of State and Simulation Approach

To simulate the mother-phase process zone, we utilized the following equation of state:

$$\Delta u - \left[a + \sqrt{\frac{2}{r}} \cos(\theta/2) \right] u + bu^3 - u^5 = 0 \tag{16.1}$$

This equation is analogous to (12.12), which is used for describing the daughter-phase zone at a stationary crack, employing dimensionless variables. A notable distinction is the positive sign in front of the potential $(2/r)^{1/2} \cos(\theta/2)$, which is necessary for the formation of the mother-phase zone within the daughter-phase matrix. We have already discussed this aspect previously in Sect. 9.1. Formation of the mother-phase zone embedded into the daughter-phase matrix occurs within the phase diagram domain below the binodal:

[1] This material is published for the first time.

© The Author(s), under exclusive license to Springer Nature Switzerland AG 2026

A. Boulbitch and A. Korzhenevskii, *Stress-Induced Local Phase Transformations Near Cracks and Dislocations*, Springer Series in Materials Science 356,
https://doi.org/10.1007/978-3-032-00945-6_16

$$a \leq \frac{3b^2}{16} \tag{16.2}$$

The Dirichlet boundary condition is defined as

$$u_{|\partial\Omega} = u_{\mathrm{d}} \tag{16.3}$$

where

$$u_{\mathrm{d}} = \sqrt{\frac{b + \sqrt{b^2 - 4a}}{2}} \tag{16.4}$$

represents the equilibrium value of the dimensionless order parameter u in the daughter phase, far from the crack tip. The nonzero boundary conditions (16.3) simplify the simulation considerably compared to the daughter-phase process zone discussed earlier. No lengthy, multi-step relaxation method is necessary in this case, and the stationary problem can be solved directly. As a result, mesh construction is straightforward, as described in Sect. 38.2.7. However, near the binodal (16.2), numerical convergence becomes challenging, hindering a detailed analysis of the mother-phase zone behavior in this region.

16.1.2 The Order Parameter Distribution

The outcomes of typical simulations are illustrated in Figs. 16.1 and 16.2 for the parameter $b = 0.1$. Figure 16.1 displays the order parameter distribution $u = u(x, y)$, while Fig. 16.2 presents its cross-sections $u(x, 0)$ along the crack line. The crack is positioned at the negative x-axis part and $y = 0$, with its tip at the origin $(0, 0)$. The tip is depicted by blue lines in Fig. 16.1, though omitted in Fig. 16.2 for clarity. It is important to note that these simulations were conducted using rescaled dimensionless coordinates. In all images, far from the tip, the distribution asymptotically approaches the level $u = u_{\mathrm{d}}$ (16.4). With b fixed at 0.1, the value of u_{d} varies across images, depending on the specific value of a used in each simulation.

As discussed in Sect. 7, the dip in the distribution never entirely disappears. This phenomenon is particularly evident in Figs. 16.1a and 16.2a. Both images are obtained at $a = -1$. Here, the zone is barely discernible yet present, the distribution $u(x, y)$ exhibiting a beak-shaped form, as observed in Fig. 16.2a. This configuration persists for all values of a below -1.

***

In summary, *the order parameter distribution in the daughter phase consistently shows a dip. Far below the binodal, this dip is small and barely noticeable, taking on a subtle beak shape. As the system approaches the binodal, the dip increases while maintaining its beak-like form. Once its depth reaches the level $u = 0$, continued movement toward the binodal causes the mother-phase zone to gradually broaden laterally, eventually extending to infinity at the binodal point.*

16.2 The Complete Phase Diagram of the Process Zone at the Crack Tip

Building on the numerical solutions presented in the previous chapter and the analysis of the daughter phase process zone detailed in Chaps. 8 and 12, *we construct a comprehensive phase diagram that delineates the regions occupied by various process zones.*[2]

***

Figures 16.1 and 16.2 consistently demonstrate a significant reduction in the order parameter distribution across all examined parameter values. Detailed numerical analysis reveals that the function $u(x, y)$ invariably retains a nonzero value within the order parameter dip, ranging from 10^{-2} to 10^{-5}. This persistent non-zero state necessitates a conventional approach to defining the threshold value of the order parameter at which a point is considered part of the mother-phase zone.

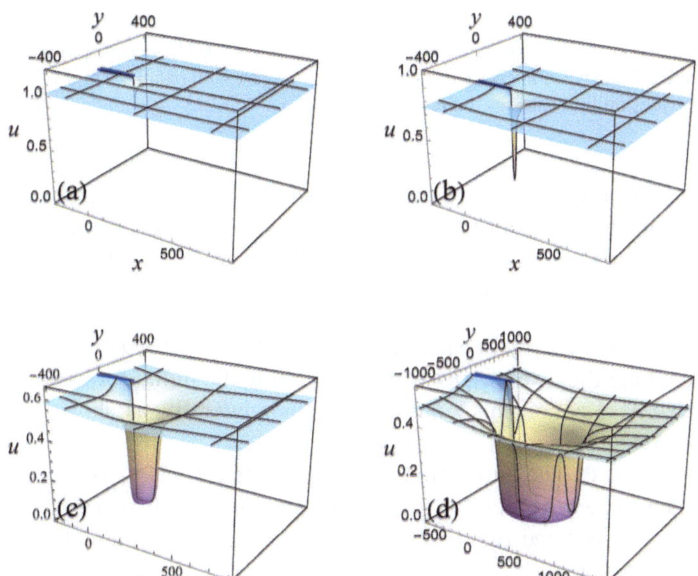

Fig. 16.1 3D order parameter distribution representing the mother-phase zone. The images are obtained by simulations of (16.1) with $b = 0.1$. Blue lines schematically indicate the crack tip. Panel (a) shows the beak-shaped distribution at $a = -1$. Panels (b–d) $a = -0.3, -0.12$ and -0.05 progressively demonstrate the mother-phase zone's lateral growth while a increases towards the binodal

[2] This material is published for the first time.

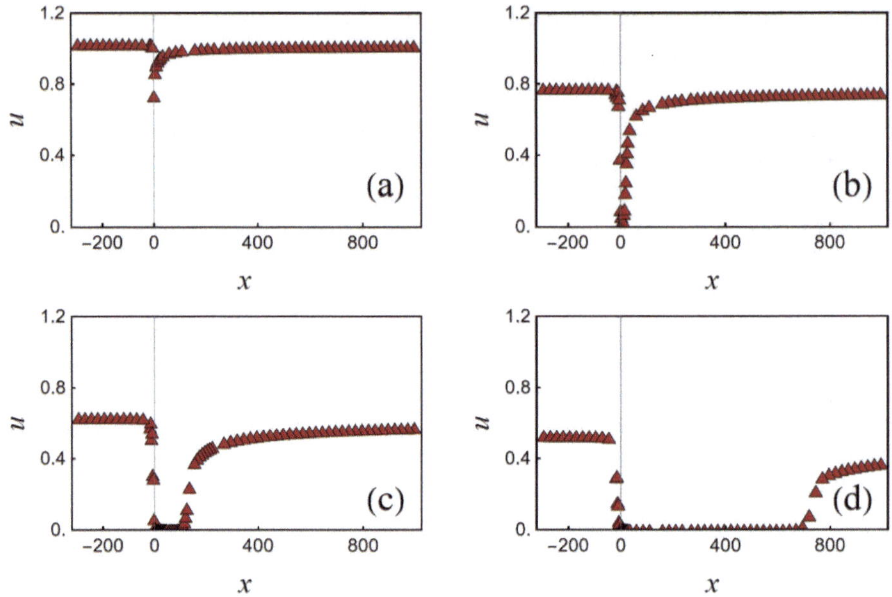

Fig. 16.2 Cross-section through the mother-phase zone at $y = 0$. They correspond to the distributions displayed in Fig. 16.1. Panels (a–d) depict various stages, with panel (b) highlighting the point where the distribution first reaches the level $u = 0$ at $a = -0.3$

Note 16.2.1 The mother-phase zone is characterized by the relative magnitude of the order parameter distribution, denoted as $|u|/u_{\mathrm{d}}$. A domain is classified as the *noticeable mother-phase zone* when $|u|/u_{\mathrm{d}}$ is less than $1/2$. Furthermore, a domain where $|u|/u_{\mathrm{d}}$ is about 10^{-2} or less is considered a *developed mother-phase zone*.

In Figs. 16.1b and 16.2b, for a parameter value of $a = -0.3$, a beak-shaped zone is observed touching the level $|u|/u_{\mathrm{d}} \approx 10^{-2}$, marking the initial formation of a developed mother-phase zone. This zone then expands laterally, as shown in Figs. 16.1 and 16.2c and d.

The Case A < 0. The simulation results provide a framework for constructing the phase diagram, with a focus on the region containing the first-order process zone. The bulk binodal, defined by (16.2), is delineated in Fig. 16.3a and b by the solid blue line marked as (*i*).

Figure 16.3a corresponds to a negative value of the striction constant A, showcasing the local binodal[3] (black dot-dashed line ii). The equation

$$a = a_{\mathrm{d}} + \frac{3\phi}{16} b^2$$

[3] Recall that we refer to the local binodal as the line on the phase diagram where the first-order process zone emerges.

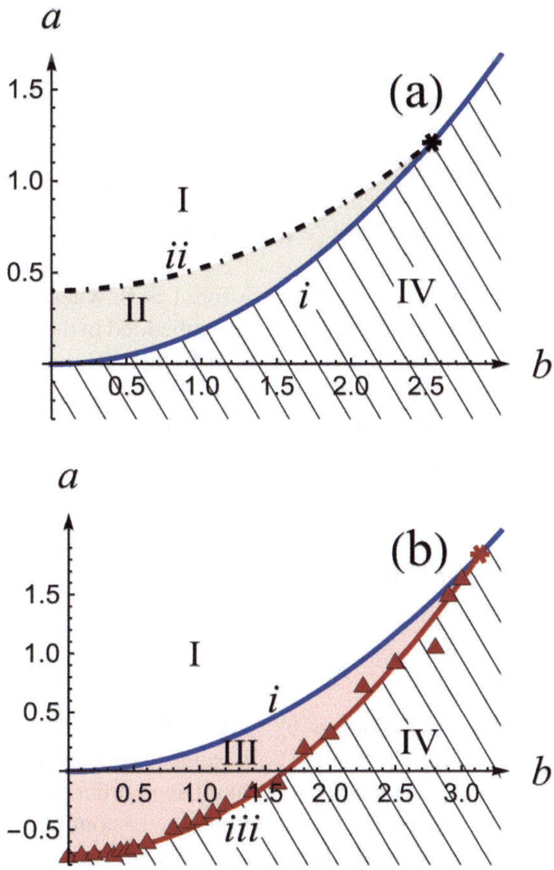

Fig. 16.3 Process zone phase diagram superimposed on the bulk phase diagram. Region I represents an area where the crystal remains in the mother phase, with the crack undressed. Region IV denotes the domain where the daughter phase occupies the entire crystal, and a small beak-shaped zone forms near the crack tip. The bulk phase transition binodal is marked by (*i*). Panel (a) illustrates the case where $A < 0$. Here, the daughter phase process zone forms along the local binodal line, given by (16.5), and indicated by (*ii*). This region is limited on the right by a critical point with coordinates (16.6), shown by the asterisk. In this scenario, the mother-phase zone does not exist. Panel (b) shows the case where $A > 0$. In this case, only the mother-phase zone exists below the bulk binodal (*i*). This zone appears within Region III, bounded from below by the local binodal (16.7), marked by (*iii*), and is limited on the right by a critical point (16.8), indicated by the asterisk. The daughter-phase zone is absent in this scenario

describes the local binodal with $a_\mathrm{d} \approx 0.4$ (11.2), and ϕ defined by (8.27), leading to $\phi \approx 0.67$ and the local binodal equation:

$$a \approx 0.4 + 0.126\, b^2 \tag{16.5}$$

At this line, the daughter-phase zone (as shown in Fig. 13.1) emerges at the crack tip. The bulk and local binodals intersect at:

$$b_{\mathrm{cr}}^{(\mathrm{d})} \approx 2.54 \tag{16.6}$$

denoted by the black asterisk in Fig. 16.3a. For $b > b_{\mathrm{cr}}^{(\mathrm{d})}$, the daughter-phase zone does not form. It only exists in Region II (shaded light brown), bounded from below by the bulk binodal (i). In the case $A < 0$, no mother-phase zone forms at the crack tip.

Case A > 0. Figure 16.3b shows the case for $A > 0$, where the daughter-phase zone ceases to exist, and the mother-phase zone embedded in the bulk daughter phase forms, as shown in Fig. 16.3b–d. This occurs in Region III, located below the bulk binodal (i) and bounded from below by a line marked with red triangles (iii), which are derived from simulations. In this region, the ratio $|u|/u_{\mathrm{d}}$ drops below 10^{-2}. We fitted this zone boundary with a parabolic equation:

$$a \approx -a_{\mathrm{m}} + 0.26\, b^2, \tag{16.7}$$

where $a_{\mathrm{m}} \approx 0.71$. This line represents the lower boundary of the mother-phase zone, which intersects the bulk binodal at:

$$b_{\mathrm{cr}}^{(\mathrm{m})} \approx 3.138, \tag{16.8}$$

as indicated by the red asterisk. For $b > b_{\mathrm{cr}}^{(\mathrm{m})}$, the mother-phase zone no longer exists.

Region IV, depicted in line-hatched in both panels, is occupied by the bulk mother phase. In panel (b), it is characterized by a beak-shaped order parameter distribution, as shown in Figs. 16.1a and 16.2a. The order parameter configuration in Region IV contrasts sharply with the more developed state in Region III, providing insight into the varying properties of the mother-phase zone across different parameter ranges.

The expressions (16.5) and (16.7) can be presented in physical coordinates. Applying the rescaling relations (12.1)–(12.6), the temperature at which the daughter-phase zone emerges, $T_{\mathrm{d}} = T_{\mathrm{d}}(K_I)$ is given by:

$$T_{\mathrm{d}} \approx T_0 + 0.04 \left[\frac{6AK_I\,(1+\nu)}{\alpha_0 E} \right]^{4/3} \left(\frac{\alpha_0}{g} \right)^{1/3} \tag{16.9}$$

Similarly, the temperature at the lower boundary of the mother-phase zone, $T_{\mathrm{m}} = T_{\mathrm{m}}(K_I)$, is:

$$T_{\mathrm{m}} \approx T_0 - 0.08 \left[\frac{6AK_I\,(1+\nu)}{\alpha_0 E} \right]^{4/3} \left(\frac{\alpha_0}{g} \right)^{1/3} \tag{16.10}$$

***

In summary, *the daughter-phase zone appears only for $A < 0$, while the mother-phase zone exists for $A > 0$. The former defines the upper limit of the zone's existence, while the latter defines the lower limit. Both zones have finite sizes, bounded on the right by critical points. Above these zones, the crack tip is bare, while below, a small beak-shaped mother-phase zone forms.*

16.3 Typical Values of the Key Parameters of the Crack Tip Zone

In this section, we present numerical estimates of critical parameters for the process zone at the crack tip, focusing on geometric characteristics $\mathbb{R}_0$, $\mathbb{R}_d$ and $\mathbb{L}_{cont}^{(max)}$, temperature intervals $T_d - T_c$ and $T_m - T_c$, and the critical velocity V_d. Since many material constants of inorganic materials generally fall within narrow ranges, we use this insight to derive typical values for the aforementioned parameters, allowing our findings to be applicable beyond the specific crystal studied.

***

16.3.1 Estimates

The estimated parameters of the process zone at the crack tip, calculated using the material constants from Tables A.1 and A.2 are summarized in Table 16.1.

***

In summary, *the estimates of the crack tip zone's parameters show the following features and peculiarities:*

Table 16.1 Estimates of the crack tip zone parameters

Property	Dimension	Estimate for displacive transitions	Estimate for order-disorder transitions	References
$T_d - T_c$*	K	$\sim 10^2$	~ 1 to 10	(10.16)
$T_m - T_c$*	K	$\sim - 10^2$	$\sim - 1$ to -10	(16.10)
$\mathbb{R}_0$	nm	~ 1	~ 1	(12.1)
$\mathbb{R}_d$**	nm	$\sim 10^4$	$\sim 10^{-1}$ to 10	(14.6)
$\mathbb{L}_{cont}^{(max)}$**	nm	$\sim 10^5$	~ 1 to 10^2	(14.25)
V_d	m/s	~ 10	$\sim 10^{-1}$	(12.7)

* $T = T_0 + 1\,\text{K}$.
** $T = T_0 + 10\,\text{K}$

1. **Distance from the Binodal**. *The parameters* $\mathbb{R}_d$ *and* $\mathbb{L}_{cont}^{(max)}$ *from Table* 16.1 *depend on* $T - T_0$*, both being proportional to* $(T - T_0)^{-2}$*, and they diverge as the temperature approaches the binodal. For these parameters, estimates were made at* $T = T_0 + 1$ K.
2. **The Zone's Size and Hysteresis**. *Both* $T_d(K_{IC}) - T_c$ *and* $T_m(K_{IC}) - T_c$ *for displacive transitions span a few hundred degrees, significantly exceeding the hysteresis width,* ΔT_{hyst} *(see Sect.* 34.1.1*). This suggests that the bifurcation temperature* $T_d(K_{IC})$*, lies far beyond the hysteresis.*
3. **The Zone's Size and Neighboring Phases** The sizes of the phase diagram regions occupied by various phases vary considerably. In terms of temperature, they can span approximately 10 K *in materials such as Bi and Te, around* $\sim 10^2$ K *in Ca, Sr, Mn, and Fe, and up to* $\sim 10^3$ K *in Be, Ba, Ti, Zr and Hf* [1]. *In some cases, the temperature intervals* $T_d - T_c$ *and* $|T_m - T_c| \sim 100$ K*, may exceed the temperature range of the mother or daughter phase on the phase diagram. For example, if a high-temperature phase precedes the mother phase in the range* $T_c < T < T_d$*, one will observe the daughter-phase zone within the mentioned high-temperature phase. The analogous situation can take place with the low-temperature phases. Namely, if below the daughter phase, there is another more low-temperature phase at* $T_c > T > T_m$*, it will exhibit a mother-phase process zone. Let us note that the situations with several high-temperature and low-temperature phases are typical rather than exceptional. For example, in* $BaTiO_3$*, we regard the cubic phase as the mother- and the tetragonal as the daughter phase. However, above the cubic phase, the phase diagram of* $BaTiO_3$ *contains the hexagonal, while below—the orthorhombic and rhombohedral phases. The temperature interval between each of the latter phases is about one hundred degrees (see Fig.* 3.1*b). This material thus falls under the situation described here.*
   Let us note that for the order-disorder transitions $T_d(K_{IC}) - T_c$ *and* $T_m(K_{IC}) - T_c$ *are on the order of a few tens degrees. That is, they either fall within the hysteresis, or exceed it only slightly.*
4. **Dependence on Stress Intensity Factor** K_I. *All parameters estimated in Table* 16.1 *depend on the stress intensity factor,* K_I*. Depending on the applied load, the latter can take various values and does not characterize the material. In contrast, its value,* K_{IC}*, at which the crack starts moving is the material's "fingerprint." For this reason, we made all estimates for* $K_I = K_{IC}$*. We estimated* $T_d - T_c$ *and* $T_m - T_c$ *at* $K_I = K_{IC}$*. It should be kept in mind that the bifurcation still occurs at* $K_I < K_{IC}$*. The temperature differences* $T_d(K_I) - T_c$ *and* $T_m(K_I) - T_c$ *are proportional to* $K_I^{4/3}$*. At* $K_I < K_{IC}$*, the bifurcation temperatures decrease, narrowing the domain of the zone's existence, though it remains considerable.*
5. **Natural Spatial Scale** $\mathbb{R}_0$. *The estimated value of* $\mathbb{R}_0$*, is quite small at* $K_I = K_{IC}$*, barely exceeding the size of a unit cell. It could be even smaller than a unit cell, implying that the continual theory may not accurately describe the zone, and results obtained by such a theory should be regarded as qualitative. However,* $\mathbb{R}_0(K_I) \sim K_I^{-2/3}$ (10.2). *Therefore, if a crystal dwells at some temperature,*

T, smaller than $T_d(K_{IC})$, the bifurcation occurs at $K_I < K_{IC}$. The latter yields $\mathbb{R}_0(K_I) > \mathbb{R}_0(K_{IC})$. In this case, there is such a value of $K_I < K_{IC}$ that $\mathbb{R}_0(K_I)$ is greater than a by at least one order of magnitude. Thus, there is a domain of applicability of the continuous approach.

6. **Lateral Size of the Mature Crack Tip Zone** $\mathbb{R}_d$. $\mathbb{R}_d$, *is considerably large due to the breadth of $T_d - T_c$ and $T_m - T_c$, and the relatively gentle slope of the potential $U \sim r^{-1/2}$. This substantial size significantly impacts crack properties, especially fracture toughness. As the temperature approaches the binodal, the size of the zone increases rapidly.*
7. **Critical Velocity** V_d. *For the transitions of the displacement type, the critical velocity V_d is typically around $\sim 10\,\mathrm{m/s}$, lower than Rayleigh's speed $V_R \sim 10^3\,\mathrm{m/s}$. Since the crack's velocity cannot surpass Rayleigh's speed, the zone exists at velocities from 0 to V_d. For materials with higher values of A and K_{IC},* the critical velocity *can exceed V_R, resulting in a process zone at any propagating crack velocity. For the order-disorder transitions the critical velocity is smaller, as one can expect: $V_d \sim 10^{-1}\,\mathrm{m/s}$. The latter implies that for such transitions the process zone can only be observed at small velocities below V_d.*

Reference

1. E.Y. Tonkov, *High Pressure Phase Transformations: A Handbook*, vols. 1–3 (Gordon and Breach, Amsterdam, 1992); E.Y. Tonkov, E.G. Ponyatovsky, *Phase Transformations of Elements Under High Pressure*, vol. 4 (CRC Press, Boca Raton, FL, 2005)

Part III
Zone Effect on the Behavior of the Generating Crack

Chapter 17
Transformation Toughness: The Field-Theoretical Perspective

Abstract This chapter applies the field-theoretical approach to analyze a transformation toughness in a crystal with a motionless crack. The loaded crack engenders a process zone at its tip. As elaborated earlier, due to the surface tension of the zone's interface, an interval close to the crack tip exists where the crack surface borders the process zone. It leads to the mechanical shielding of the crack tip, thereby giving rise to transformation toughness. The field-theoretical method empowers us to establish its correlation with the solid's position on its phase diagram. Based on the outcomes of our simulations described earlier in the book, we have derived an expression for effective fracture toughness using two distinct approaches: (i) leveraging the elastic Green's function and (ii) utilizing weight functions. Building upon these findings, we have formulated an accurate analytical approximation that characterizes transformation toughness, expressing it in terms of experimentally attainable parameters related to the phase diagram, including the hysteresis width, the slope of the binodal line, and the transformation strain (Boulbitch and Korzhenevskii 2021).

An unequivocal correlation has been established between the formation of the process zone and the enhancement of fracture toughness, a phenomenon referred to as *transformation toughness*. Observations of transformation toughness have been documented in various materials, including TRIP steels [1, 2], stishovite [14], zirconia ceramics [15–19], and ceramic composites containing zirconia components [20]. This has sparked considerable interest in unraveling the origins of transformation toughness.

17.1 Comparison of Mechanical and Field-Theoretical Approaches

This introductory section briefly reviews the historically first approach—still widely used today—that describes the crack tip process zone and its impact on fracture toughness, referred to here as the 'mechanical approach.'

***

© The Author(s), under exclusive license to Springer Nature Switzerland AG 2026

A. Boulbitch and A. Korzhenevskii, *Stress-Induced Local Phase Transformations Near Cracks and Dislocations*, Springer Series in Materials Science 356,
https://doi.org/10.1007/978-3-032-00945-6_17

17.1.1 Mechanical Approach

The mechanical approach has played a pioneering role in theoretical investigations concerning zone-induced transformation toughness. It operates on the assumption that

1. The zone differs from the matrix in terms of its elastic properties,
2. A spontaneous strain is present within the zone,
3. Moreover, the zone-matrix interface position is controlled by equality to a specific value of some invariant constructed of stress tensor components.

Within these assumptions, the zone's behavior is treated as a problem involving a moving boundary [9, 21–23].

The first to elucidate transformation toughness within a simplified mechanical framework was Antolovich [1]. By analogy to plasticity, he introduced the energy of the long-lasting wake into the fracture energy, qualitatively explaining the growth of fracture toughness. Subsequent studies followed Antolovich's approach [16–19].

In the context of materials undergoing a phase transition with a dilatant transformation strain, McMeeking and Evans proposed a *mechanical shielding mechanism* [55] that accounts for the compression occurring within the wake, opposing crack opening. Budiansky and colleagues analyzed this mechanism in composites with particles exhibiting stress-induced transformation [24]. Rose, employing Hutchinson's approach [25], and Stump and Budiansky, in a detailed study, examined the process zone's divergence, referred to as *resonance* [26, 27]. These studies employed the phase boundary criterion $\sigma_{\alpha\alpha} = \sigma_{\mathrm{m}}$, where $\sigma_{\alpha\alpha}$ represents the spur of the stress tensor, and σ_{m} is regarded as a material constant. Under these assumptions, the process zone takes on a cardioid shape, not contributing to transformation toughness. The latter is only impacted by the wake [24, 55]. Other papers [29, 30] analyzed models incorporating both dilatant and shear stresses into the criterion. Budiansky, Truskinovsky, and Simha developed a model focusing on shear strain within spherical ZrO_2 inclusions embedded in an elastic matrix, considering the tetragonal to monoclinic transition as a proper ferroelastic transition [28, 31].

In the aforementioned studies [1, 16, 17, 19, 24, 26, 27, 55], two key assumptions prevailed. Firstly, the researchers assumed that (i) crack propagates, thus forming the wake, though the propagation was considered quasi-static. Secondly, (ii) they considered phase transformations as irreversible phenomena. The latter hypothesis was vital for this group of studies, as it is equivalent to the statement that the wake, once formed, persists indefinitely. This long-lasting wake forms contact with the crack surfaces behind its tip, giving rise to mechanical shielding, a significant contributor to transformation toughness, as elucidated in [55]. For a comprehensive review up to the year 2003, readers can refer to the paper by Kelly and Rose [9].

More recent research has been focused on the phase transformations occurring at the crack tip in shape memory alloys, such as NiTi and various steels. Notably, Freed et al., in their work [32], studied shape memory alloys, considering the influence of temperature on the boundary criterion. Baxevanis and collaborators explored the path-dependent nature of the J-integral and analyzed the transformation's impact on the ratio of far-field to crack-tip energy release rates in their works [33, 34]. Ma, Korsunsky, and McMeeking offered insights into transformation toughening within anisotropic solids in their study [35]. Lexellent and coauthors analyzed NiTi, focusing on the tension-compression asymmetry, as detailed in [36]. Several recent papers have addressed the crack-tip transformation in NiTi material [57]. Maletta and Young developed a model for the NiTi process zone that accounts for stress triaxiality in their research [38]. Krupp et al. analyzed the extent of martensite formation ahead of the crack tip in metastable austenite steel, examining its dependence on crack length in [39]. Vatne et al. investigated the crack tip bcc-fcc transformation in Fe in their work [40]. Additionally, studies have explored toughening mechanisms in bulk metallic glasses, as explored in [41].

It is worth noting that the criteria for phase boundaries hypothesized in most of the papers mentioned above were inspired by those used in plasticity. Some of these assumptions and criteria will be discussed in more detail in Sect. 17.4.1. Note that the plasticity-like boundary conditions applied in these studies assume physical mechanisms distinct from those occurring during phase transformations. Consequently, such boundary conditions are either physically unfounded or describe a scenario different from the one under consideration.

Boundary conditions based on the physically accurate requirement of a global free energy minimum were proposed by Roitburd [42, 43] and further developed by Grinfeld [44], Boulbitch, and Korzhenevskii [3]. However, the resulting criterion is mathematically complex and, to the best of our knowledge, has not yet been applied to the description of the crack tip zone.

17.1.2 The Field-Theoretical Approach

Defects such as dislocations and cracks are recognised as strong stress concentrators. These same mechanisms also lead to localised, stress-induced phase transformations—so-called process zones—at dislocation and crack tips. Such transformations are highly nonlinear and present considerable theoretical challenges. A further important aspect involves linking mechanical behaviour with crystal thermodynamics. Realistic modelling of process zone behaviour requires a description grounded in the crystal phase diagram. The field-theoretical approach inherently meets these needs by integrating Landau theory of phase transformations with conventional mechanics.

In this framework, the process zone is described following [13], by the field of the order parameter, $\eta = \eta(\mathbf{R})$. As in classical Landau theory [45], the order parameter is constructed from combinations of atomic displacements averaged over a small yet macroscopic volume. It thus represents internal degrees of freedom within the solid, characterising the distinction between the parent and product crystal structures. The function $\eta(\mathbf{R})$) satisfies the well-established Ginzburg–Landau equation of state. Within this approach, both the process zone boundary and the reversibility of transformation naturally emerge from solutions to this equation.

The field-theoretical method was initiated by Nabutovskiĭ and Shapiro [46], who described a superconducting 'filament' at a dislocation. One of us first applied this method to crack tip zones in [47]. It was subsequently developed in [3–8, 10–13], as well as in the paper [48]. Several recent studies have incorporated the phase-field method into this framework. It has also been applied to analyse process zones in ferroelectrics [49]. In [50], this combination was used to determine strain and stress fields and the zone configuration at a crack tip. Zhao et al. examined polydomain zone structure during early crack propagation by merging the field-theoretical and phase-field approaches [51]. Zhu et al. used the same combination to investigate intergranular microcrack nucleation in polycrystalline tetragonal zirconia [52]. Levitas and collaborators explored zone wetting effects at crack surfaces [53] and capillary effects stemming from interactions between phase-field gradients and the strain field [54].

***

In summary, *plasticity-like criteria lack a solid physical foundation when applied to phase transformations. The physically well-established criteria proposed by Roitburd, Grinfeld, Boulbitch, and Korzhenevskii differ qualitatively from plasticity-like boundary conditions.*

Furthermore, the concept of phase transformation irreversibility—central to the mechanical approach—is relatively uncommon (discussed in detail in Section 17.4.1.5). *In most solids, the interface between stable and metastable phases propagates at a finite velocity, decreasing the volume of the metastable phase. These arguments suggest an alternative framework for transformation toughness. The latter, referred to as the field-theoretical approach enable one to describe the process zone and relate this description to the crystal thermodynamics and kinetics.*

17.2 Effective Stress Intensity Factor

This section investigates the screening effect of the setback in the crack tip process zone, illustrated in Fig. 13.2*b (Sect.* 13.1). *It leads to a perturbation in the stress intensity factor applied to the crack depending on a dimensionless parameter,* ζ_0, *which we express in quadratures. To verify accuracy, we apply two mathematical methods: one based on elastic Green's functions and the other on weight functions. Numerical evaluation provides the relationship* $\zeta_0 = \zeta_0(a)$, *for which we also derive an analytical approximation.*

***

17.2.1 Bueckner's Approach Revisited

The transformation from a zoneless crack to one that possesses a zone results in the emergence of transformation strain. This strain augments $\varepsilon_{ij}^{(0)}$ (4.7) and consequently gives rise to an additional stress distribution around the crack tip. This additional stress influences the load experienced by the crack, thereby perturbing the stress intensity factor. The net stress intensity factor K_I can be expressed as follows:

$$K_I = K_I^{(\mathrm{ap})} + \delta K_I \tag{17.1}$$

(see 1.3a). Here, δK_I is the perturbation induced by the zone of an applied stress intensity factor $K_I^{(\mathrm{ap})}$. In this approach, we determine δK_I in a self-consistent manner.

Let us note that (3.16) can be rewritten as:

$$\sigma_{\alpha\beta} = \sigma_{\alpha\beta}^{(0)} + A\eta^2\delta_{\alpha\beta}$$

where

$$\sigma_{\alpha\beta}^{(0)} = \frac{E\nu}{(1-2\nu)(1+\nu)}\varepsilon_{\gamma\gamma}\delta_{\alpha\beta} + \frac{E}{(1+\nu)}\varepsilon_{\alpha\beta}$$

is the standard Hook's law representation of stress. Correspondingly, we rewrite the equation of mechanical equilibrium, $\partial\sigma_{\alpha\beta}/\partial x_\beta = 0$, as:

$$\frac{\partial\sigma_{\alpha\beta}^{(0)}}{\partial x_\beta} = f_\alpha$$

where

$$f_\alpha = -A\frac{\partial\eta^2}{\partial x_\alpha} \quad (\alpha = 1, 2) \tag{17.2}$$

can be interpreted as an in-plane force, $\mathbf{f} = (f_1, f_2, 0) \equiv (f_x, f_y, 0)$.

One can view this as if the zone generates the force (17.2) at all its points. To handle such forces, we can implement the approach introduced by Bueckner, which allows us to convert the load applied on the body into a force distributed along the crack's surface [56] (see also [58]). This approach involves two steps:

First Step:

Consider a hypothetical crack-free body to which bulk forces are applied. Determine its stressed state, referred to as the first stressed state.

Second Step:

Calculate the stress of the first stressed state at the points of the crack faces and use it to determine the perturbation of the stress intensity factor.

Now, leveraging the results presented in Sects. 13.1 and 14.2, let us switch to the calculation of the perturbation of the stress intensity factor using Bueckner's algorithm.

17.2.2 Application of the Green Function in Bueckner's Method

17.2.2.1 Step 1: Defining the First Stressed State

In the initial step of Bueckner's approach, we define the first stressed state within a hypothetical crack-free body, which still contains the process zone. To achieve this, we use the Green's function. Specifically, we use it to determine the stress at point $z = x + iy$ induced by a local force $f_x + if_y$ applied at point $Z = X + iY$ as derived in [58]. Here i is the imaginary unit. The expressions are as follows:

$$\Phi(z) = \frac{f_x + if_y}{8\pi(1-\nu)} \times \frac{1}{z - Z}$$

$$\Psi(z) = \frac{(3-4\nu)(f_x - if_y)}{8\pi(1-\nu)} \times \frac{1}{z - Z} - \frac{\overline{Z}(f_x + if_y)}{8\pi(1-\nu)} \times \frac{1}{(z - Z)^2} \tag{17.3}$$

In these equations $\Phi(z)$ and $\Psi(z)$ are Kolosov-Mushelishvili's potentials (as defined in [59]), and the bar over the symbols (e.g., $\overline{Z}$, or $\overline{\Phi'(z)}$) denotes their complex conjugation, where $\overline{Z} = X - iY$.

Note 17.2.1 Importantly, all coordinates and variables (x, y, z, X, Y, Z, f_x and f_y) in this section are dimensional.

The relation of (17.3) to the process zone one achieves by equating the complex force, $f_x + if_y$, to the expression (17.2) as follows:

$$f_x + if_y = -A\left[\frac{\partial\eta^2(X,Y)}{\partial X} + i\frac{\partial\eta^2(X,Y)}{\partial Y}\right]$$

The stress at point z can be computed using the standard relation from the plane theory of elasticity. We are particularly interested in σ_{yy} which is given by [59]:

$$\sigma_{yy} = \mathrm{Re}\left[2\Phi(z) + z\overline{\Phi'(z)} + \Psi(z)\right]. \tag{17.4}$$

By evaluating this expression at the point lying on the crack contour $z = x$ (where $x < 0$ and $y = 0$) and integrating over the domain Ω, we obtain the stress $\sigma_{yy}(x)$ of the first stressed state. To simplify the writing of formulas, we represent it as the sum of two terms:

$$\sigma_{yy}(x) = \sigma_{yy}^{(1)}(x) + \sigma_{yy}^{(2)}(x),$$

where

$$\sigma_{yy}^{(1)}(x) = -\frac{A}{4\pi(1-\nu)}\iint\limits_{\Omega}\left\{\frac{(X-x)\left[(1+2\nu)Y^2-(1-2\nu)(X-x)^2\right]}{\left[(X-x)^2+Y^2\right]^2}\frac{\partial\eta^2(X,Y)}{\partial X}\right\}dXdY \tag{17.5}$$

$$\sigma_{yy}^{(2)}(x) = -\frac{A}{4\pi(1-\nu)}\iint\limits_{\Omega}\left\{\frac{Y\left[(3-2\nu)Y^2+(1-2\nu)(X-x)^2\right]}{\left[(X-x)^2+Y^2\right]^2}\frac{\partial\eta^2(X,Y)}{\partial Y}\right\}dXdY \tag{17.6}$$

These expressions represent the first stressed state, thus, laying the foundation for the second step of Bueckner's algorithm.

17.2.2.2 Step 2: Perturbation of the Stress Intensity Factor

This step uses the first stressed state to compute the perturbation δK_I, as defined in (1.3a):

$$\delta K_I = -\sqrt{\frac{2}{\pi}}\int_0^{\infty}\frac{\sigma_{yy}(-x)}{\sqrt{x}}dx \tag{17.7}$$

To simplify the integration, we transition from dimensional parameters (order parameter, η, coordinates x, X, y and Y with length dimensions) to dimensionless parameters u, ξ, x' and y':

$$\eta(X,Y) = \eta_0 u(x',y');\quad x = -\mathbb{R}_0\xi;\quad X = \mathbb{R}_0 x';\quad Y = \mathbb{R}_0 y' \tag{17.8}$$

The factors η_0 and $\mathbb{R}_0$ are determined by (12.3) and (12.1).

Using (17.5) and (17.6), we express the perturbation δK_I as follows:

$$\delta K_I = -A\eta_0^2\mathbb{R}_0^{1/2}\zeta_0 \tag{17.9}$$

The dimensionless parameter ζ_0 is an integral expressed in terms of dimensionless variables ξ, x', and y':

$$\zeta_0 = \int\limits_0^{\infty}\frac{d\xi}{\sqrt{\xi}}\iint\limits_{\Omega}dx'dy'\left\{Q_1(\xi,x',y')\frac{\partial u^2(x',y')}{\partial x'} + Q_2(\xi,x',y')\frac{\partial u^2(x',y')}{\partial y'}\right\} \tag{17.10}$$

Where:

$$Q_1(\xi,x',y') = -\frac{(x'+\xi)\left[y'^2(1+2\nu)-(x'+\xi)^2(1-2\nu)\right]}{(2\pi)^{3/2}(1-\nu)\left[(x'+\xi)^2+y'^2\right]^2} \tag{17.11}$$

and

$$Q_2(\xi, x', y') = -\frac{y'\left[y'^2(3-2\nu) + (x'+\xi)^2(1-2\nu)\right]}{(2\pi)^{3/2}(1-\nu)\left[(x'+\xi)^2 + y'^2\right]^2} \tag{17.12}$$

These equations lead us to the perturbed stress intensity factor K_I in terms of $K_I^{(\mathrm{ap})}$, which was initially applied to the crack:

$$K_I = K_I^{(\mathrm{ap})} - A\eta_0^2 \mathbb{R}_0^{1/2} \zeta_0 \tag{17.13}$$

Therefore, the problem ultimately boils down to calculating the single dimensionless parameter ζ_0 (17.10). Below, we estimate this integral numerically. Once we have ζ_0, we can determine the effective stress intensity factor K_I using (17.13) and, with its value, ascertain the fracture toughness.

17.2.2.3 Numerical Calculation of the Integral ζ_0

To perform the integration in (17.10), we require knowledge of the derivatives $\partial u^2(x', y')/\partial x'$ and $\partial u^2(x', y')/\partial y'$. This necessitates having the function $u(x', y')$, which we have already obtained through simulations outlined in Annex Sect. 38, with the results[1] detailed in Sect. 13.1. These simulations involved solving (12.13) using COMSOL 5.5, allowing us to determine the derivatives $\partial u^2/\partial x$ and $\partial u^2/\partial y$ required for the integral expression of ζ_0 (17.10).

As illustrated in Fig. 17.1 we observe the lateral behavior of these derivatives for a specific case of $a = 0.24$. Similar distributions were obtained for a range of values, including $a = 0.2, 0.22$, 0.23 and 0.24. These distributions maintain the same shapes while varying in size. They exhibit sharp crests localized along the zone boundary, with one part of each crest in the upper semi-space and the other in the lower. Consequently, portions of each distribution are positive while others are negative. A closer examination at a smaller scale reveals that each distribution exhibits a bell-shaped character, as demonstrated in Fig. 13.11a.

The sawtooth appearance of these distributions is associated with errors introduced during the numerical derivation of $\partial u^2/\partial x$ and $\partial u^2/\partial y$ on the mesh from the numerical data for u.

Subsequently, we imported the data for $\partial u^2/\partial x$ and $\partial u^2/\partial y$ into the software Mathematica where we performed interpolation. This integral (17.10) we solve using the interpolation functions instead of $\partial u^2/\partial x$ and $\partial u^2/\partial y$. During the calculation we assigned $\nu = 1/3$. In this specific case, the functions $Q_1(\xi, x, y)$ and $Q_2(\xi, x, y)$ (17.11) and (17.12) take the following simplified forms:

$$\begin{aligned} Q_1(\xi, x, y) &= -\frac{(x+\xi)\left[5y^2-(x+\xi)^2\right]}{(2\pi)^{3/2}\left[(x+\xi)^2+y^2\right]^2} \\ Q_2(\xi, x, y) &= -\frac{y\left[7y^2+(x+\xi)^2\right]}{(2\pi)^{3/2}\left[(x+\xi)^2+y^2\right]^2} \end{aligned} \tag{17.14}$$

Using the interpolation functions for $\partial u^2/\partial x$ and $\partial u^2/\partial y$ in combination with $Q_1(\xi, x, y)$ and $Q_2(\xi, x, y)$ (17.14), we numerically computed the integral (17.10) utilizing Mathematica's

[1] We recommend skipping this section on the first reading.

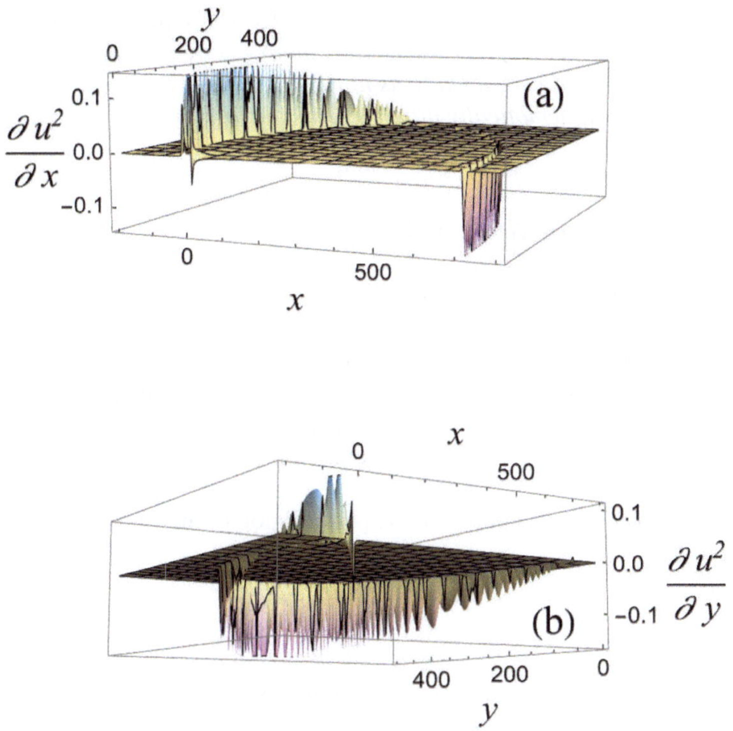

Fig. 17.1 Lateral behavior of the derivatives of the squared order parameter. Panel (a) shows $\partial u^2/\partial x$, and panel (b) displays $\partial u^2/\partial y$, for $a = 0.24$ and $b = 1$

built-in numerical routine NIntegrate. It's worth noting that this integral exists, but due to the singularity of $Q_1(\xi, x, y)$ and $Q_2(\xi, x, y)$ at $x = -\xi$ and $y = 0$, achieving numerical convergence requires a specific strategy.

We employed a strategy involving point sampling based on the 3D Cartesian rule and applied the Gauss-Kronrod approach with additional Gauss points. The selected options included: Points$\rightarrow$ 3, MinRecursion$\rightarrow$ 3, MaxRecursion$\rightarrow$ 7, Exclusions$\rightarrow$ $\{0, 0, 0\}$. In the integral with respect to ξ, we adopted a finite upper limit equal to the coordinate ξ of the left boundary of the domain Ω instead of infinity. This adjustment does not affect the result significantly because at this ξ value, the derivatives $\partial u^2/\partial x$ and $\partial u^2/\partial y$ are nearly zero.

With these parameters, we achieved an error within a few percent and reasonable calculation times, typically below ten minutes, contingent on the size of the integration domain. The result is demonstrated in Fig. 17.4.

***

It's important to acknowledge *that the estimation of the integral* (17.10), (17.11) and (17.12) *is susceptible to numerical errors. These errors are partly attributed to the singular nature of the kernnels $Q_1(\xi, x', y')$ and $Q_2(\xi, x', y')$. Additionally, determining the derivatives $\partial u^2(x', y')/\partial x'$ and $\partial u^2(x', y')/\partial y'$ from the numerical* *data for $u(x', y')$ introduces its own set of errors. For validation purposes, alongside the Green's function method, we concurrently apply the weight function method. The latter neither employs diverging kernels nor derivatives of the order parameter and is, in this respect, mathematically different from the former.*

17.2.3 Weight Functions in Field Theory

17.2.3.1 Weight Functions Method Revisited

The weight functions method, initially introduced by Buechner [8] found application in defining the perturbation of the stress intensity factor by the process zone and the zone-wake complex within the mechanical approach, as demonstrated by McMeeking and Evans [55]. Here, we adapt this method for use in the field-theoretical framework.

In the weight functions approach, the perturbation of the stress intensity factor is expressed as:

$$\delta K_I = \sigma_n \oint_{\mathfrak{C}} (\mathbf{H} \cdot \mathbf{n}) dS \tag{17.15}$$

where σ_n represents the normal stress at the zone boundary [55]. In terms of the order parameter one expresses it as

$$\sigma_n = A\eta^2.$$

To apply the weight functions method within the field-theoretical approach, we need to approximate the order parameter distribution $\eta = \eta(x, y)$ with a step-function. We assume it to be strictly zero outside of the zone and take a constant value inside, derived from the solution of a homogeneous problem:

$$\eta = \eta_0 u_{\text{roof}} \tag{17.16}$$

where u_{roof} represents the height of the roof β. With reasonable accuracy, one can replace u_{roof} with unity (see Sect. 13.2.2.3). As always, aiming to regard the manifested first-order case, we consider $b = 1$.

The integral (17.15) is computed along the closed contour $\mathfrak{C} = \mathfrak{C}_1 \cup \mathfrak{C}_2$. The contour $\mathfrak{C}_1$ follows the line with the length $\mathbb{L}_{\text{inv}}$ (referred to as the *dimensional contact length*) where the zone contacts crack surfaces. The contour $\mathfrak{C}_2$ traces the boundary between the mother and daughter phases and can be characterized by the dimensional radius-vector $\mathbb{R} = \mathbb{R}(\theta)$, as shown in Fig. 17.2. We transform $\mathbb{L}_{\text{inv}}$ and $\mathbb{R}$ with the dimensions of length into dimensionless variables l_{inv} and r:

$$\mathbb{L}_{\text{inv}} = \mathbb{R}_0 \times l_{\text{inv}}; \quad \mathbb{R}(\theta) = \mathbb{R}_0 \times r(\theta) \tag{17.17}$$

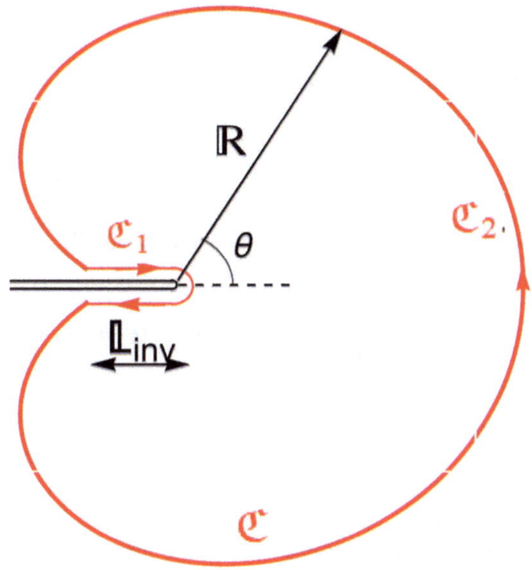

Fig. 17.2 Integration contour. Contour $\mathfrak{C}$ of integration of (17.15) consists of two subcontours, $\mathfrak{C}_1$ along the crack and $\mathfrak{C}_2$ along the zone boundary described by $\mathbb{R} = \mathbb{R}(\theta)$. The contact length $\mathbb{L}$ where the zone meets the crack surfaces is indicated

where $\mathbb{R}_0$ is the rescaling factor (see relation (12.1)).

In terms of the dimensionless variables l_{inv} and $r(\theta)$, the weight functions vector $\mathbf{H}$, the normal unit vector $\mathbf{n}$ and the arc element dS along $\mathfrak{C}_2$ can be expressed as:

$$\mathbf{H} = \frac{1}{2\sqrt{2}\pi(1-\nu)\mathbb{R}_0^{1/2}\sqrt{r(\theta)}}(h_1, h_2) \tag{17.18}$$

where

$$h_i = \begin{cases} \cos(\theta/2)\,[2\nu + \sin(\theta/2)\sin(3\theta/2)]\,, & i = 1 \\ \sin(\theta/2)\,[-2\nu - \cos(\theta/2)\cos(3\theta/2)]\,, & i = 2 \end{cases}$$

and

$$\mathbf{n} = \left(\frac{\sin(\theta)r'(\theta) + \cos(\theta)r(\theta)}{\sqrt{r'^2(\theta) + r^2(\theta)}}, -\frac{\cos(\theta)r'(\theta) - \sin(\theta)r(\theta)}{\sqrt{r'^2(\theta) + r^2(\theta)}} \right) \tag{17.19}$$

and finally

$$dS = R_0\sqrt{r'^2(\theta) + r^2(\theta)}d\theta \tag{17.20}$$

We represent δK_I as a sum $\delta K_I = \delta K_I^{(1)} + \delta K_I^{(2)}$ corresponding to the portions $\mathfrak{C}_1$ and $\mathfrak{C}_2$ of the contour $\mathfrak{C}$. The first contribution can be calculated straightforwardly as:

$$\delta K_I^{(1)} = -\frac{4A\eta_0^2 \mathbb{R}_0^{1/2}}{\sqrt{2\pi}} \sqrt{l_{\text{inv}}} \tag{17.21}$$

The second contribution takes the form:

$$\delta K_I^{(2)} = \frac{A\eta_0^2 \mathbb{R}_0^{1/2}}{2\sqrt{2\pi}(1-\nu)} I \tag{17.22}$$

where I is the integral along the contour $\mathfrak{C}_2$:

$$I = \frac{1}{2\sqrt{2\pi}(1-\nu)} \int_0^{\pi} \left\{ r'(\theta) p_1(\theta) + r(\theta) p_2(\theta) \right\} \frac{d\theta}{\sqrt{r(\theta)}} \tag{17.23}$$

where

$$p_i(\theta) = \begin{cases} \sin(\theta/2)\,[(5-8\nu)\cos(\theta) - 4\nu + 1]\,, & i = 1 \\ \cos(\theta/2)\,[(7-8\nu)\cos(\theta) + 4\nu - 5]\,, & i = 2 \end{cases}$$

By combining these results, we arrive at the expression (17.9) with

$$\zeta_0 = \zeta_1 + \zeta_2; \quad \zeta_1 = \frac{2^{3/2}}{\pi^{1/2}} \sqrt{l_{\text{inv}}}; \quad \zeta_2 = -\frac{I}{\pi\sqrt{2}(1-\nu)} \tag{17.24}$$

Once the contour $\mathfrak{C}_2$ is obtained, the dimensionless parameter ζ_0 can be calculated. This parameter determines the magnitude of the zone effect on fracture toughness, making it critical to determine its absolute value, a task we undertake in the following section.

17.2.3.2 Integration Within the Weight Functions Approach

We recommend skipping this section on the first reading. Based on the COMSOL simulations, we acquired the distribution of $(\nabla u)^2 = (\partial u/\partial x)^2 + (\partial u/\partial y)^2$. This distribution consistently exhibits a sharp crest along the zone boundary, characterized by a bell-shaped cross-section. Evidently, $(\nabla u)^2$ is everywhere positive. Figure 17.3a provides an illustrative example of such a distribution for $a = 0.24$, $b = 1$. It's worth noting that $(\nabla u)^2$ distributions at different a values share qualitatively similar configurations and differ only in scale.

The sawtooth shape seen in the distribution is attributed to numerical differentiation errors. The crest point heights vary along the zone contour. We selected points in close proximity to the crest in Cartesian coordinates by specifying a threshold value for $(\nabla u)^2$, determined through trial and error within the range of 0.005 to 0.02. These points accurately represent the zone boundary contour. We imported this data into Mathematica as lists and transformed them to express the boundary points in cylindrical coordinates $r = r(\theta)$, which are depicted as dots in Fig. 17.3b.

We further divided the interval $0 \le \theta \le \pi$ into three overlapping intervals: $0 \le \theta \le \theta_1 + 0.1$, $\theta_1 - 0.1 \le \theta \le \theta_2 + 0.1$ and $\theta_2 - 0.1 \le \theta \le \pi$. Within each interval, we fitted the

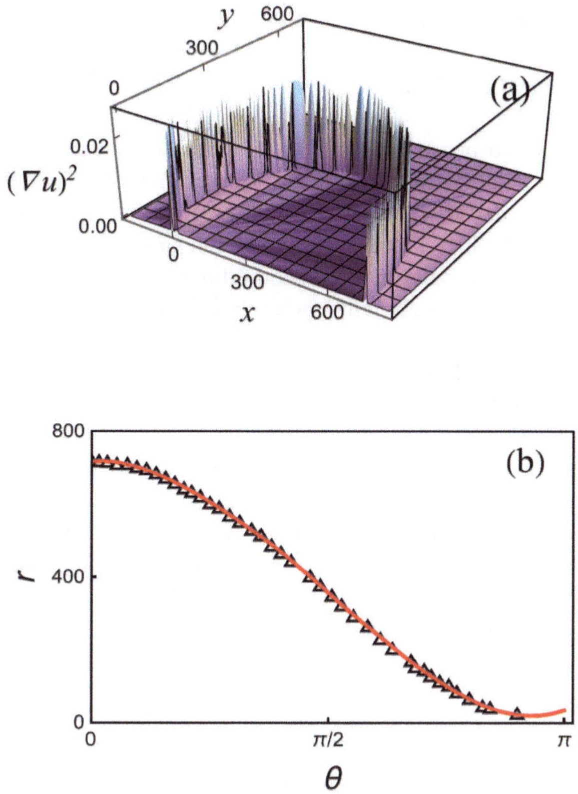

Fig. 17.3 Definition of the zone boundary based on numerical data. **a** The lateral distribution of $(\nabla u)^2$ for $a = 0.24$, $b = 1$ **b** The coordinates of the points of its crest transformed to cylindrical coordinates $r = r(\theta)$. The dots indicate the boundary points obtained form the simulation data, while the solid line shows its fitting

curves using a fourth-order polynomial in terms of the angle θ, employing Mathematica's standard FindFit routine. We adjusted the boundaries θ_1 and θ_2 through trial and error to achieve continuous and piecewise smooth curves. Importantly, the presence of two kink points introduced negligible error into the final result since integration is not sensitive to that. The fitting outcome is depicted by the solid line in Fig. 17.3b. Similar results were obtained for all a values: 0.2, 0.22, 0.23, 0.24 and 0.25.

We incorporated the analytical functions representing $r(\theta)$ obtained through fitting into the integral I (17.23) and estimated the integral numerically using Mathematica's standard NIntegrate routine. In this instance, the integration process was straightforward.

17.2.4 The ζ_0 Behavior

17.2.4.1 Numerical Analysis

Utilizing the described procedures, we have successfully determined the behavior of the dimensionless factor ζ_0 as a function of parameter a, denoted as $\zeta_0(a)$. This behavior is comprehensively illustrated in Fig. 17.4. In this Figure, data points represent ζ_0 at $b = 1$ and distinct a values, derived from simulation studies utilizing both the Green function and weight functions methodologies. Remarkably, these analyses yielded results that align closely with each other, demonstrating consistency across methods.

We denote the binodal position as $a_b = 3/16$ indicated by the dashed vertical line in Fig. 17.4. It is discernible from the graph that ζ_0 experiences an uptick as it approaches the binodal (3.32).

This particular observation marks a significant departure from prior research findings. Notably, in the context of this study, preceding research endeavors by McMeeking and Evans [55], Budiansky [24], along with other scholars [9], consistantly obtained a ζ_0 value of zero. Hence, the manifestation of a substantially non-zero ζ_0 emerges as a distinct outcome rooted in the field-theoretical approach, the genesis of which will be explored in subsequent discussions.

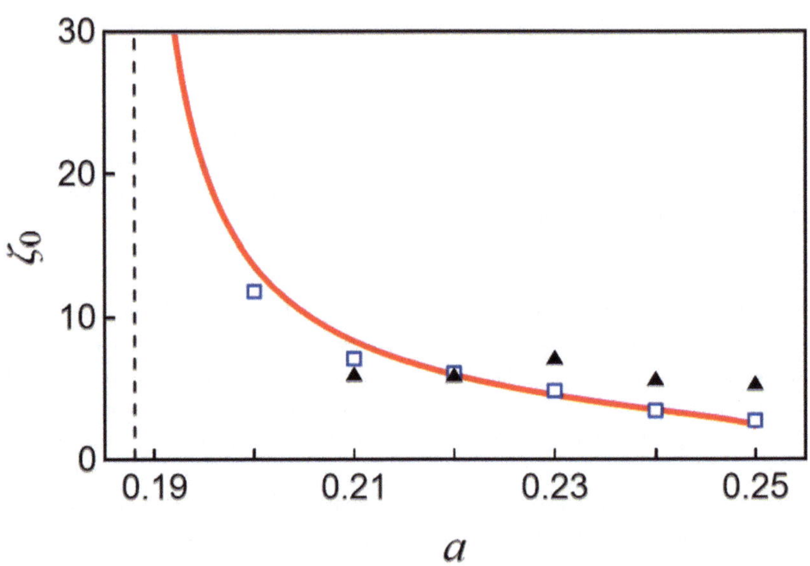

Fig. 17.4 Variation of ζ_0 with a for Fixed $b = 1$. The filled triangles illustrate results garnered from the simulation employing the Green function technique, while the open squares represent data calculated through the weight functions method. The unbroken line shows the behavior of the analytical function $\zeta_0(a)$ as dictated by (17.25). Furthermore, the dashed line signifies the binodal position at $a = 3/16$

17.2.4.2 Analytical Formulation for the ζ_0

To decipher the nuance highlighted earlier, we initially observe that ζ_1 (the primary component of ζ_0 elucidated via the weight function method (17.24) exhibits a proportionality to $l_{\text{inv}}^{1/2}$. Remember that l_{inv} signifies the dimensionless contact length, a topic thoroughly explored in Sect. 13.1.

Subsequent numerical analysis illustrates that, consistently, the ratio ζ_2/ζ_1 maintains a small magnitude, approximately 0.2. This allows us to fairly accurately approximate ζ_0 as $\zeta_0 \approx \zeta_1/0.8 \sim l_{\text{inv}}^{1/2}$. In Sect. 14.2.2, we discussed the analytical derivation of l_{inv} as a function of variables a and b (14.12).

Applying the formulas detailed in references (17.24), (17.16) and (14.12), we arrive at the following expression for ζ_0:

$$\zeta_0 \approx 3.22 \frac{\left[\left(b^2-4a\right)^{1/2}+b\right]\left[b^2-2a+b\left(b^2-4a\right)^{1/2}\right]^{1/8}}{\left(16a-3b^2\right)\left[8a-b^2-b\left(b^2-4a\right)^{1/2}\right]^{1/4}} \tag{17.25}$$

The solid curve in Fig. 17.4 graphically presents ζ_0 (17.25) at $b = 1$, exhibiting a good agreement with our numerical evaluations.

We transition from dimensionless parameters to the original physical variables, using the transformation defined in (12.4, 12.6) applying it to (17.25). This renders ζ_0 in terms of the original material constants, as:

$$\zeta_0 \approx 3.28 \left[\frac{A(1-2\nu)(1+\nu)}{Eg^{1/4}} K_I\right]^{1/6} \left(\frac{\gamma}{\beta^2}\right)^{1/8} s\left(\frac{\alpha\gamma}{\beta^2}\right) \tag{17.26}$$

Here, for the sake of brevity, we introduce a dimensionless variable s, defined as a function of $\mathfrak{z}$:

$$s(\mathfrak{z}) = \frac{\left[1+(1-4\mathfrak{z})^{1/2}\right]\left[1+(1-4\mathfrak{z})^{1/2}-2\mathfrak{z}\right]^{1/8}}{(16\mathfrak{z}-3)^{1/2}\left[8\mathfrak{z}-1-(1-4\mathfrak{z})^{1/2}\right]^{1/4}} \tag{17.27}$$

With $\mathfrak{z}$ defined as:

$$\mathfrak{z} = \alpha\gamma/\beta^2,$$

the expression (17.26) is defined within the range $\alpha_{\mathfrak{b}} < \alpha \leq \alpha_{\text{up}}$, i.e., below the upper spinodal but above the binodal boundary. This leads us to express these boundaries in terms of α, β and γ, defining $\alpha_{\mathfrak{b}} = 3\beta^2/16\gamma$ and $\alpha_{\text{up}} = \beta^2/4\gamma$.

Importantly, the factor $s = s(\mathfrak{z})$ is solely reliant on the dimensionless combination $\mathfrak{z} = \alpha\gamma/\beta^2$, indicating its dependence on the position of the solid phase diagram within the specified $\mathfrak{z}$ range:

$$\frac{3}{16} < \frac{\alpha\gamma}{\beta^2} \leq \frac{1}{4}$$

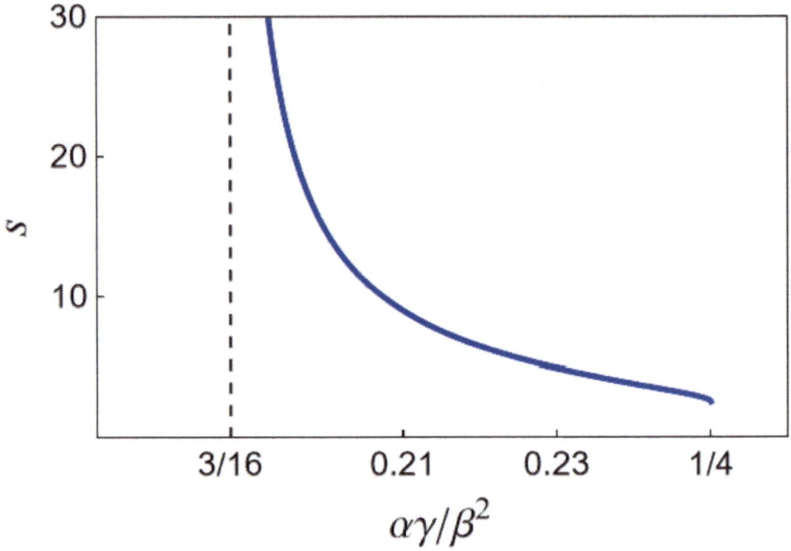

Fig. 17.5 Behavior of the dimensionless factor $s(\alpha\gamma/\beta^2)$. The factor $s(\alpha\gamma/\beta^2)$ is influenced by the dimensionless combination of Landau potential coefficients, $\alpha\gamma/\beta^2$. The dashed-vertical line marks the position of the binodal

Here, $s(\mathfrak{z})$ transitions from a minimum value, $s_{\min} \approx 2.59$, at the upper spinodal at $\mathfrak{z} = 1/4$ to an infinite value at the binodal ($\mathfrak{z} = 3/16$).

Figure 17.5 vividly showcases this behavior. We also delineate the binodal temperature, T_0, such that $\alpha_0(T_0 - T_c) = 3\beta^2/16\gamma$, permitting us to express $\mathfrak{z}$ as:

$$\mathfrak{z} = \frac{3}{16}\frac{T - T_c}{T_0 - T_c} \tag{17.28}$$

showing that $s = s\left[(T - T_c)\,/\,(T_0 - T_c)\right]$. This establishes that s is a function of the temperature difference relative to the Curie temperature, T_c.

As we approach the binodal $\alpha_b = 3\beta^2/16\gamma$, the expression for ζ_0 (17.26) simplifies to:

$$\zeta_0 \approx 0.67\frac{|\beta|^{5/4}\gamma^{1/8}}{[\alpha_0\,(T - T_0)]^{3/4}}\left[\frac{A(1-2\nu)(1+\nu)}{Eg^{1/4}}\right]^{1/6} K_I^{1/6} \tag{17.29}$$

This reveals a divergence in ζ_0 in close proximity to the binodal ζ_0 scaling as $(T - T_0)^{-3/4}$, as demonstrated in Fig. 17.4.

If a tricritical point where $\beta = 0$ exists in the phase diagram, then as $\beta \to 0$, ζ_0 exhibits a divergence, scaling as $\zeta_0 \sim \beta^{-1/4}$. Its minimal value is reached at the upper spinodal α_{up}, given by:

$$\zeta_0^{(\mathrm{up})} \approx 3.01 \left(\frac{\gamma}{\beta^2}\right)^{1/8} \left[\frac{A(1-2\nu)(1+\nu)}{Eg^{1/4}}\right]^{1/6} K_I^{1/6} \tag{17.30}$$

***

In summary, *the key qualitative finding of this section is that the setback of the process zone's interface induces mechanical screening over a specific distance from the crack tip. This screening reduces the effective stress intensity factor acting on the crack compared to the applied one. As the system approaches the binodal, this reduction diverges following a* $(T - T_0)^{-3/4}$ *relationship.*

17.3 The Transformation Toughness K_{IC}

17.3.1 Analytical Framework for Transformation Toughness

17.3.1.1 Identifying the Measured Fracture Toughness Value in Fracture Experiments

It is worth noting that the material constant, K_{IC}, is not directly measured during a fracture experiment. Instead, what is measured is the applied stress intensity factor, denoted as $K_I^{(\mathrm{ap})}$, that is exerted on the solid. The exact relationship to fracture toughness can vary based on the specifics of the experiment. We represent this measured value of fracture toughness as $K_{IC}^{(\mathrm{m})}$, where the superscript "m" signifies "measured". This measured value, $K_{IC}^{(\mathrm{m})}$, is observed to escalate once the process zone comes into existence.

The initiation of crack propagation occurs when the perturbed stress intensity factor K_I (refer to (17.1)) equates to the material constant K_{IC}. Consequently, replacing in (17.13) $K_I \to K_{IC}$ and $K_I^{(\mathrm{ap})} \to K_{IC}^{(\mathrm{m})}$ provides the derivation of the measured fracture toughness in terms of the material constant K_{IC} and ζ_0:

$$K_{IC}^{(\mathrm{m})} = K_{IC} + A\eta_0^2 \mathbb{R}_0^{1/2} \zeta_0 \tag{17.31}$$

Utilizing the analytical expression detailed in (17.26) for ζ_0, we can derive $K_{IC}^{(\mathrm{m})}$:

$$K_{IC}^{(\mathrm{m})} = K_{IC} + 3.41\, s \left(\frac{\alpha\gamma}{\beta^2}\right) [(1-2\nu)(1+\nu)]^{1/2} \left(\frac{A^{12} g}{\beta^2\gamma^3}\right)^{1/8} \sqrt{\frac{K_{IC}}{E}} \tag{17.32}$$

The second term in (17.32) describes the zone impact on the fracture toughness.

17.3.2 Relative Perturbation of the Fracture Toughness

We have derived an analytical expression for the transformation toughness, as described in the reference (17.32). The relative perturbation of fracture toughness, denoted as $(K_{IC}^{(m)} - K_{IC})/K_{IC}$, is represented by the following equation:

$$\frac{K_{IC}^{(m)} - K_{IC}}{K_{IC}} = s\left(\frac{\alpha\gamma}{\beta^2}\right)\delta \qquad (17.33)$$

where δ is a dimensionless parameter expressed as follows:

$$\delta = 3.41\left[\frac{(1-2\nu)(1+\nu)}{K_{IC}E}\right]^{1/2}\left(\frac{A^{12}g}{\beta^2\gamma^3}\right)^{1/8} \qquad (17.34)$$

and s is defined by the expression (17.27). In this context, the variable s depends solely on the position within the phase diagram, while δ relies on the specific values of material constants. For any given point on the phase diagram, the parameter δ (17.34) determines the absolute magnitude of the perturbation in fracture toughness. Therefore, it is of utmost importance to accurately estimate this parameter, a topic that will be addressed in Sect. 17.3.3.

17.3.3 Estimates of the Transformation Toughness Perturbation

Let us estimate the perturbation in transformation toughness (17.33). We begin with general relations based on the key equations and parameters involved.

In the pursuit of deriving these estimates, we begin by establishing order-of-magnitude relationships, guided by our study of a first-order process zone. Since (4.12) characterizes the first-order transformation, it is expected that, under regular circumstances, all of its terms should be of similar magnitude. If any terms substantially differ from the others, we can safely disregard the smaller ones. A similar principle applies to the potential (4.14). Consequently, we can derive the following estimate:

$$\alpha_0 \Delta T_{\mathrm{h}} \eta^2 \sim |\beta|\, \eta^4 \sim \gamma\eta^6 \qquad (17.35)$$

Here $\Delta T_{\mathrm{h}} = T_{\mathrm{up}} - T_0$ representing the temperature difference between the upper spinodal and the binodal.

Next, let us consider the internal spatial scale, $\mathbb{R}_0$ (12.1):

$$\mathbb{R}_0 \sim \left(\frac{gE}{AK_{IC}}\right)^{2/3} \qquad (17.36)$$

Within the narrow transition region of the kink, the expression $g\,(\nabla\eta)^2$ also takes on a magnitude akin to that of the terms in (17.35). We can approximate it as: $g\,(\nabla\eta)^2 \sim g\eta^2/\mathbb{R}_0^2$ yielding the following relation:

$$\frac{g\eta^2}{\mathbb{R}_0^2} \sim |\beta|\,\eta^4 \tag{17.37}$$

In most cases, we have knowledge of the slope of the phase transition line $\partial T_0/\partial p$. As discussed in Sect. 3.3.3, the striction constant A can be roughly associated with $\partial T_0/\partial p$ as follows:

$$|A| \sim \alpha_0 E \left|\frac{\partial T_0}{\partial p}\right| \tag{17.38}$$

For the transformation strain ε_0, the estimate is expressed as:

$$\varepsilon_0 \sim \frac{A\eta^2}{E} \tag{17.39}$$

Utilizing the estimates (17.35)–(17.39), we arrive at a straightforward estimation for the parameter δ:

$$\delta \sim \frac{E\varepsilon_0}{\Delta T_{\mathrm{h}}}\left|\frac{\partial T_0}{\partial p}\right| \tag{17.40}$$

It's noteworthy the results (17.33), (17.34), and (17.40) are applicable if $\partial T_0/\partial p < 0$. In such instances, the process zone of the daughter phase forms within the matrix of the mother phase above the temperature T_0 of the bulk phase transition (see Fig. 9.1).

Additionally, it's worth noting that the product $E\varepsilon_0(\partial T_0/\partial p)$ represents the shift of the binodal temperature, denoted as ΔT_0, under the influence of hydrostatic pressure $p_0 \sim E\varepsilon_0$ as follows: $\Delta T_0 = E\varepsilon_0(\partial T_0/\partial p)$. Thus, we can express δ as:

$$\delta \sim \frac{\Delta T_0}{\Delta T_{\mathrm{h}}} \tag{17.41}$$

17.3.3.1 Numerical Estimates

When we consider typical values for various parameters involved in structural phase transitions, we find that the transformation strain, ε_0, typically falls in the range of $\varepsilon_0 \sim 10^{-3}$ to 10^{-2}, while the Young's modulus, denoted as E, typically registers at $E \sim 10^{11}$ Pa (see Table 34.0.1). The slope of the phase transition line $T_0(p)$ on the phase diagram can vary, but its typical magnitude is $k_0 = \partial T_0/\partial p \sim 1 \div 10$ K/ kbar, which is equivalent to $10^{-8} \div 10^{-7}$ K/ Pa. Furthermore, the hysteresis width is typically within the range of $\Delta T_{\mathrm{h}} \approx 1 \div 10$ K, as documented by [61]. With these considerations in mind, we can provide an estimate for δ, which typically ranges from $\delta \sim 0.1$ to $\delta \sim 10^2$.

Considering that the factor s is greater than or equal to 0.92, these estimates underscore that the measured fracture toughness, $K_{IC}^{(\mathrm{m})}$, can indeed attain values comparable to the original K_{IC} and, in some cases, substantially exceed it.

***

In summary, *we show that the fracture toughness deep within the mother phase, where no process zone forms, should be considered an intrinsic material constant. The formation of a zone reduces the observed fracture toughness due to mechanical screening around the crack tip, caused by the setback of the zone's interface. This screening does not require a wake and is therefore independent of crack propagation. The perturbation in fracture toughness depends on the thermodynamic properties of the crystal and its position within the phase diagram.*

17.4 Relationship to Previous Results

17.4.1 The Mechanical and Field-Theoretical Approaches Compared

In this section, we present our qualitative findings and compare them with key aspects of the mechanical approach.

***

17.4.1.1 The Mechanical Approach: Defining the Zone Boundary Criterion

In the mechanical framework, the determination of the process zone boundary relies on a criterion that equates the stress to a critical value, treated as a material constant. This concept has been extensively discussed by McMeeking and Evans, Budiansky, Rose, Stump [24, 26, 27, 55], and others, as reviewed by Kelly and Rose [9]. Such criteria are formulated by analogy to plasticity theory, plausibly due to the lack of alternative approaches at the time. Plasticity theory, in turn, links the formation of a plastic zone to phenomena such as the generation of dislocations or the activation of their mobility. Both processes are triggered when the shear stress[2] surpasses a specific threshold, a mechanism described by Friedel [62] and Hirth and Lothe [37].

Within this approach, the elastic mechanical problem is analogous to that of an "*elastic washer*" made from a material different from the surrounding matrix and inserted into a matrix orifice under a preload. In this analogy, the "washer" represents the process zone. The principal distinction between the "washer" problem and the process zone lies in the fact that the configuration of the process zone is not

[2] It can be an invariant combination of the stress tensor components.

known *a priori*. Instead, it must be derived as part of the solution by applying the boundary criterion outlined earlier in this section. From a mathematical perspective, this dependence on an unknown configuration places the problem in the category of so-called moving boundary problems. This complexity, in turn, makes the description of the process zone within the mechanical framework particularly challenging.

In the context of heterophase solids, the phase boundary criterion becomes notably more intricate, entailing an interplay of stress field, transformation strain, and boundary surface tension. Consequently, the criteria applied in [9, 24, 26, 27, 55] fail to adequately describe the phase boundary phenomenon.

Roitburd, as documented in [42, 43], along with Grinfel'd [44], detailed a criterion for sharp, coherent[3] phase boundaries within crystalline materials. This criterion is based on the requirement that the phase boundary adopts a configuration minimizing the total free energy of the crystal, a fundamental principle of equilibrium in physical systems. Thus, it is physically justified. In principle, this criterion complements the mechanical approach but introduces an additional integral equation imposed on the zone boundary, rendering the theory challenging to handle. Furthermore, Roitburd and Grinfel'd focused on plane phase boundaries and, therefore, did not address the boundary surface tension. To account for its influence we have extended the Roitburd-Grinfel'd approach, as detailed in [3]. While this extension offers valuable insights, it also introduces a complex integral equation to characterize the phase boundary configuration. This equation poses significant challenges in terms of tractability.

17.4.1.2 Essence of the Field-Theoretical Approach

In this book, we follow the field-theoretical approach, a fundamentally different method for tackling the boundary criterion problem. This approach re-expresses the process zone dilemma by formulating it in terms of the order parameter. The order parameter is subjected to the nonlinear Ginzburg-Landau equation of state (referenced as (4.12) in Sect. 4.1). This equation characterizes the spatial distribution of the order parameter within an inhomogeneous spatial environment.

Within the context of this approach, the solution $\eta(\mathbf{r})$ exhibits a kink, and *we describe the zone boundary as the configuration of this kink.* Multiple images vividly illustrate such kink configurations, as seen in Figs. 13.1, 13.3, 13.4, 13.5a, 13.6a–c, 13.7, and others.

It's worth noting that the field-theoretical approach inherently accommodates the zone boundary surface tension, attributable to the term $\sim g\,(\nabla\eta)^2$ referenced as (3.8).

In the model developed in this book, surface tension plays a central role in determining the zone setback and, consequently, the transformation toughness. From (17.33) and (17.34) one sees that this relationship is succinctly summarized as:

$$\frac{(K_{IC}^{(\mathrm{m})} - K_{IC})}{K_{IC}} \sim g^{1/8}.$$

[3] The reference is to boundaries without misfit dislocations.

In other words, the transformation toughness depends directly on the constant g controlling the zone boundary surface tension.

17.4.1.3 Peculiarities in Boundary Criteria for Various Solids

It's important to highlight that the zone boundaries in monocrystals, ceramics, and composites exhibit significant distinctions.

In Crystals Crystals are natural candidates for applying the field-theoretical approach. However, they often contain defects, such as dislocations and their various combinations, which generate additional stress fields that contribute to the formation of the process zone. Consequently, the theory for an idealized pure crack is quantitatively applicable only if its distance to other defects greatly exceed the size of the crack tip zone. The field-theoretical approach applies well to single crystals, provided the previous condition holds. In addition to them, one can apply this approach to polycrystals and crystals with mosaic structure, provided that the zone size, $\mathbb{R}_{\mathrm{d}}$, is smaller or comparable to the characteristic dimension of the crystallite $\mathbb{D}_{\mathrm{cr}}$. In the opposite limit $\mathbb{R}_d \gg \mathbb{D}_{\mathrm{cr}}$, the crystal exhibits the features of a ceramics.

In Ceramics In ceramics, the formation of zone boundaries is influenced by stress fields arising from both the crack tip and the surrounding confined grains. These grains vary in size, shape, and orientation, introducing randomness into the stress distribution. This stochastic nature is shaped by the material's history, making it challenging to determine the exact configuration of the zone boundaries. When the zone's dimensions significantly exceed the grain size, the field-theoretical approach becomes applicable upon averaging.

In Composites Combining Grains Exhibiting and Not Exhibiting Transformation In the case of composites, the boundary can pass outside of the grains undergoing the phase transformation. In such instances, established mechanical approach criteria [9], as well as such as those formulated in [3, 42–44], do not directly apply. Consequently, selecting an appropriate boundary criterion for these materials becomes a complex task, often leaning towards a phenomenological approach.

Within the field-theoretical approach the description of the ceramics and composites face the same difficulties.

17.4.1.4 When the Two Approaches Coincide

In certain materials characterized by a *highly manifested first-order transition*, the order parameter experiences a sharp discontinuity at the transition point, with the order parameter jump approaching its saturated value, η_{s}. Here, the subscript "s" stands for "saturated." The $\eta = \eta(T)$ relationship closely resembles a step function: $\eta \approx \eta_{\mathrm{s}}$ at $T \le T_0$ and $\eta = 0$ at $T > T_0$. In such cases, employing the Landau theory of phase transitions offers little advantage. Instead, the transition can be comprehensively characterized by the discontinuity η_{s} at the binodal point T_0, or equivalently by the transformation strain ε_0.

Another pivotal determinant of the zone is the criterion governing its phase boundary. If this criterion accounts for the zone's surface tension, the mechanical and field-theoretical approaches should align.

17.4.1.5 Reversibility Versus Irreversibility of Phase Transformations

In numerous research papers, such as [1, 16, 17, 19, 24, 26, 27, 55] and many others noted in the review of Kelly and Rose [9], the phase transformation into a metastable daughter phase is often assumed to be irreversible. It's important to understand that the metastable daughter phase inherently corresponds to a less energetically favorable minimum of the free energy compared to the mother phase. Consequently, the metastable phase exists perpetually out of equilibrium. In contrast, in true equilibrium conditions, the daughter phase would vanish, rendering the transformation reversible.

Nevertheless, from a practical standpoint, at least two mechanisms contribute to making the transition irreversible.

1. Pinning-Induced Irreversibility
The first mechanism involves transformation-induced plastic deformation. Zone formation can generate substantial shear stress at its boundary, potentially rearranging existing dislocations and even creating new ones. These effects have been reported in materials like NiTi [63] and in a nanostructured Fe alloy [64]. As the zone advances, these plastic defects can effectively "pin" the daughter phase in place, creating an irreversible wake. Such a wake is not metastable, but stable. It is stabilized by the stress field generated by the dislocations that pin it. This stabilized wake enhances the material's fracture toughness. Notably, the dislocations pinning the wake themselves contribute significantly to the stress intensity factor, with an impact comparable to that of the wake. However, the role of pinning defects and their effect on fracture toughness has yet to be thoroughly examined.

2. Irreversibility Due to Long Characteristic Time
In cases where the transformation follows an order-disorder or reconstructive pathway, the time, τ, required to transform into the stable phase can be considerable, even in the absence of pinning. If τ significantly exceeds the time required for fracture, the screening approach described in the works of [24, 55], and others becomes applicable.

The most well-known material exhibiting transformation toughness is ZrO_2, which involves a tetragonal parent phase and a monoclinic daughter phase. As reported by [28], this transformation is *reversible*. Consequently, this kind of irreversibility of the wake cannot account for the transformation toughness phenomenon in this material.

Quasi-Static Crack Propagation
Early approaches [16–19], as well as the work of McMeeking and Evans, Rose, Budiansky, and their coauthors [24, 26, 27, 29, 30, 55], emphasized the necessity of quasi-static crack propagation. This requirement stemmed from the understanding

that the formation of the wake was integral to explaining transformation toughness. It was hypothesized that the wake emerges following the propagation of the crack-zone complex. Consequently, it was assumed that the crack is not entirely motionless but propagates at an exceedingly low velocity. This corresponds to $V \ll V_G$.

As detailed in Chap. 13, in cases of reversible transformations, the wake vanishes completely during crack propagation at a velocity below V_G. The metastable wake forms only when the crack velocity exceeds Gordon's speed. Therefore, the concept of a persisting metastable wake following quasi-statically propagating cracks is not applicable. In this context, only the previously discussed pinning-stabilized wake may be relevant in the case of a highly defective crystal.

17.4.1.6 Fracture Toughness Beyond the Hysteresis Region

Traditionally, the transformation toughness phenomenon has been associated exclusively with the hysteresis region of the phase diagram, where the daughter phase remains metastable. This belief originates from the explanation related to the formation of a persistent metastable wake and its energetic [1, 16–19] or screening effect [24, 26, 27, 29, 30, 55]. The hysteresis region is critical, as it is the only region where the metastable phase can emerge. As we have already discussed, these arguments have scope for a limited number of materials but do not apply to most solids.

However, we present compelling evidence in this chapter that the transformation toughness phenomenon extends beyond the hysteresis region for a motionless as well as for a propagating crack. Indeed, the transformation toughness is generated by the setback and associated contact length, l_{inv} (Fig. 13.2). Its physical origin lies in the surface tension at the interface of the process zone. The presence of the surface tension is unrelated to the position on the phase diagram inside and outside the hysteresis region but only to the zone presence.

It is worth noting that the bifurcation line (dot-dashed brown line and brown squares in Fig. 22.6) lies above the upper spinodal on the phase diagram at least, at small b. Consequently, the transformation toughness phenomenon manifests across the entire region between the bifurcation and the binodal lines. At the same time, only the region between the upper bulk spinodal and bulk binodal lines is considered to be part of the hysteresis. However, the bifurcation line may lie above the upper bulk spinodal line—that is, outside the hysteresis.

Furthermore, in this chapter, our focus has predominantly been on first-order phase transitions. Nevertheless, in the case of second-order transitions, the transformation toughness phenomenon also arises due to surface tension effects.

17.4.1.7 'Resonance' Phenomenon

Under specific conditions, the size of the process zone exhibits divergence, a phenomenon initially noted by Rose [26] and referred to as a *resonance* and further explored by Stump and Budiansky [27]. They argued that resonance occurs when a parameter $\sim \sigma_T/\sigma_m$ reaches a critical value, where σ_m represents the stress spur at the zone boundary, and σ_T is the transformation stress in the unconstrained solid.

Our approach offers insight into the physical origin of such resonances, revealing that they occur as the solid approaches the binodal line on the phase diagram (3.32).

17.5 Chapter Summary

From the solution for the order parameter distribution at the crack tip, we derived an effective stress intensity factor acting on the crack, influenced by both the externally applied load and the internal stress from the process zone.

Our findings challenge common assumptions by revealing that transformation toughness can arise even in stationary cracks, without requiring a quasistatic propagation regime. We identified that transformation toughness originates from the surface tension at the process zone interface, a previously unrecognized mechanism. Furthermore, our results show that transformation toughness is influenced by the material's proximity to the binodal in its phase diagram: the closer the material, the stronger the effect. Specifically, transformation toughness is directly proportional to the slope of the first-order phase transformation line in the phase diagram and the spontaneous transformation stress, while inversely proportional to the hysteresis width.

Regarding the formation of an irreversible metastable wake during quasistatic motion, we demonstrated that this is impossible if the crack velocity is below Gordon's speed. However, in highly defective crystals, such a wake could form due to stabilization by defects. In these cases, the contribution of defects to the stress intensity factor is expected to be comparable to that of the wake and must be accounted for. Additionally, such a wake may arise in crystals with an extremely long relaxation time for the phase transition.

***

References

1. S.D. Antolovich, Trans. Metall. Soc. AIME **242**, 2371 (1968)
2. S.D. Antolovich, D. Fahr, Eng. Fract. Mech. **4**, 133 (1972)
3. A. Boulbitch, A.L. Korzhenevskii, Phys. Rev. B **96**, 054106 (2017)
4. A. Boulbitch, P. Tolédano, Phys. Rev. Lett. **81**(4), 838 (1998)
5. A. Boulbitch, P.E. Pumpyan, Sov. Phys. Crystallogr. **35**(2), 156 (1990)
6. A. Boulbitch, A.L. Korzhenevskii, Phys. Rev. Lett. **107**, 085505 (2011)
7. A. Boulbitch, A.L. Korzhenevskii, Eur. Phys. J. B **89**, 261 (2016)
8. A. Boulbitch, A.L. Korzhenevskii, Phys. Rev. E **93**, 063001 (2016)
9. P.M. Kelly, L.R.F. Rose, Prog. Mater. Sci. **47**, 463 (2002)
10. A. Boulbitch, Math. J. **20**, 1–21 (2018)

11. A. Boulbitch, A.L. Korzhenevskii, Phys. Rev. **E101**, 033003 (2020); idem Phys. Rev. **E101**, 033004 (2020)
12. A. Boulbitch, A.L. Korzhenevskii, Phys. Rev. E **108**, 014114 (2023)
13. A. Boulbitch, A.L. Korzhenevskii, Phys. Rev. E **103**, 033001 (2021)
14. K. Yoshida et al., Sci. Rep. **5**, 10993 (2015)
15. R.C. Garvi, Hannink, R.T. Pascoe, Nature **258**, 703 (1975); L.K. Lenz, A.H. Heuer, J. Am. Ceram. Soc. **65**, 192 (1982); R.P. Indel, R.W. Rice, D. Lewis, J. Am. Ceram. Soc. **65**, 108 (1982); F.F. Lange, J. Mater. Sci. **17**, 240 (1982)
16. T.K. Gupta, *Fracture Mechanics of Ceramics*, ed. by R.C. Bradt (Plenum Press, New York, 1978), pp. 877–889
17. T.K. Gupta, F.F. Lange, J.H. Behtold, J. Mater. Sci. **13**, 1464 (1978)
18. N. Claussen, Z. Werkstofftechnik **13**, 138 (1982)
19. P.F. Becher, V.J. Tennery, *Fracture Mechanics of Ceramics*, vol. 6, ed. by R.C. Bradt (Plenum, New York, 1983), p. 383
20. K. Tsukuma, M. Shimada, J. Mater. Sci. **20**, 1178 (1985); A.G. Evance, J. Am. Ceram. Soc. **73**, 187 (1990)
21. I. Birkby, R. Stevens, Key Eng. Mater. **122–124**, 527 (1996)
22. R.I. Todd, M.P.S. Saran, Transformation toughening, in *Materials Processing Handbook*, vol. 20, ed. by J.R. Groza (CRC Press, Boca Raton, FL, 2007), pp.1–20
23. A.G. Evans, R.M. Cannon, Acta Metall. **34**, 761 (1986)
24. B. Budiansky, J.W. Hutchinson, J.C. Lambropoulos, Int. J. Solids Struct. **19**, 843 (1983)
25. J.W. Hutchinson, Harvard University Report, Division of Applied Science, DEAP S-8 (1974)
26. L.R.F. Rose, J. Mech. Phys. Solids **19**, 337 (1986)
27. D.M. Stump, B. Budiansky, Acta. Met. **37**, 3297–3304 (1989)
28. B. Budiansky, L. Truskinovsky, J. Mech. Phys. Solids **41**, 1445 (1993)
29. D.M. Stump, Phil. Mag. **A64**, 879 (1991)
30. G.T.M. Stam, E.V.D. Giessen, P. Meijers, Int. J. Solids Struct. **31**, 1923–1948 (1994)
31. N. Simha, L. Truskinovsky, Acta Metall. Mater. **42**, 3827 (1994)
32. Y. Freed, L. Banks-Sills, J. Mech. Phys. Solids **55**, 2157 (2007)
33. T. Baxevanis, Y. Chemisky, D.C. Lagoudas, Smart Mater. Struct. **21**, 903 (2012)
34. T. Baxevanis, A.F. Parrinello, D.C. Lagoudas, Int. J. Plast. **50**, 158 (2013)
35. L. Ma, A.M. Korsunsky, R.M. McMeeking, J. Appl. Mech. Trans. ASME **80,** 051001 (2013)
36. C. Lexcellent, F. Thiebaud, Scripta Mater. **59**, 321 (2008); C. Lexcellent, M.R. Laydi, V. Taillebot, Int. J. Fract. **169**, 1 (2011)
37. J.P. Hirth, J. Lothe, *Theory of Dislocations* (Krieger Publishing Company, Malabar, Florida, US, 1991)
38. C. Maletta, M.L. Young, J. Mater. Eng. Perform. **20**, 597 (2011)
39. U. Krupp et al., *Materials Structure & Micromechanics of Fracture*, vol. 465, ed. by P. Sandera (Trans Tech Publications, Durnten-Zurich, Switzerland, 2011), p. 55
40. I.R. Vatne, A. Stukowski, C. Thaulow, E. Ostby, J. Marian, Mater. Sci. Eng. A **560**, 306 (2013)
41. J. Corteen, M. Rainforth, I. Todd, Scr. Mater. **65**, 524 (2011)
42. A.L. Roitburd, Sov. Phys. Usp. **17**, 326 (1974)
43. A.L. Roitburd, Sov. Phys. Solid State **26**, 1229 (1984)
44. M.A. Grinfel'd, Lett. Appl. Eng. Sci. **19**, 1031 (1981); idem Izvestia Earth Phys. **18**, 28 (1982)
45. L.D. Landau, E.M. Lifshitz, *Statistical Physics*, 3rd edn. (Pergamon Press, Oxford, 1980)
46. V.M. Nabutovskiĭ, V.Y. Shapiro, Sov. Phys. JETP **48**(3), 480 (1978)
47. A.A. Bulbich, J. Mater. Sci. **27**, 1070 (1992)
48. C.F. Nigro, C. Bjerken, P.A. Olsson, Int. J. Fract. **209**, 91 (2017)
49. T. Sluka, K.G. Webber, E. Colla, D. Damjanovic, Acta Mater. **60**, 5172 (2012); H. Yu, J. Wang, S. Kozinov, M. Kuna, Acta Mater. **154**, 334 (2018)
50. M. Mamivand, M.A. Zaeem, H.E. Kadiri, Acta Mater. **64**, 208 (2014)
51. T. Zhao, J. Zhu, J. Luo, Eng. Fracture Mech. **159**, 155 (2016)
52. J. Zhu, J. Luo, Mater. Sci. Eng. **A701**, 69 (2017)
53. V.I. Levitas, H. Jafarzadeh, G.H. Farrahi, M. Javanbakhtd, Nanoscale **11**, 22243 (2019)

54. V.I. Levitas, H. Jafarzadeh, G.H. Farrahi, M. Javanbakhtd, Int. J. Plast. **111**, 1 (2018)
55. R.M. McMeeking, A.G. Evans, J. Am. Ceram. Soc. **65**, 242 (1982)
56. H.F. Brueckner, Trans. ASME **80**, 1225 (1958)
57. A. Falvo, F. Furgiuele, A. Leonardi, C. Maletta, J. Mater. Eng. Perform. **18**, 679 (2009); C. Maletta, F. Furgiuele, Acta Mater. **58**, 92 (2010); C. Maletta, E. Sgambitterra, F. Furgiuele, Fatigue Fract. Eng. Mater. Struct. **36**, 903 (2013); S. Hazar, W. Zaki, Z. Moumni, G. Anlas, Int. J. Plast. **67**, 26 (2015)
58. Section 57, p. 218 of the book [2]
59. N.I. Mushelisvili, *Some Basic Problems of the Mathematical Theory of Elasticity*. Reprint of the Second English edition, translated by J.R.M. Radok (Leyden, The Netherlands, Noordoff International Publishing, 1977)
60. H.F. Bueckner, Z. Angew, Math. Mech. **50**, 529 (1970)
61. E.Y. Tonkov, *High Pressure Phase Transformations: A Handbook*, vols. 1–3 (Gordon and Breach, Amsterdam, 1992); E.Y. Tonkov, E.G. Ponyatovsky, *Phase Transformations of Elements Under High Pressure*, vol. 4 (CRC Press, Boca Raton, FL, 2005)
62. J. Friedel, *Dislocations* (Pergamon Press, Oxford, 1964)
63. T. Ungar, J. Frenzel, S. Gollerthan, G. Ribarik, L. Balogh, G. Eggeler, J. Mater. Res. **32**, 4433 (2017)
64. L.K. Huang, W.T. Lin, B. Lin, F. Liu, Acta Mater. **118**, 306 (2016)

Chapter 18
Crack Velocity Jumps

Abstract We examine the dynamics of a crack as it progresses through a thin elastic plate, with a particular focus on the influences of the process zone at the crack tip. This zone contributes to energy dissipation, exerting a viscous force on the crack tip that significantly influences its propagation patterns. As we further analyze the escalation of the driving force–the principal control parameter in this scenario–we observe notable variations in the crack propagation velocity (Boulbitch and Korzhenevskii 2016).

For a long time, the speed, V, of a brittle crack has been believed to be a continuous, monotonic function of the driving force, G [1]. However, recent experiments have challenged this notion, revealing fast crack propagation [2] with a terminal crack speed significantly lower than the Rayleigh velocity, V_R [2, 3]. Moreover, discontinuities in the relationship between crack velocity and driving force have been observed in various materials, such as silicon [4, 5] and limestone [6]. To date, no satisfactory explanation for these discontinuities has been provided. One possible mechanism for such a behavior if the crack tip process zone's formation and diappearance. This mechanism first studied in [7] we present below.

18.1 Equation for Crack Motion

To describe the concerted motion of the crack-zone complex, we combine the dynamic equation for the order parameter with Freund's equation, which governs crack propagation. We incorporate the drag force resulting from energy dissipation within the process zone into Freund's equation. The first step involves rendering this system dimensionless, followed by a comprehensive analysis of the resulting system.

***

© The Author(s), under exclusive license to Springer Nature Switzerland AG 2026

A. Boulbitch and A. Korzhenevskii, *Stress-Induced Local Phase Transformations Near Cracks and Dislocations*, Springer Series in Materials Science 356,
https://doi.org/10.1007/978-3-032-00945-6_18

18.1.1 Modification of Freund's Equation

In linear elastic fracture mechanics, the steady propagation of a crack is described by the inertialess equation derived by Freund for an infinite body [1]. For a more detailed information, please refer to Sect. 1.1.

However, Freund's equation (1.9) overlooks the process zone, not to mention its internal dynamics. The latter gives rise to energy dissipation, manifested as the configurational quasi-viscous friction force, f_{v} (3.51), (3.58) and (6.44) acting on the crack tip. Let us focus on the case of a crack in the elastic plate with the thickness $\mathbb{L}$, and denote this force per unit thickness as $\Delta\Gamma$:

$$\Delta\Gamma = f_{\mathrm{v}}/\mathbb{L}$$

To account for this friction force, $\Delta\Gamma$ should be added to the right-hand side of Freund's equation (1.9):

$$(1 - V/V_{\mathrm{R}})G = \Gamma + \Delta\Gamma \tag{18.1}$$

Assuming that the crack tip propagates with velocity V along the Ox axis and switching to the co-moving frame $(x - Vt, y)$, and using (3.51) and (3.58), the friction force can be expressed as:

$$\Delta\Gamma = \kappa V \int \left(\frac{\partial\eta}{\partial x}\right)^2 d^2r \tag{18.2}$$

The system of (4.10), (18.2), and (18.1) yields:

$$\begin{cases} g\Delta\eta + \kappa V \frac{\partial\eta}{\partial x} - \left[\alpha - B_{\mathrm{c}}\frac{\cos(\theta)}{\sqrt{r}}\right]\eta - \beta\eta^3 = 0 \\ \left(1 - \frac{V}{V_{\mathrm{R}}}\right) G = \Gamma + \kappa V \int \left(\frac{\partial\eta}{\partial x}\right)^2 d^2r \end{cases} \tag{18.3}$$

In this study, our focus is solely on describing the second-order process zone. Consequently, assuming $\beta > 0$, we omit the term $\gamma\eta^5$ in the (4.10). Let us consider the bulk mother phase ($\alpha > 0$; $\eta = 0$) and the daughter-phase process zone localized at the crack tip ($x = 0$, $y = 0$) and vanishing away from it. As usual, we apply the boundary condition:

$$\eta(\infty) = 0$$

corresponding to no daughter phase away from the crack tip.

These equations (18.3), along with the definitions of G, Γ (1.10), and B_{c} (9.2), constitute a self-consistent system with unknowns V and $\eta(\mathbf{r})$, describing the concerted, steady propagation of the crack and the process zone. We present its solution in the subsequent section.

18.1.2 Equation Describing Crack Motion

18.1.2.1 Reduction of System (18.3) into One Equation

By leveraging the findings from the Sect. 10.1, we can explicitly determine the friction force (18.2). After substituting the expression for the order parameter (6.34) into the second equation, as outlined in (18.3), and conducting the integration, we arrive at a single equation (18.1), which characterizes the motion of the crack. In this equation, $\Delta\Gamma$ represents a piecewise smooth function:

$$\Delta\Gamma = \begin{cases} 0, & \alpha > \alpha_\mathrm{d} \text{ or } V \geq V_\mathrm{d} \\ \frac{\kappa}{\beta} V \left(\alpha_\mathrm{d} - \alpha - \frac{\kappa^2 V^2}{4g}\right) S, & \alpha \leq \alpha_\mathrm{d} \text{ and } V < V_\mathrm{d} \end{cases} \tag{18.4}$$

Here, the function $S = S(V)$ is defined as:

$$S(V) = \frac{J_2}{J_4} I_2 \tag{18.5}$$

This function incorporates integrals J_2 and J_4 (6.37) and a dimensionless kinetic integral I_2 (6.45). The integrals J_2, J_4 and I_2 are contingent on the crack velocity, and for $V \neq 0$, analytical calculation is not feasible. Figure 18.1 presents the outcome of a numerical evaluation of $S(V)$.

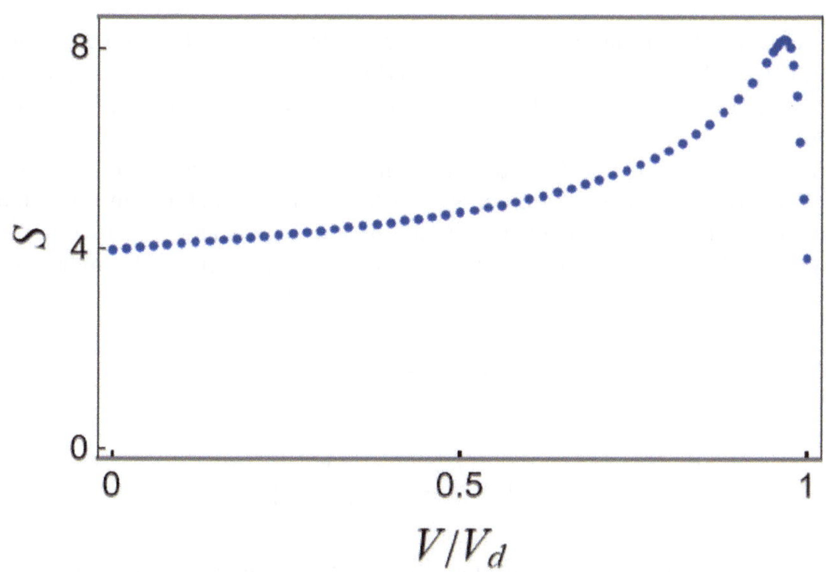

Fig. 18.1 Numerically evaluated dependence of S on V. Its average value is $S_0 \approx 4.9$

It is important to emphasize that the derivation of (18.1) relies on the solutions found in (6.34) and (6.36). These solutions are valid only in the immediate vicinity of the bifurcation point. Consequently, (18.1) holds true primarily within this limited region of applicability. Despite its narrow scope, an analysis of (18.1) will provide insight into the behavior of the crack in the presence of a dynamic process zone.

The nonlinearity inherent in (18.1) is evident in two aspects: (i) the nonlinear relationships within the factor $\alpha_d - \alpha - \kappa^2 V^2/4g$ as follows from (6.27) and that of $S(V)$ (Fig. 18.1), and (ii) its piecewise-smooth character. This nonlinearity leads to a bifurcation in its solutions.

18.1.2.2 Rescaling to Dimensionless Parameters

Equation (18.1) is cumbersome since it includes many dimensional parameters. In this section, we first introduce several parameters to rewrite this equation, making it more conducive for analysis.

Initially, we define parameters α_c and V_c as follows:

$$\alpha_c = \frac{2^{2/3}}{\pi^{2/3} g^{1/3}} \left[\frac{|A|\,(1-2\nu)(1+\nu)}{E} \right]^{4/3} \times K_{IC}^{4/3} \tag{18.6}$$

$$V_c = \frac{2^{4/3} g^{1/3}}{\pi^{1/3}\kappa} \left[\frac{|A|\,(1-2\nu)(1+\nu)}{E} \right]^{2/3} \times K_{IC}^{2/3} \tag{18.7}$$

The subscript "c" in these parameters indicates their dependence on K_{IC}, unlike α_d and V_d (6.27) and (10.22), which depend on K_I. Thus, α_c and V_c depend only on the material constants: g, A, E, ν, κ, and K_{IC}, making them material constants themselves.

Freund's equation (1.9) depends solely on two dimensionless parameters: V/V_R and Γ/G. However, once the process zone forms at the crack tip, the equation of crack motion becomes dependent on a multitude of parameters. By utilizing (18.6) and (18.7), (18.1) necessitates seven dimensionless parameters for formulation:

$$\mathbb{A} = \frac{\alpha}{\alpha_d}; \quad \mathbb{A}_c = \frac{\alpha}{\alpha_c}; \quad v = \frac{V}{V_d}; \quad v_{dR} = \frac{V_d}{V_R}; \quad v_R = \frac{V}{V_R}; \quad v_{cR} = \frac{V_c}{V_R} \tag{18.8}$$

and

$$\mathbb{G} = 1 - \frac{\Gamma}{G} \equiv 1 - \left(\frac{K_{IC}}{K_I} \right)^2 \tag{18.9}$$

The latter is referred to as the *inverse driving force*. Among the seven parameters (18.8) and (18.9), three dimensionless velocities, v, v_{dR}, and v_R, are subjected to the identity:

$$v \times v_{\mathrm{dR}} = v_{\mathrm{R}} \tag{18.10}$$

Therefore, out of these parameters, only six are independent, which adds complexity to the problem.

It's worth mentioning that $\mathbb{A}$ and $\mathbb{A}_c$ define the actual position on the phase diagram:

$$\mathbb{A} = \frac{T - T_c}{T_{\mathrm{d}}(K_I) - T_c}; \quad \mathbb{A}_c = \frac{T - T_c}{T_{\mathrm{d}}(K_{IC}) - T_c} \tag{18.11}$$

where $T_{\mathrm{d}}(K_I)$ is the emergence temperature at the actual value of the stress intensity factor, K_I, while $T_{\mathrm{d}}(K_{IC})$ is the value of T_{d} at $K_I = K_{IC}$. Note that at $K_I = K_{IC}$, the crack is still motionless. At $T = T_c$, $\mathbb{A}_c = 0$, while $T = T_{\mathrm{d}}(K_{IC})$ yields $\mathbb{A}_c = 1$.

Furthermore, rather than using the driving force, G (1.10), we introduce the dimensionless inverse driving force, $\mathbb{G}$. It is equal to zero when $G = \Gamma$ (i.e., where crack propagation starts) and becomes unity as $G \to \infty$. Thus, crack propagation occurs within the conveniently analyzed interval $0 \le \mathbb{G} \le 1$.

It's important to observe that parameters $v_{c\mathrm{R}}$, $\mathbb{A}_c$ and v_{dR} depend only on K_{IC}, making them invariant with respect to the crack load, K_I. For this reason, in the following discussion, we address crack dynamics in terms of these parameters on one hand and the K_I-dependent parameter $\mathbb{G}$ on the other hand.

For convenience, let us also list the expressions of v, v_{dR}, $\mathbb{A}$ (18.8) as well as α_{d} G and K_I in terms of invariant parameters and $\mathbb{G}$:

$$v = (1 - \mathbb{G})^{1/3} \frac{V}{V_c}; \quad \mathbb{A} = \mathbb{A}_c (1 - \mathbb{G})^{2/3}; \ v_{\mathrm{dR}} = \frac{v_{c\mathrm{R}}}{(1 - \mathbb{G})^{1/3}} \tag{18.12}$$

$$\alpha_{\mathrm{d}} = \frac{\alpha_c}{(1 - \mathbb{G})^{2/3}}; \qquad G = \frac{\Gamma}{1 - \mathbb{G}}; \qquad K_I = \frac{K_{IC}}{\sqrt{1 - \mathbb{G}}} \tag{18.13}$$

18.1.2.3 Derivation and Solution of the Dimensionless Equation of Motion[1]

Equation of Motion in Terms of the Rescaled Parameters. In terms of v and A the configurational friction force $\Delta\Gamma$ can be expressed as follows:

$$\Delta\Gamma = \begin{cases} \Delta\Gamma_0 P(v, \mathbb{A}), & 1 - v^2 - \mathbb{A} > 0 \\ 0 & 1 - v^2 - \mathbb{A} \le 0 \end{cases} \tag{18.14}$$

where

$$\Delta\Gamma_0 = \frac{\kappa v_{c\mathrm{R}} \alpha_c}{\beta} V_R \tag{18.15}$$

[1] The following sections are technical. We suggest to skip them during the first reading.

is the amplitude of $\Delta\Gamma$ with dimensions of friction force per unit plate thickness. Thus, the ratio $\Delta\Gamma/\Delta\Gamma_0$ is dimensionless. Furthermore, $P(v, A)$ is a dimensionless function:

$$P(v, \mathbb{A}) = v\left(1 - v^2 - \mathbb{A}\right) S(v) \tag{18.16}$$

where S is determined by (18.5). Numerical evaluation of the integrals J_2, J_4 and I_2(6.37) and (6.45) involved in S (18.5) enables us to obtain the ratio $\Delta\Gamma(v, A)/\Delta\Gamma_0$ (18.14) numerically. Figure 18.2a displays the dependence of the dimensionless ratio $\Delta\Gamma/\Delta\Gamma_0$ on the parameters A and v (18.8).

Observe that $\Delta\Gamma$ initially increases and then decreases as a function of the crack tip velocity, V, a trend that occurs for any value of A. Figure 18.2b illustrates this behavior for three values of A: 0.3, 0.5 and 0.7.

Note 18.1.1 Notably, the decreasing portion of $\Delta\Gamma(V)$ inevitably leads to motion instability. We will investigate these phenomena in greater detail later in this chapter.

By substituting the expression from (18.14) into (18.1) and applying the replacement rules from (18.8), we arrive at the following equation of the crack motion in terms of the rescaled parameters:

$$\mathbb{G} - v_{\mathrm{R}} = k\begin{cases} P(v, \mathbb{A}), & 1 - v^2 - \mathbb{A} > 0 \\ 0, & 1 - v^2 - \mathbb{A} \le 0 \end{cases} \tag{18.17}$$

This equation includes a dimensionless parameter, k, which is expressed as a product of material constants:

$$k = \frac{\Delta\Gamma_0}{\Gamma} \equiv \frac{\kappa\alpha_c V_c E}{\beta(1-\nu^2)K_{IC}^2} \tag{18.18}$$

This parameter controls the impact of the process zone dynamics on crack motion.

Equation (18.17) describes crack propagation based on the inverse driving force.

Semi-Analytical Approach. Equation of motion (18.17) contains the function $S = S(v)$. If one knew its analytical form, the whole problem would be solvable analytically. This function is, however, a combination of three integrals: J_2, J_4 and I_2 (6.37) and (6.45), which can only be evaluated numerically. To address this issue, we solve part of the problem numerically and then approximate the solution with a suitable analytical expression for further analytical progress.

To implement this approach, we fitted the numerical function $P(v, A)$ to a simple polynomial as follows:

$$P(v, \mathbb{A}) \approx c_1 v - c_3 v^3 \tag{18.19}$$

for various values of $A_i \in [0, 1]$ where $i = 1, 2, ..., 20$. The fitting procedure yields two lists $c_1 = c_1(A_i)$ and $c_3 = c_3(A_i)$. The latter were also fitted to polynomials:

$$\begin{aligned} c_1(\mathbb{A}) &\approx 4.85 - 5.57\mathbb{A} + 0.73\mathbb{A}^2 \\ c_3(\mathbb{A}) &\approx 4.49 - 0.36\mathbb{A} \end{aligned} \tag{18.20}$$

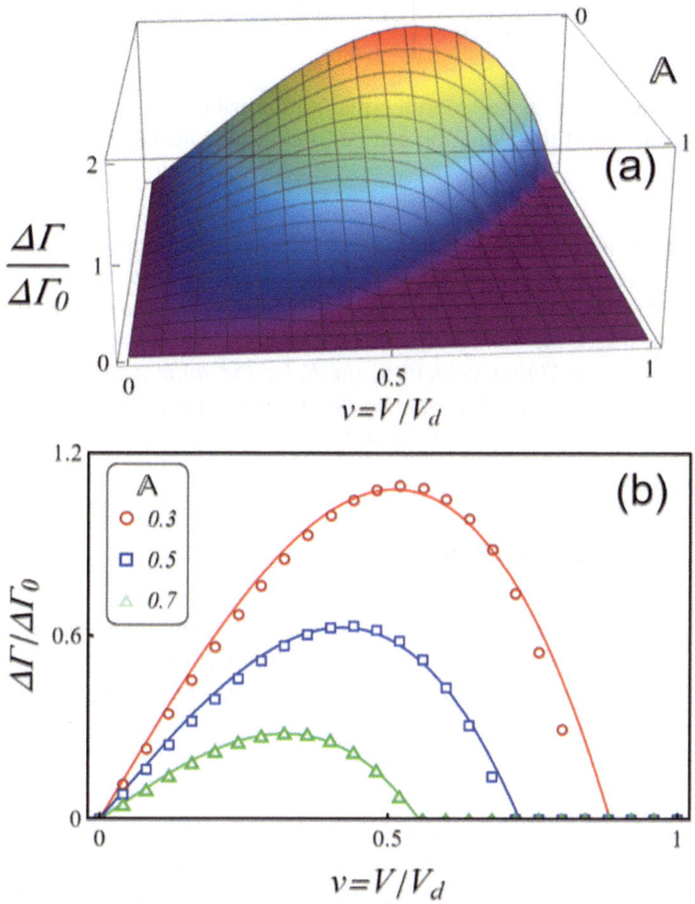

Fig. 18.2 Behavior of $\Delta\Gamma/\Delta\Gamma_0$. **a** The configurational friction force per unit plate thickness, $\Delta\Gamma$, exerted on the crack tip versus $\mathbb{A} = \alpha/\alpha_d$ and $v = V/V_d$. Here $\Delta\Gamma_0$ is the friction force amplitude (18.16), the ratio $\Delta\Gamma/\Delta\Gamma_0$ being dimensionless. **b** The cross-sections of $\Delta\Gamma/\Delta\Gamma_0$ by the planes $\mathbb{A} = 0.3$, 0.5 and 0.7. The results obtained by the numerical integration are shown by dots, while their fitting to (18.19) and (18.20) are indicated by the solid lines

The comparison between the numerical results and their fitting is shown in Fig. 18.2b. Using this procedure, we obtain an analytical approximation of the function $P(v, A)$. Now, standard analytical approaches can be applied for its further analysis.

The function $P(v, A)$ exhibits a nonlinear dependence on v, as depicted in Fig. 18.2b.

Using (18.19) and (18.20), we simplify the equation of crack motion (18.17) to the following concise form:

$$\mathbb{G} - v_{\mathrm{R}} = kv \times \begin{cases} c_1 - c_3 v^2, & 1 - v^2 - \mathbb{A} > 0 \\ 0, & 1 - v^2 - \mathbb{A} \leq 0 \end{cases} \tag{18.21}$$

This nonlinear, piecewise-smooth equation describes steady crack propagation. A detailed analysis of this equation can be found in the next section.

***

In summary, *we derived a nonlinear, piecewise smooth equation of motion for the crack-process zone complex in a rescaled form, dependent on only six independent parameters. At low crack velocities, the equation's right-hand side is linear in the crack velocity, where the configurational resistance force resembles typical viscous friction. However, at higher velocities, the behavior shifts dramatically: once the velocity reaches a critical value causing the order parameter to vanish, the process zone disappears, eliminating the friction force.*

18.2 Characteristic Plane

Crack propagation regimes are effectively analyzed within the characteristic plane $(\mathbb{G}, \mathbb{A}_c)$*, which represents crack dynamics. In this plane, only the region where* $\mathbb{A}_c > 0$ *and* $0 \leq \mathbb{G} \leq 1$ *is relevant, as each point within this domain fully defines the state of crack motion.*

***

Let us take a good look at the $(\mathbb{G}, \mathbb{A}_c)$ plane. The inverse driving force, $\mathbb{G} = 1 - \Gamma/G$ (18.9), determines the force propelling the crack. The crack's velocity, V, depends upon $\mathbb{G}$. When $\mathbb{G} = 0$ on the left side of the characteristic plane, $V = 0$; when $\mathbb{G} = 1$ on the right side, it corresponds to the terminal velocity, $V = V_R$.

The parameter $\mathbb{A}_c = \alpha/\alpha_c$ determines the temperature and fixes the solid's position in its phase diagram.

In the $(\mathbb{G}, \mathbb{A}_c)$-plane, any directed line stretching from $\mathbb{G} = 0$ to $\mathbb{G} = 1$ represents a trajectory that shows the evolution of crack motion as K_I increases from K_{IC} to infinity. Hence, any crack propagation scenario corresponds to a certain trajectory on the characteristic plane, referred to as the *crack evolution trajectory*.

Below, we discuss the simplest type of trajectories, the horizontal ones, as more complex trajectories are beyond the scope of this book.

18.2.1 Dynamics of the Dressed Crack

Generally, the characteristic plane is divided into two regions: Region I, where $1 - v^2 - \mathbb{A} < 0$, indicates no process zone at the crack tip, making the tip *undressed*;

and Region II, where $1 - v^2 - \mathbb{A} > 0$, indicates the tip *dressed* with a process zone. The process zone emerges or disappears when the condition below holds:

$$1 - v^2 - \mathbb{A} = 0 \tag{18.22}$$

provided that v is one of the solutions of the equation of motion (18.17). Let us note that (18.22) represents the bifurcation condition (6.27) in a slightly different notation.

The Trivial Solution. Let us examine the solution within Region I first. When $1 - v^2 - \mathbb{A} < 0$, the equation of motion (18.21) is linear, yielding the solution:

$$v_{\mathrm{R}} = \mathbb{G} \tag{18.23}$$

In different notation, it takes the form:

$$V = \left(1 - \frac{\Gamma}{G}\right) V_{\mathrm{R}} \tag{18.24}$$

which is Freund's equation solution (see (1.9) and (1.11)), as expected.

The Bifurcation Line. To derive the explicit expression for the bifurcation line, we use (18.23). This involves applying the second and third relations (18.12). The resulting equation, is as follows:

$$\mathbb{A}_c = \frac{1}{(1 - \mathbb{G})^{2/3}} - \left(\frac{\mathbb{G}}{v_{c\mathrm{R}}}\right)^2 \tag{18.25}$$

In the $(\mathbb{G}, \mathbb{A}_c)$ plane, equation (18.25) defines a line. When the crack is initially undressed, it undergoes dressing as the crack evolution trajectory crosses this line from above. Consequently, we refer it to as the *dressing boundary*.

First, it's important to highlight that (18.25) is derived from (6.27). However, a fundamental distinction exists between these two equations. While (6.27) is expressed in terms of the rescaled velocity, $v = v(G)$, the equation of the bifurcation line, (18.25), relies exclusively on the independent inverse driving force $\mathbb{G}$ and a single parameter, $v_{c\mathrm{R}}$, where $v_{c\mathrm{R}} = V_c / V_{\mathrm{R}}$ (as indicated in (18.8)). This latter parameter represents a composite of material constants and can rightfully be regarded as a material constant in its own right.

Second, in the preceding sections, we explored the bifurcation of (6.1) while assuming that the defect velocity, V, is an independent control parameter. Under such circumstances, the bifurcation exhibits a soft nature, with both direct and reverse bifurcations occurring under the same conditions. However, the self-consistent system of (18.3) introduces added complexity. In this system, the condition for the reverse (undressing) bifurcation differs from the dressing bifurcation condition (18.25).

We will examine the undressing condition in the upcoming section.

18.2.2 *Analysis of the Equation of Motion*

18.2.2.1 Analytical Approach[2]

When $1 - v^2 - A > 0$, the expression (18.21) takes the form of a cubic equation. Of the three solutions it offers, our focus is on identifying a real positive solution that tends toward zero as G approaches zero. This specific behavior is manifested in the following solution:

$$v = \frac{(i\sqrt{3} - 1)(kc_1 + v_{\mathrm{dR}})}{2^{2/3}\mathbb{M}} - \frac{(1 + i\sqrt{3})\mathbb{M}}{6 \times 2^{1/3}kc_3} \tag{18.26}$$

Here, M is defined as:

$$\mathbb{M} = \left\{3kc_3\left[81\gamma^2c_3^2k^2 - 12kc_3(kc_1 - v_{\mathrm{dR}})^3\right]^{1/2} - 27\gamma c_3^2k^2\right\}^{1/3}$$

The parameters c_1 and c_3 are functions of A and are given by the relation (18.20). Although expressed in terms of complex numbers, it is easy to make sure that the solution (18.26) is real.

Let us define the critical value of A as A_{d}, at which the process zone either emerges or disappears:

$$1 - v^2 - \mathbb{A}_{\mathrm{d}} = 0 \tag{18.27}$$

This condition holds provided v is one of the solutions, (18.23) or (18.26), of the equation of motion (18.21).

Dressing Condition. If the crack is initially undressed, we can determine the equation that defines the line in the characteristic plane where the process zone emerges by substituting the undressed solution $v = G/v_{\mathrm{dR}}$ into (18.27). This substitution yields:

$$A_{\mathrm{d}} = 1 - (G/v_{\mathrm{dR}})^2$$

This equation characterizes the point at which the process zone comes into play in the crack propagation.

Undressing Condition. When the speed of an initially dressed crack increases, the transition to an undressed state occurs at $A < 1 - (G/v_{\mathrm{dR}})^2$.

Figure 18.3 illustrates the graphical solution of the equation of motion (18.21). In Fig. 18.3a, one can see the curve representing its right-hand part. (b), (c), and (d) depict the straight lines representing the left-hand part of (18.21), corresponding to incrementally increasing values of G.

We will analyze the scenario where a crack starts initially dressed at $G = 0$ and then G gradually increases. At the position shown by (b), (18.21) has a single solution (indicated by a purple dot at the intersection of the lines (a) and (b) in Fig. 18.3. This

[2] The following sections are technical. We suggest that the reader skip them during the first reading.

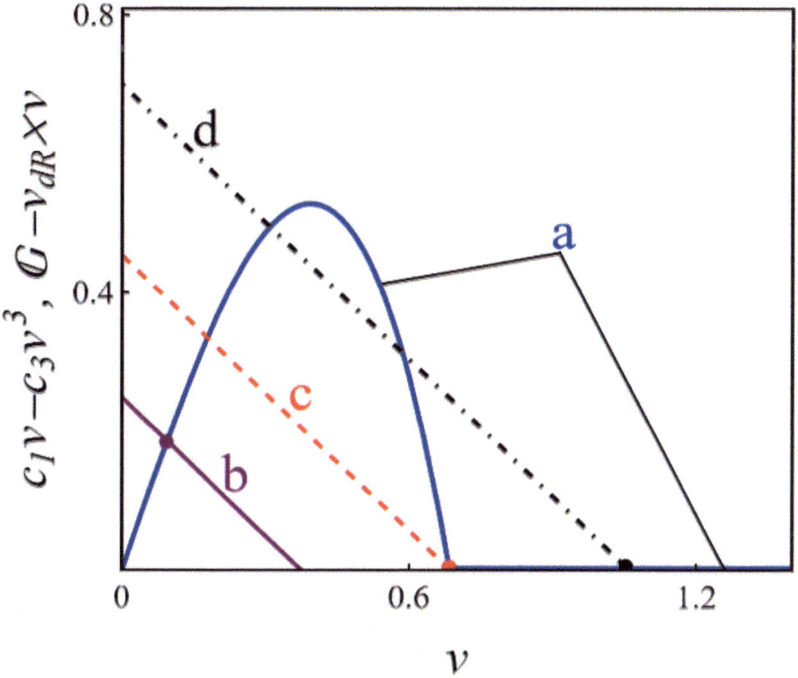

Fig. 18.3 Graphical solution of the equation of motion (18.21). **a** shows the piecewise smooth right-hand side of the equation of motion: $c_1\omega - c_3\omega^3$ for $1 - v^2 - \mathbb{A} < 0$ and 0 in the opposite case. **b**, **c** and **d** show the left-hand side of the equation of motion for different $\mathbb{G}$. Dots at the intersection of the curve and the straight lines indicate the solution of (18.21) actually taking place. The case of an initially dressed crack corresponds to $\mathbb{G}$ increasing from zero. For small $\mathbb{G}$ one finds a single point of intersection of the lines (a) and (b) corresponding to the non-Freund solution. The line (c) corresponds to the transition from the non-Freund to the Freund solution at the intersection of the line (c) with the abscissa. For larger $\mathbb{G}$, shown by the line (d), of the three possible solutions of (18.21), only the one lying on the abscissa takes place

is the nonlinear solution (18.26) of the cubic equation (18.21) describing the dressed crack.

As G increases to the point where the corresponding line takes position (c) in Fig. 18.3, a second solution emerges. For position (d), three solutions exist. Two of them are solutions of the cubic equation, while the third is given by (18.23) and corresponds to the undressed crack.

In situations with multiple solutions, we need a criterion to determine which one occurs in reality. We hypothesize that the correct criterion favors the solution that offers minimal resistance to crack propagation. According to this criterion, both (c) and (d) yield the surviving solution, which is (18.23).

Thus, the transition from the dressed to the undressed state occurs in the scenario indicated by (c) in Fig. 18.3.

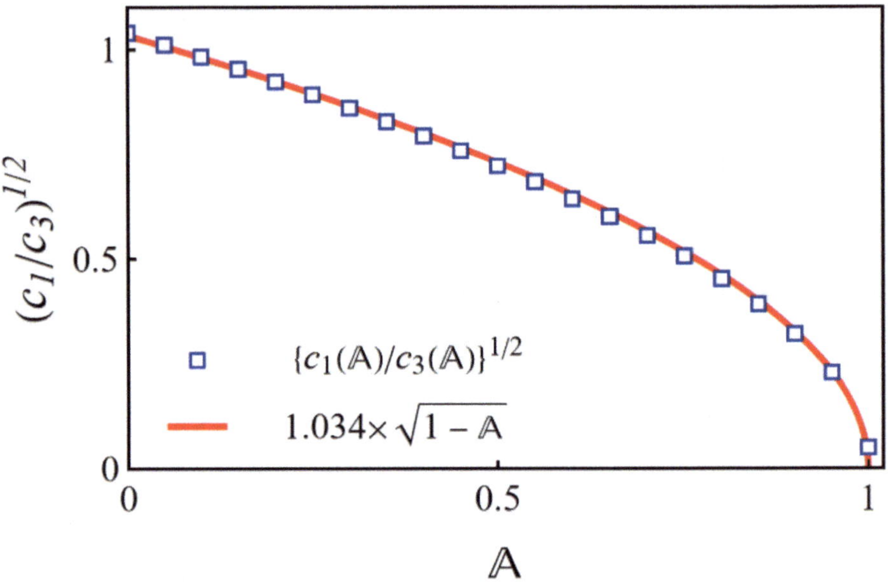

Fig. 18.4 Illustration of approximation (18.28). The ratio $[c_1(\mathbb{A})/c_3(\mathbb{A})]^{1/2}$ is shown by open squares, while $1.034(1-\mathbb{A})^{1/2}$ is displayed by the solid line

Undressing Equation. Now, let us discover the condition that governs this bifurcation.

The surviving solution satisfies two key relationships: $G = v_{\mathrm{dR}}v$ and $c_1(A) - c_3(A)v^2 = 0$ (as shown in Fig. 18.3c), which leads to:

$$\mathbb{G} = v_{\mathrm{dR}}\left[\frac{c_1(\mathbb{A})}{c_3(\mathbb{A})}\right]^{1/2}$$

This equation defines the undressing line. The expression $[c_1(\mathbb{A})/c_3(\mathbb{A})]^{1/2}$ can be accurately approximated as follows:

$$\sqrt{\frac{c_1}{c_3}} \approx 1.034(1-A)^{1/2} \tag{18.28}$$

as demonstrated in Fig. 18.4.

Using now (18.12), we can express the undressing condition as follows:

$$\mathbb{A}_c \approx \frac{1}{(1-\mathbb{G})^{2/3}} - 0.94\left(\frac{\mathbb{G}}{v_{c\mathrm{R}}}\right)^2 \tag{18.29}$$

One can see that the undressing condition (18.29) only slightly differs from the dressing condition (18.25).

***

In summary, *we analyzed the nonlinear, piecewise-smooth equation of motion for the crack-zone complex and established distinct conditions for its dressing and undressing bifurcations. Each condition is represented by a curve in the* $(\mathbb{G}, \mathbb{A}_c)$ *plane.*

18.3 Crack Propagation Scenarios

In this section, we use the (G, A_c) *plane to outline four qualitatively distinct scenarios of crack propagation.*

***

An analysis of the crack motion equation delineates three primary scenarios of crack-process zone propagation. The first scenario occurs when at $K_I = K_{IC}$, the temperature, T, exceeds $T_{\rm d}$ (Fig. 18.5a).

Let $\mathbb{G}_0$ denote the point where the crack evolution trajectory crosses the line (18.25) illustrated in Fig. 18.5a, shown in solid blue. The crack tip remains undressed when $\mathbb{G} < \mathbb{G}_0$. Consequently, the crack propagation adheres to the solution (1.11) of the Freund equation (1.9). At the juncture $\mathbb{G} = \mathbb{G}_0$ a process zone emerges at the crack tip, initiating a configurational friction force. For $\mathbb{G} > \mathbb{G}_0$, the crack velocity notably diminishes, subsequently followed by further acceleration (Fig. 18.5b).

Figure 18.6a presents another scenario where the entire crack evolution trajectory resides within domain II, implying the crack is continually dressed. This scenario fosters a solution that consistently exhibits a non-Freund type, and lies below the velocity delineated by Freund's solution (1.11).

The scenarios depicted in Figs. 18.5 and 18.6 invariably occur when the boundary line separating regions I and II undergoes a monotonic growth.

In cases of non-monotonic boundary behaviors, a third potential scenario emerges, as illustrated in Fig. 18.7a. Here, we examine a crack evolution trajectory initiating from region II, pertaining to an initially dressed crack. Its propagation adheres to the (18.21), with a velocity lesser than that prognosticated by Freund's solution (1.11).

Upon crossing the boundary (18.29) at point $\mathbb{G}_{01}$, the process zone vanishes, and the undressed crack advances with velocity given by (1.11), witnessing a substantial acceleration in crack speed. With a further increment of the inverse driving force, the trajectory re-enters region II crossing the line (18.25) at $\mathbb{G} = \mathbb{G}_{02}$. Hence, the crack tip dresses once more, obeying (18.21), leading to a decline in velocity. This results in a $V = V(\mathbb{G})$ relation with one Freund's segment in $\mathbb{G}_{01} < \mathbb{G} < \mathbb{G}_{02}$ and two non-Freund's segments at $0 \leq \mathbb{G} \leq \mathbb{G}_{01}$ and $\mathbb{G}_{02} < \mathbb{G} < 1$.

This scenario necessitates that both lines (18.25) and (18.29), showcase a non-monotonic behavior, occurring at $v_{c\rm R} \lesssim 0.7$.

It's worth noting the existence of a fourth possible crack propagation regime, feasible when the initiation point of the crack evolution trajectory is slightly elevated

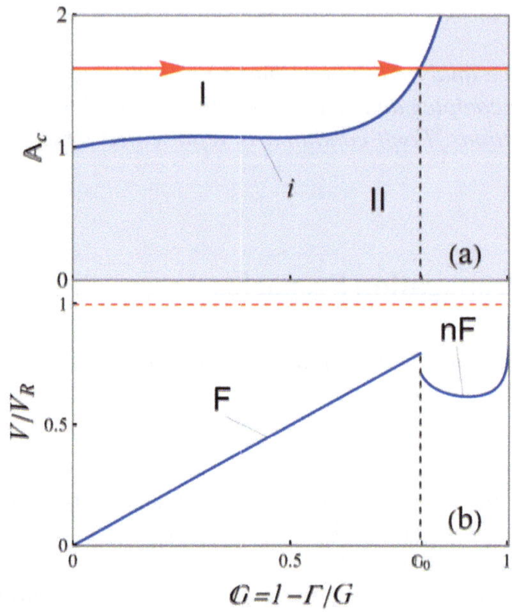

Fig. 18.5 Characteristic Plane (a) and corresponding plot of the dimensionless crack velocity V/V_R, versus inverse driving force $\mathbb{G}$ **(b)**. The diagram was generated with $v_c = 0.7$ and $k = 0.3$. In (a) the characteristic plane, $(\mathbb{G}, \mathbb{A}_c)$, is divided into parts (I) where the crack is undressed and (II) where it is dressed. The red arrow indicates the crack evolution trajectory. (i) indicates the bifurcation line (18.25). Panel (b) depicts the V/V_R dependence on $\mathbb{G}$ related to the crack evolution trajectory shown in panel (a). "F" signifies the velocity segment according to Freund's solution (1.11) while "nF" highlights the part characterized by the non-Freund solution given by (18.26). The horizontal dashed line marks the Rayleigh velocity level

above $\mathbb{A}_c = 1$, yet beneath the maximum point of the boundary separating regions I and II in Fig. 18.7a. Here, the stationary crack is initially undressed, dresses as soon as the crack propagation trajectory crosses the boundary for the first time. The latter takes place at a small crack speed. Further, the crack undresses again at a marginally escalated speed. Similar to preceding regimes, the crack dresses once more at high velocity when the trajectory crosses the boundary for the third time. The ranges of $\mathbb{A}_c$ and v_{cR} where this regime could manifest are quite restricted, implying a low likelihood of this scenario unfolding.

An essential observation to highlight is the transition appearances between propagation regimes, resembling jumps as depicted in Fig. 18.7b. However, owing to the soft nature of the bifurcation in proximity to boundaries $\mathbb{G}_{01}$ and $\mathbb{G}_{02}$ the crack transitions from dressed to undressed states and back gradually. Consequently, velocity variations near these points remain continuous, albeit potentially steep.

***

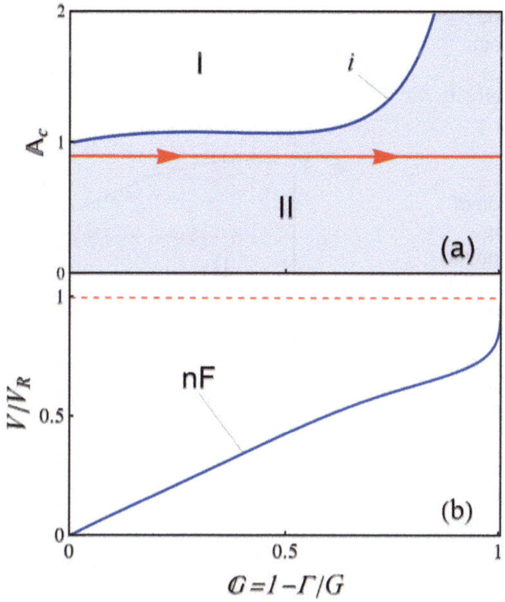

Fig. 18.6 Characteristic Plane and the $V = V(G)$ dependence. The characteristic plane (a) segregates into sections devoid of (I) and with the transformation zone at the crack tip (II). (i) marks the transition line (18.25). The red arrow demonstrates the crack evolution trajectory nestled within domain II. In panel (b), the V/V_R dependence on $\mathbb{G}$ corresponds to the crack evolution trajectory depicted in panel (a), uniformly exhibiting a non-Freund nature. The horizontal dashed line signifies the Rayleigh velocity level. This diagram was produced with $\nu_c = 0.7$ and $k = 0.16$

In conclusion, *the nature of crack propagation can be effectively analyzed using the characteristic plane, which reveals four possible scenarios. These scenarios differ based on (i) the position of the initial point of the crack evolution trajectory in the characteristic plane—either above or below the dressing line—and (ii) the number of times the trajectory intersects the dressing or undressing line. When the trajectory crosses the bifurcation line, the process zone either emerges at or disappears from the crack tip, altering the resistance force exerted on it. This results in deceleration after the zone emerges or an increase in velocity upon its disappearance.*

18.4 Simulations

This section addresses the simulation of the system of equations describing crack-zone complex propagation, with the aim of validating the scenarios predicted through analytical methods.

***

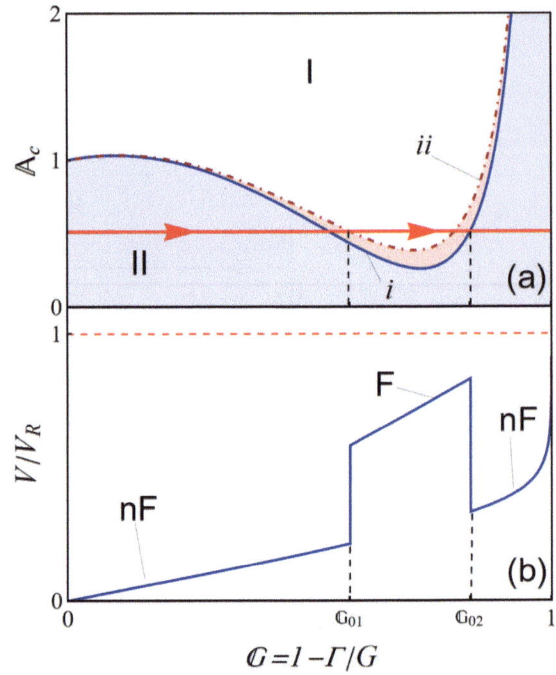

Fig. 18.7 Characteristic plane with non-monotonic phase boundary and corresponding crack velocity progression. In (a), the characteristic plane demonstrates a non-monotonic variation of the boundary separating regions I and II based on the inverse driving force, $\mathbb{G}$. The crack evolution trajectory crosses the domain boundaries (I and II) twice. Panel (b) represents the V/V_R relation to $\mathbb{G}$ aligning with the crack evolution trajectory shown in panel (a). The graph was generated with $v_c = 0.5$ and $k = 0.4$

Simulating this system of differential and integral equations is quite intricate. To perform the simulations, we rescaled the system using a specialized approach and employed a computational trick that enabled COMSOL to solve the equations. Details of the rescaling method and the numerical simulation procedure are provided in Annex D. Below we focus on the obtained results.

The simulation results for $v_{c\mathrm{R}} = 0.31$ and $q_2 = 0.9$[3] are summarized in Fig. 18.8. In panel (a), we present the results obtained with $\mathbb{A}_c = 1.9$, where the crack evolution trajectory mostly resides within domain I. It only enters domain II at a relatively high value of $\mathbb{G}$ (approximately $\mathbb{G} = \mathbb{G}_0 \approx 0.85$), as seen in Fig. 18.5. For $\mathbb{G} \leq \mathbb{G}_0$ the results align with the analytical solution $V/V_R = \mathbb{G}$, displaying Freund behavior. Conversely, for $\mathbb{G} > \mathbb{G}_0$ the solution exhibits the non-Freund type.

Figure 18.8b obtained at $\mathbb{A}_c = 0.5$, illustrates the scenario where the crack evolution trajectory remains entirely within domain II, with the crack tip consistently dressed, as shown in Fig. 18.6. In this case, the solution $V(\mathbb{G})/V_R$ is uniformly of the non-Freund type.[4]

Figure 18.7c demonstrates a situation with a non-monotonic I-II boundary. For $\mathbb{G} < \mathbb{G}_{01}$ and $\mathbb{G} > \mathbb{G}_{02}$, the solutions exhibit the non-Freund type, while Freund's

[3] Definition of the latter parameter and its relation to the equations of motion (18.3) one finds in (37.2) in Annex D.

[4] The numerical process converges poorly for $\mathbb{G}$ close to unity. For this reason, we disregarded a few points in the close vicinity of $\mathbb{G} = 1$.

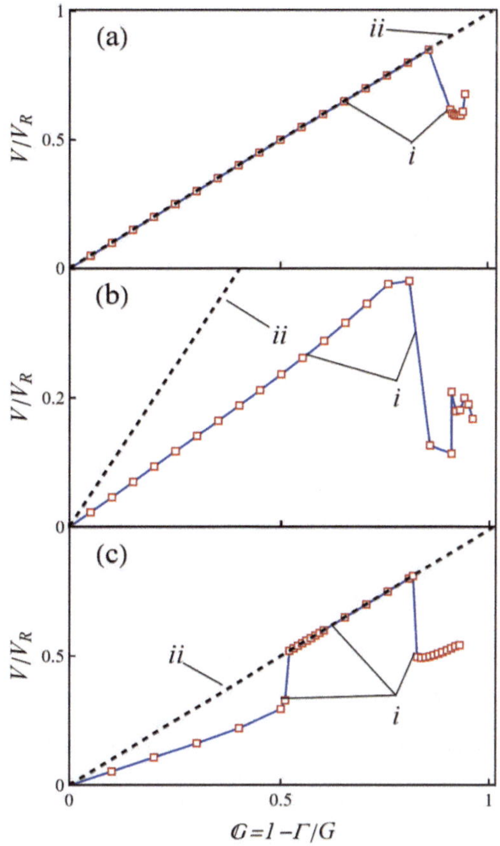

Fig. 18.8 V/V_R versus inverse driving force $\mathbb{G} = 1 - \Gamma/G$: simulation results. Open squares represent simulation results, while the solid line (i) provides visual guidance. The dashed line (ii) represents Freund's solution $V/V_R = \mathbb{G}$. All the images are generated with $v_{cR} = 0.31$ and $q_2 = 0.9$. (a) corresponds to $\mathbb{A}_c = 1.9$, (b) to $\mathbb{A}_c = 1.01$ and (c) to $\mathbb{A}_c = 0.5$. The constants $q_{1,2}$ are defined in (37.2) (Annex D)

solution occurs at $\mathbb{G}_{01} < \mathbb{G} < \mathbb{G}_{02}$. In our simulations, we found $\mathbb{G}_{01} \approx 0.52$ and $\mathbb{G}_{02} \approx 0.85$.

Figure 18.9 illustrates the distributions of the rescaled order parameter u as a function of dimensionless coordinates x_1 and y_1 taken from the simulation with $v_{cR} = 0.31$, $q_2 = 0.9$,[5] and $\mathbb{A}_c = 1.01$ reported in Fig. 18.8c.

Figure 18.9a exhibits the case $\mathbb{G} = 0$ corresponding to a motionless crack. In Fig. 18.9b the value $\mathbb{G} = 0.5$ is slightly below $\mathbb{G}_{01}$ where a jump in velocity occurs (Fig. 18.8c).

The order parameter distribution after the jump (at $\mathbb{G} = 0.52$) is shown in Fig. 18.9c. In this case, the order parameter value within the distribution is so small that it is not visible within the image's scale. Finally, Fig. 18.9d displays the order parameter just after the second downward jump ($\mathbb{G} \geq \mathbb{G}_{02} \approx 0.85$).

It's worth noting that the distributions in Fig. 18.9b–d are stretched backward compared to Fig. 18.9a, consistent with our analytical predictions. However, the dis-

[5] See Annex D.

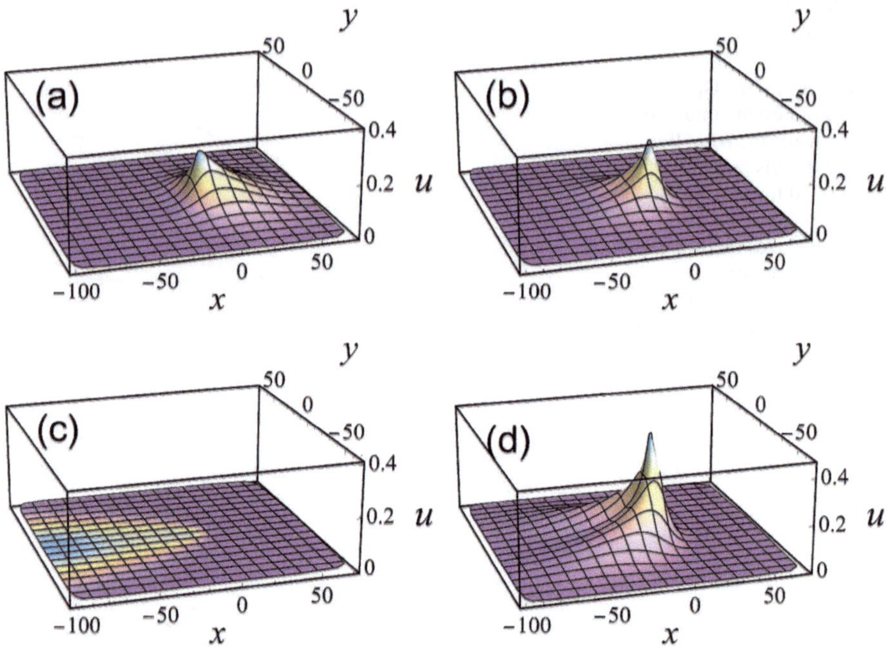

Fig. 18.9 Simulated order parameter distribution. The distribution $u = u(x_1, y_1)$ with $\mathbb{A}_c = 1.01$, $v_c = 0.31$ and $q_2 = 0.9$ corresponding to the scenario shown in Fig. 18.8b. Panel (a) corresponds to the motionless crack, ($\mathbb{G} = 0$), panel (b) to $\mathbb{G} = 0.5$ (before the first transition), panel (c) to $\mathbb{G} = 0.52$ (after the first transition) and panel (d) to $\mathbb{G} = 0.85$ (after the second transition). The constants $q_{1,2}$ are defined by (37.2) in Annex D)

tributions significantly differ from those shown in Fig. 10.2 depicting our analytic solution (10.21). This discrepancy arises because the asymptotic solution obtained from the bifurcation is accurate only in the vicinity of the bifurcation point, represented by the boundary line between domains I and II in Figs. 18.5, 18.6 and 18.7a. At a finite distance from this line, the bifurcation theory's solution provides a poor approximation.

***

In conclusion, *the numerical solutions indeed confirm the qualitative results predicted analytically, thereby validating these predictions.*

18.5 The Chapter Summary

In this chapter, we examine the dynamics of the crack–process zone complex. We derived a viscous-like configurational force that resists crack motion and incorporated it into Freund's equation of crack propagation. By solving this equation in conjunction with one describing the dynamics of the order parameter, we identified the region in the characteristic plane where the process zone exists. The latter enabled us to describe four distinct regimes of crack propagation based on the material's position within this diagram.

The first regime occurs when the material is outside the region of process zone existence. Here, the crack tip remains undressed up to very high driving forces, which are difficult to achieve experimentally. In this case, crack dynamics is largely as predicted by Freund's equation.

In the second regime, the crack evolution trajectory consistently remains within the process zone region of the characteristic plane. Here, the crack behavior everywhere deviates significantly from predictions of linear elastic fracture mechanics.

In the third and fourth regimes, the crack evolution trajectory intermittently enters and exits the process zone region, causing crack dynamics to alternate between Freund-type and non-Freund-type behavior, followed by the velocity jumps during these transitions.

***

References

1. L.B. Freund, *Dynamic Fracture Mechanics* (Cambridge University Press, Cambridge, 1998)
2. J. Fineberg, M. Marder, Phys. Rep. **313**, 1 (1999)
3. J.E. Field, Contemp. Phys. **12**, 1 (1971)
4. J.A. Hauch et al., Phys. Rev. Lett. **82**, 3823 (1999)
5. M.J. Buehler et al., Phys. Rev. Lett. **99**, 165502 (2007)
6. A. Bertram, J.F. Kalthoff, Key Eng. Mater. **251–252**, 423 (2003)
7. A. Boulbitch, A.L. Korzhenevskii, Phys. Rev. E **93**, 063001 (2016)

Chapter 19
Wedging-Induced Stick–Slip Crack Propagation

Abstract In this chapter, we explore the critical role of the process zone at the crack tip, which serves as an additional velocity-dependent channel for energy dissipation during fracture events. Incorporated into the fracture energy, denoted as $\Gamma = \Gamma(V)$, this zone contribution grants the fracture energy a negative slope, represented mathematically as $d\Gamma(V)/dV < 0$ fostering a positive feedback mechanism. When combined with the crack's inertial properties manifesting within the wedging process, this phenomenon precipitates self-oscillations. This exploration aims to shed new light on the intricate dynamics at play during material fracturing, potentially unveiling innovative approaches to understanding and manipulating fracture mechanics (Boulbitch and Korzhenevskii 2011).

19.1 Introduction

Unstable crack growth in the form of stick-slip motion has been widely observed in various materials such as glasses, ceramics, and brittle polymers [1]. Linear elastic fracture mechanics attributes this type of motion to a reduction in a solid's resistance to crack growth. This reduction is characterized by a negative derivative of the velocity-dependent fracture energy, given as $d\Gamma(V)/dV < 0$. This behavior is widely believed to be associated with non-linear processes that are localized in the process zone and are heavily influenced by the properties of this zone. Several mechanisms have been proposed as potential contributors to this phenomenon, including molecular relaxation, transition from isothermal to adiabatic regimes, crystallization at the crack tip, and microplasticity [1, 2]. While most of these mechanisms have been discussed qualitatively, attempts to mathematically model microplasticity have determined them to be intrinsically ill-posed [3].

Advancements in experimental techniques for studying rapid crack propagation have led to their application in examining various materials [4, 5]. The observed similarities in dynamic fracture behaviors across materials with different microstructures, like brittle polymers, glasses, and certain crystals, suggest the possibility of a universal model for describing them. This idea has prompted the development of several

© The Author(s), under exclusive license to Springer Nature Switzerland AG 2026
A. Boulbitch and A. Korzhenevskii, *Stress-Induced Local Phase Transformations Near Cracks and Dislocations*, Springer Series in Materials Science 356,
https://doi.org/10.1007/978-3-032-00945-6_19

phase-field models [6], which successfully replicate phenomena like microbranching and oscillatory instabilities in straight cracks as noted in [5].

Recent studies have hypothesized that the characteristic length scale, which indicates the distance at which a crack profile deviates from linear elastic fracture mechanics, is much larger than scales associated with any dissipative mechanisms [7]. The findings of a weakly non-linear model have shown agreement with data from brittle gels [8].

This non-linearity, as discussed in [8] can be linked to acoustic phonons. In solids with non-primitive lattice cells, the non-linearity of non-acoustic degrees of freedom (e.g., related to optical phonons) may be observed. When this occurs, it becomes a significant factor that deviates from the predictions of linear elastic fracture mechanics, potentially leading to the formation of a transformational process zone near the crack tip and introducing a specific characteristic scale.

In this chapter, we discuss a physical mechanism that underlies the non-monotonic behavior of $\Gamma = \Gamma(V)$. This mechanism is associated with the process zone at the crack tip, leading to self-oscillations during its propagation [18].

19.2 Wedging Dynamics

This section presents the derivation of equations modeling the wedging dynamics in the propagation of the crack-zone complex. We build upon the thin-wedge theory by Barenblatt and Cherepanov [9] *and the dynamic wedging equations developed by Barenblatt and Salganik* [10, 11], *integrating these with the process zone theory described in Chap.* 6 *of this book. This framework offers a comprehensive basis for understanding the interaction between wedging dynamics and the process zone at the crack tip.*

***

Figure 19.1 depicts a schematic representation of a solid material undergoing cleavage facilitated by a thin wedge. In this context, we evaluate the scenario wherein the solid manifests a transformational process zone at the tip of the cleavage crack. As we will elucidate below, this process zone contributes significantly to initiating the stick-slip regime of crack propagation.

In linear elastic fracture mechanics, crack propagation in an infinite elastic sheet is delineated by the inertialess equation proposed by Freund, as referenced in (1.9). This stands in stark contrast to phenomena observed in the double cantilever beam configuration as documented in various studies [12–15], and during wedging events as detailed in [10, 11], where crack propagation indeed showcases inertia.

To further understand this, we dive into the wedging dynamics of a solid, employing the approach formulated by Barenblatt and Salganik [10, 11]. This approach presents the crack propagation equation as:

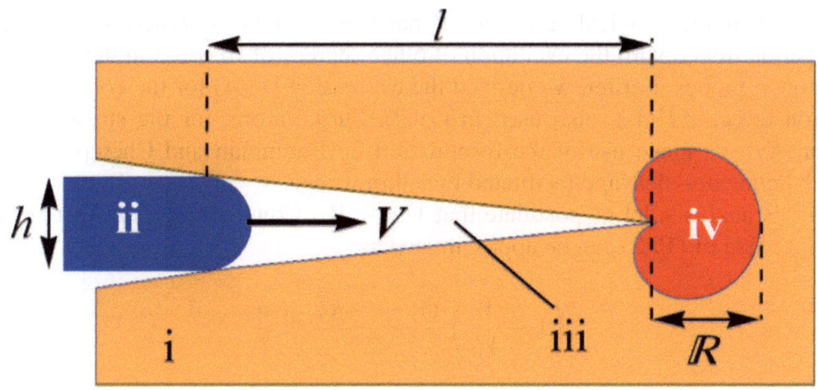

Fig. 19.1 Cleavage of a solid exhibiting a process zone by a thin wedge. A solid (i) is penetrated by a thin wedge (ii) characterized by thickness h and velocity V. The wedge facilitates opening of the crack (iii) with a length denoted by l, resulting in generation of a process zone (iv) at its tip which exhibits the size $\mathbb{R}$

$$m\frac{d^2l}{dT^2} = \left(1 - \frac{V}{V_R}\right)\frac{(1-\nu^2)K_I^2}{E} - \Gamma \tag{19.1}$$

Here, m represents the effective mass of the crack, T denotes time, and ν is the Poisson's ratio. Moreover, l stands for the distance between the contact points of the wedge and the faces of the crack and the crack tip, as illustrated in Fig. 19.1. During wedging, this distance is identified as the crack length [9].

Equation (19.1) effectively describes the cleavage process in structureless elastic solids that do not exhibit a transformation process zone.

19.2.1 Equation of Motion of the Crack-Zone Complex During Wedging

While progressing through this book, readers should have familiarized themselves with the concept that solids undergoing phase transformation often develop a transformational process zone at the crack tip. Once formed, this zone induces a configurational friction force, $\Delta\Gamma$ (18.4), stemming from energy dissipation during its dynamics. $\Delta\Gamma$ must be incorporated into the fracture resistance, Γ. In situations involving the formation of the transformational process zone, the equation governing the cleavage dynamics morphs into the form:

$$m\frac{d^2l}{dT^2} = \left(1 - \frac{V}{V_R}\right)\frac{(1-\nu^2)K_I^2}{E} - \Gamma - \mathfrak{s}V - \Delta\Gamma(V) \tag{19.2}$$

Here, a phenomenological parameter, $\mathfrak{s}$, has been introduced to account for energy dissipation linked with the excitation of other degrees of freedom in the solid, such as phonon modes. Earlier, we derived the expression (18.4) for the configurational friction force, $\Delta\Gamma(V)$, also used in (19.2). Furthermore, for the stress intensity factor, K_I, we make use of the formulation by Barenblatt and Cherepanov (1.7) which delineates cleavage facilitated by a thin wedge.

In this analysis, let us postulate that $V << V_{\mathrm{R}}$. Consequently, the first term on the right side of (19.2) can be approximated as:

$$\left(1 - \frac{V}{V_{\mathrm{R}}}\right)\frac{(1-\nu^2)K_I^2}{E} \approx \frac{\mathfrak{q}}{l}$$

where

$$\mathfrak{q} = \frac{Eh^2}{2\pi(1-\nu^2)} \tag{19.3}$$

Substituting these values into (19.2), we get a modified equation of motion as:

$$m\frac{d^2l}{dT^2} = \mathcal{H}(l, V) \tag{19.4}$$

Here, we introduced a function $\mathcal{H}(l, V)$, such that above the bifurcation point where $\xi \equiv 0$, $\mathcal{H} = \mathcal{H}_0$ is given by:

$$\mathcal{H}_0(l, V) = \frac{\mathfrak{q}}{l} - 2\Sigma - \mathfrak{s}V \tag{19.5}$$

with Σ denoting the fracture energy. Below the bifurcation (i.e., at $\xi \neq 0$) we find:

$$\mathcal{H}(l, V) = \mathcal{H}_0(l, V) - \frac{\kappa S_0}{\beta}V \times \mathcal{M}\left(\frac{k_1}{l^{2/3}} - \alpha - \frac{\kappa^2}{4g}V^2\right) \tag{19.6}$$

where

$$k_1 = \frac{1}{g^{1/3}}\left[\frac{A(1-2\nu)h}{\pi(1-\nu)}\right]^{4/3} \tag{19.7}$$

$S(V)$ represents the integral (18.5), while $S_0 \approx 4.9$ is its average value. Further, $\mathcal{M}(x)=x \times \Theta(x)$, where $\Theta(x)$ signifies the Heaviside theta-function. The final term in (19.6) elucidates the contribution of the process zone to the dynamic fracture resistance, represented by the non-analytical, piecewise-smooth function $\mathcal{H}$.

19.2.2 Steady Wedging

Assuming V_0 to be a constant velocity at which the wedge steadily penetrates into the split, forming the crack with the length $l = L$ in its front one relates them by the equation $\mathcal{H}_0(L, V_0) = 0$ yielding:

$$L = \frac{\mathfrak{q}}{2\Sigma + \mathfrak{s}V_0} \tag{19.8}$$

In this case, V_0 represents one of the parameters controlled by experimentalist, complementary to the crystal temperature. Consequently, we aim to study the wedging process in the plane (T, V_0) or in the equivalent (α, V_0)-plane.

The controlling condition for the formation of the process zone is given by

$$\alpha = \alpha_d(V).$$

Utilizing the expression for $\alpha_d(V)$ (10.24) derived earlier in Sect. 10.3. In conjunction with (19.8) and (1.7), we establish the bifurcation condition, $\alpha = \alpha_d(V_0)$, which one can regards as a line in the (α, V_0) plane:

$$\alpha_d(V_0) = \frac{1}{g^{1/3}}\left[\frac{2A^2(1+\nu)(1-2\nu)^2}{\pi E(1-\nu)}\right]^{2/3}(\mathfrak{s}V_0 + 2\Sigma)^{2/3} - \frac{\kappa^2 V_0^2}{4g} \tag{19.9}$$

This equation delineates a specific trajectory in the (V_0, α) plane, where below this line, $\alpha \leq \alpha_d(V_0)$, the crack tip is dressed.

***

In summary, *this section presents the equation of motion for the crack tip, a solution describing its steady motion regime, and the bifurcation condition. Together, these findings establish a foundation for exploring the self-oscillatory regime of wedging dynamics.*

19.3 Self-oscillatory Wedging

In this section, we apply the van der Pol–Andronov technique to describe the self-oscillatory regime of wedging dynamics.

***

In the steady regime, the crack tip and the wedge share identical velocities, V_0. In contrast, now we assume that the velocity of the crack tip, $V = V(t)$, is the sum of the constant wedge velocity, V_0, and the relative speed of the crack tip with respect to the wedge, $dl(t)/dt$.

$$V = V_0 + \frac{dl}{dt} \tag{19.10}$$

We seek a solution to the wedging equation (19.4) with the velocity V.

19.3.1 Dimensionless Formulation of the Wedging Equation

To further explore this, let us express the crack length in the form:

$$l(t) = L\,[1 + z(\tau)]\,; \qquad \tau = \mathfrak{q}^{1/2} m^{-1/2} L t \tag{19.11}$$

where L denotes the steady-state crack length as elucidated in (19.8), z and τ are the dimensionless length and time.

During the steady regime, we find that $z(\tau) = 0$ and $l(t) = L$.

For cases where $\alpha < \alpha_d$ the condition stipulating zero resistance force, $\mathcal{H}(l, V)$, in the steady regime is given by

$$\Gamma - \frac{\mathfrak{q}}{L} + \mathfrak{s} V_0 + \frac{\kappa V_0}{\beta}\left(\frac{k_1}{L^{2/3}} - \alpha\right) - \frac{\kappa^3 V_0^3}{4 g \beta} = 0 \tag{19.12}$$

This equation implicitly dictates the crack length, $L = L(V_0)$, during steady cleavage.

Incorporating the dimensionless variables from (19.11), we can restate the equation describing the crack tip motion (19.4) as:

$$z''(\tau) + z(\tau) = \Xi(z, z') \tag{19.13}$$

Here $z' = dz(\tau)/d\tau$ and $z''(\tau) \equiv d^2 z(\tau)/d\tau^2$. The nonlinear function $\Xi(z, z')$ has a complex expression. We will articulate it separately in the subsequent section.

Structure of the Function $\Xi(z, z')$.

Above the Bifurcation Point. Above the bifurcation point (specifically at $\xi \equiv 0$), the right-hand side of (19.13), denoted by $\Xi(z, z') = \Xi_0(z, z')$, assumes the form:

$$\Xi_0(z, z') = z(\tau) - 1 + \frac{1}{1 + z(\tau)} - \frac{\mathfrak{s} L}{m^{1/2}\mathfrak{q}^{1/2}} z'(\tau) \tag{19.14}$$

An expansion of the function $\Xi(z, z')$ in series in respect to z and z', coupled with the utilization of (19.8) and (19.12) confirms that Ξ_0 converges strictly to zero.

Below the Bifurcation Point. In contrast, below the bifurcation point (i.e., at $\xi \neq 0$), the function evolves to $\Xi(z, z') = \Xi_0(z, z') + \Delta\Xi$, with $\Delta\Xi$ denoted as the aggregate of three components: $\Delta\Xi = \Delta\Xi_1 + \Delta\Xi_2 + \Delta\Xi_3$ outlined below:

$$\Delta\Xi_1 = \frac{3LS_0V\kappa^3}{4gm\beta}\left[z'(\tau)\right]^2 + \frac{LS_0\mathfrak{q}^{1/2}\kappa^3}{4gm^{3/2}\beta}\left[z'(\tau)\right]^3 - \frac{k_1S_0L^{1/3}\kappa V}{\mathfrak{q}\beta\left[1+z(\tau)\right]^{2/3}} \tag{19.15}$$

$$\Delta\Xi_2 = \frac{LS_0\kappa}{m^{1/2}\mathfrak{q}^{1/2}\beta}\left(\alpha + \frac{3V^2\kappa^2}{4g}\right)z'(\tau) + \frac{k_1S_0L^{1/3}\kappa V}{\mathfrak{q}\beta} \tag{19.16}$$

$$\Delta\Xi_3 = -\frac{k_1S_0L^{1/3}\kappa}{m^{1/2}\mathfrak{q}^{1/2}\beta}\frac{z'(\tau)}{\left[1+z(\tau)\right]^{2/3}} \tag{19.17}$$

Upon examination, it is evident that the function $\Xi(z, z')$ in the proximity of the point $z(\tau) = 0$, there is a region where $|z(\tau)| < 1$ and $\left|\Xi(z, z')\right| < 1$. Within this domain, the application of the van der Pol-Andronov method, cited in [16] proves valid for analyzing equation (19.13).

19.3.2 Van der Pol-Andronov Approach

In the framework of the van der Pol-Andronov approach (19.18), we transition from the variables (z, z') to the polar ones (ρ, ϑ) as dictated by the following relations:

$$\begin{cases} z(\tau) = \rho\cos(\vartheta - \tau) \\ z'(\tau) = -\rho(\tau)\sin(\vartheta - \tau) \end{cases} \tag{19.18}$$

Here, the slowly varying "radius," $\rho(\tau) \geq 0$, and the "phase", $\vartheta(\tau)$, emerge as the new variables. Under the assumption $z < 1$, it follows that $\rho(\tau) < 1$. Consequently, the equation of motion (19.13) reformulated in terms of the new variables takes the following form:

$$\begin{cases} \frac{d\rho}{d\tau} = -\Xi\left[\rho\cos(\tau - \vartheta), -\rho\sin(\tau - \vartheta)\right]\sin(\tau - \vartheta) \\ \rho\frac{d\vartheta}{d\tau} = \Xi\left[\rho\cos(\tau - \vartheta), -\rho\sin(\tau - \vartheta)\right]\cos(\tau - \vartheta) \end{cases} \tag{19.19}$$

Central to the Van der Pol-Andronov approach is the strategy of averaging the right-hand parts of equations in (19.19) over the period 2π, concerning the variable $\zeta = \vartheta - \tau$:

$$\rho' = \mathbb{F}(\rho) = -\frac{1}{2\pi}\int_0^{2\pi}\Xi(\rho\cos\zeta, -\rho\sin\zeta)\sin\zeta d\zeta \tag{19.20}$$

and

$$\vartheta' = \mathbb{O}(\rho) = \frac{1}{2\pi\rho}\int_0^{2\pi}\Xi(\rho\cos\zeta, -\rho\sin\zeta)\cos\zeta d\zeta \tag{19.21}$$

where $\rho' \equiv d\rho/d\tau$ and $\vartheta' = d\vartheta/d\tau$. For a comprehensive exposition along with proofs, we direct the readers to the books [16, 17].

We emphasize that the prerequisite for applying the Van der Pol-Andronov approach is the requirement that the function $\mathbb{F}(\rho)$ is small.

19.3.3 Evaluation of the Functions $\mathbb{F}(\rho)$ *and* $\mathbb{O}(\rho)$

The content of the next few sections is technical and can be omitted during the first reading.

19.3.3.1 Analysis of the Function $\mathbb{F}(\rho)$

To advance our computations, it is beneficial to represent the function $\mathbb{F}(\rho)$, appearing in the right-hand side of (19.20), as a summation of three individual terms, delineated as:

$$\mathbb{F}(\rho) = \mathbb{F}_0(\rho) + \mathbb{F}_1(\rho) + \mathbb{F}_2(\rho)$$

where the terms are defined as follows:

$$\mathbb{F}_0(\rho) = -\int_0^{2\pi} \varXi_0 \sin\zeta \frac{d\zeta}{2\pi}$$

$$\mathbb{F}_1(\rho) = -\int_0^{2\pi} \{\Delta\varXi_1 + \Delta\varXi_2\} \sin\zeta \frac{d\zeta}{2\pi}$$

and

$$\mathbb{F}_2(\rho) = -\int_0^{2\pi} \varXi_3 \sin\zeta \frac{d\zeta}{2\pi}$$

The calculation of $\mathbb{F}_0(\rho)$ and $\mathbb{F}_1(\rho)$ is straightforward yielding the results:

$$\mathbb{F}_0(\rho) = -\frac{\mathfrak{s}L}{2\,m^{1/2}\mathfrak{q}^{1/2}}\rho \tag{19.22}$$

$$\mathbb{F}_1(\rho) = \frac{S_0 L\kappa}{2\,m^{1/2}\mathfrak{q}^{1/2}\beta}\left[\left(\alpha + \frac{3\kappa^2 V^2}{4g}\right)\rho + \frac{3\mathfrak{q}\kappa^2}{16gm}\rho^3\right] \tag{19.23}$$

Meanwhile, $\mathbb{F}_2(\rho)$ is given by

$$\mathbb{F}_2(\rho) = \frac{k_1 S_0 L^{1/3}\kappa}{m^{1/2}\mathfrak{q}^{1/2}\beta}\mathbb{U}(\rho)$$

and is expressed through the function $U(\rho)$, which is formulated as:

$$\mathbb{U}(\rho) = \frac{\rho}{2\pi}\int_0^{2\pi} \frac{\sin^2\zeta d\zeta}{(1+\rho\cos\zeta)^{2/3}} \tag{19.24}$$

The integral (19.24) can be evaluated exactly, representing it as a summation of hypergeometric functions ${}_2F_1\,(x_1, x_2; x_3, x_4)$ given by:

$$U(\rho) = -\frac{1}{16\rho(1-\rho^2)^{2/3}}\sum_{i=1}^{4} U_i$$

$$\mathbb{U}_1 = 3(1+\rho)^{2/3}(1-\rho)_2F_1\left(-\frac{1}{2}, \frac{2}{3}; 1, \frac{2\rho}{\rho-1}\right)$$

$$\mathbb{U}_2 = 3(1-\rho)^{2/3}(1+\rho)_2F_1\left(-\frac{1}{2}, \frac{2}{3}; 1, \frac{2\rho}{1+\rho}\right)$$

$$\mathbb{U}_3 = (1+\rho)^{5/3}(2\rho-3)_2F_1\left(\frac{1}{2}, \frac{2}{3}; 1, \frac{2\rho}{\rho-1}\right)$$

$$U_4 = (1-\rho)^{2/3}(2\rho^2+\rho-3)_2F_1\left(\frac{1}{2}, \frac{2}{3}; 1, \frac{2\rho}{1+\rho}\right)$$

The exact solution, however, is rather lengthy. We thus resort to a numerical calculation within the interval $\rho \in [0, 0.7]$, approximating it by a simple cubic polynomial:

$$\mathbb{U}_{\mathrm{app}}(\rho) \approx 0.1(5\rho - 0.09\rho^2 + \rho^3) \tag{19.25}$$

Figure 19.2 compares the exact result, $U(\rho)$, (the blue dots) with its approximation, $U_{\mathrm{app}}(\rho)$ (19.25), (the solid red line). One can see that the approximation is accurate at $\rho < 1$, which is exclusively utilized in the subsequent analysis.

Utilizing the approximation (19.25) the value of $\mathbb{F}_2(\rho)$ is derived as:

$$\mathbb{F}_2(\rho) \approx 0.1\frac{k_1 S_0 L^{1/3}\kappa}{m^{1/2}q^{1/2}\beta}\left(5\rho - 0.09\rho^2 + \rho^3\right) \tag{19.26}$$

Aggregating the terms (19.22), (19.23) and (19.24) the function $F(\rho)$ is identified as:

$$\mathbb{F} = \frac{LS_0\kappa}{2(mq)^{1/2}\beta}\rho\begin{cases} -\frac{\mathfrak{s}\beta}{\kappa S_0}, & \alpha/\alpha_{\mathrm{d}} > 1 - V_0^2/V_{\mathrm{d}}^2 \\ \mathbb{R}_1 + \mathbb{R}_2\rho - \mathbb{R}_3\rho^2, & \alpha/\alpha_{\mathrm{d}} \le 1 - V_0^2/V_{\mathrm{d}}^2 \end{cases} \tag{19.27}$$

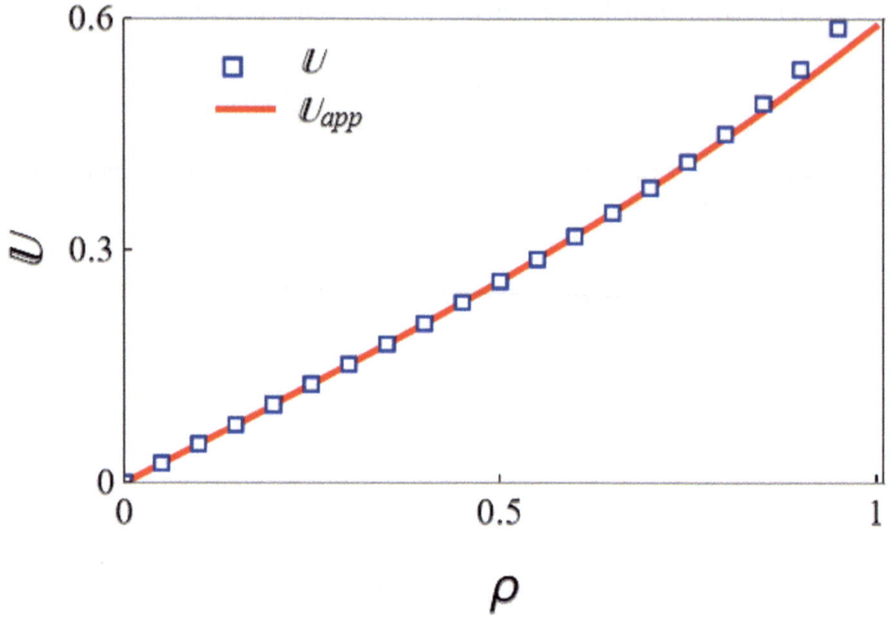

Fig. 19.2 Illustration of the polynomial approximation (see (19.25)). The exact value of the integral $\mathbb{U} = \mathbb{U}(\rho)$ is denoted by the squares and its approximation (19.25) is illustrated by the solid line

where

$$\mathbb{R}_1 \approx \left(\alpha - \frac{k_1}{L^{3/2}} + \frac{3\kappa^2 V_0^2}{4g} - \frac{\mathfrak{s}\beta}{\kappa S_0}\right) \tag{19.28}$$

$$\mathbb{R}_2 \approx 0.018\frac{k_1}{L^{3/2}} \tag{19.29}$$

and

$$\mathbb{R}_3 \approx 0.2\left(\frac{k_1}{L^{3/2}} - 0.94\frac{q\kappa^2}{gm}\right) \tag{19.30}$$

19.3.3.2 The Function $\mathbb{O}(\rho)$

Utilizing direct integration techniques, the function $\mathbb{O}(\rho)$ can be delineated as presented in reference (19.21)

$$\mathbb{O}(\rho) = \frac{1}{2} + \frac{1}{\rho^2} - \frac{1}{\rho^2\left(1-\rho^2\right)^{1/2}} + \Delta\mathbb{O}(\rho) \tag{19.31}$$

The term $\Delta\mathbb{O}(\rho)$ is further expressed as:

$$\Delta\mathbb{O}(\rho) = -\frac{1}{12\pi q\beta}\frac{1}{\rho} + \frac{k_1 L^{1/3} S_0 V\kappa}{2\rho}\sum_{i=1}^{4}\mathbb{O}_i \tag{19.32}$$

$$\mathbb{O}_1 = -\frac{9\,(1-\rho)^{1/3}}{1+\rho}\,{}_2F_1\left(-\frac{1}{2},\frac{2}{3};2,\frac{2\rho}{\rho-1}\right)$$

$$\mathbb{O}_2 = \frac{\rho-9}{(1-\rho)^{2/3}\,(1+\rho)}\,{}_2F_1\left(\frac{1}{2},\frac{2}{3};2,\frac{2\rho}{\rho-1}\right)$$

$$\mathbb{O}_3 = -\frac{9\,(1+\rho)^{1/3}}{(1-\rho)}\,{}_2F_1\left(-\frac{1}{2},\frac{2}{3};2,\frac{2\rho}{1+\rho}\right)$$

$$\mathbb{O}_4 = \frac{9+\rho}{(1+\rho)^{2/3}\,(1-\rho)}\,{}_2F_1\left(\frac{1}{2},\frac{2}{3};2,\frac{2\rho}{1+\rho}\right)$$

Above the bifurcation, $\Delta\mathbb{O} \equiv 0$, is identically zero, while it assumes a non-zero value below the bifurcation.

***

In summary, *using the van der Pol–Andronov approach, we reformulated the problem as a system of second-order ordinary differential equations, making it more suitable for qualitative analysis.*

19.4 Regimes of Wedging Dynamics

In this section, we conduct a qualitative analysis of the derived equations and construct their phase portrait, uncovering multiple self-oscillatory regimes.

***

19.4.1 Mapping Principles of Phase Portraits

Let us note that the right-hand sides of the system of equations referenced in (19.20) and (19.21) are independent of ϑ. Equation (19.20) delineates a one-dimensional nonlinear ordinary differential equation given by

$$\rho' = \mathbb{F}(\rho) \tag{19.33}$$

The qualitative analysis of this equation is straightforward: its dynamics is entirely dictated by the roots of the function $\mathbb{F}(\rho)$ which establish its stationary points. One categorizes the phase portraits of (19.13) in (z, z') plane by mapping them to the stationary points of the initial equation (19.33) along the one-dimensional ray $\rho \geq 0$.

19.4.1.1 Analysis of Stationary Points $\rho = 0$

Specifically, the attractive stationary point at $\rho = 0$ in (19.33) aligns with the stable focus of (19.13) in the (z, z') plane. In the same vein, a repulsive center at $\rho = 0$ corresponds to the unstable focus in the (z, z') plane.

19.4.1.2 Analysis of Stationary Points $\rho \neq 0$

An attractive point at $\rho \neq 0$ in (19.33) signifies a stable cycle in (19.13), whereas any repelling point at $\rho \neq 0$ denotes the unstable cycle within the (z, z')-plane [17].

19.4.1.3 Solving Equation (19.21)

Given that $\mathbb{O} = \mathbb{O}(\rho)$ and is ϑ-independent, the solution to (19.21) is solely dictated by the solution $\rho = \rho(\tau)$, of the first equation, expressed as

$$\vartheta(\tau) = \int_0^{\tau} \mathbb{O}\left[\rho(\tau')\right] d\tau'$$

This formulation is simplified when considering the limiting solutions at the attractive stationary points, ρ_{st}, of (19.33). It manifests a solution depicting rotation at a constant angular velocity, $\mathbb{O}\left(\rho_{\mathrm{st}}\right)$:

$$\vartheta(\tau) = \mathbb{O}\left(\rho_{\mathrm{st}}\right) \tau$$

19.4.2 The Bifurcation Diagram and Self-oscillation Crack Propagation

19.4.2.1 Division of the (V_0, α)-Plane into Regions with Different Regimes of the Crack Motion

We now address the analysis of the situation under consideration. For $\alpha > \alpha_{\mathrm{d}}(V)$ as described by (19.9) the crack remains undressed and the function $\mathbb{F}(\rho)$ manifests as linear, conforming to the primary line of (19.27), as depicted in Fig. 19.3 (i). This showcases a singular root at $\rho = 0$ corresponding to the attractive focal point in the (z, z') plane.

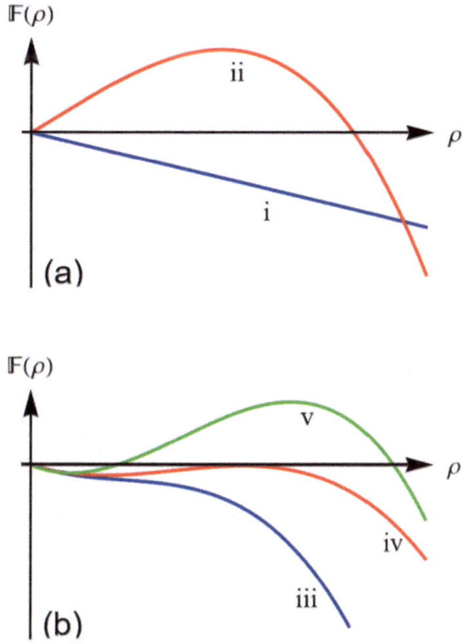

Fig. 19.3 Representative forms of the function $\mathbb{F} = \mathbb{F}(\rho)$. Panel (a) represents the "undressed" crack tip scenario (i) and the "dressed" tip case with $\mathbb{R}_1 > 0$ (ii). Meanwhile, panel (b) illustrates the "dressed" crack tip cases with $\mathbb{R}_1 < 0$. Curve (iii) aligns with $\Delta = 4\mathbb{R}_1\mathbb{R}_3 + \mathbb{R}_2^2 < 0$, curve (iv) aligns with the $\Delta = 0$ case, and curve (v) corresponds to $\Delta > 0$

Conversely, for $\alpha < \alpha_d(V)$, the crack tip is in a "dressed" state. Under the assumption that $\mathbb{R}_3 > 0$ (19.30), the function $\mathbb{F}(\rho)$ is illustrated in Fig. 19.3a, denoted by (ii). When $\mathbb{R}_1 > 0$, the function $\mathbb{F}(\rho)$ presents dual roots as shown by (ii) in Fig. 19.3a: $\rho = 0$ signifying an unstable focus, and $\rho \neq 0$ indicating a stable cycle in the (z, z') plane.

Furthermore, if $\mathbb{R}_1 < 0$ the root count is dictated by the sign of the parameter $\Delta = 4\mathbb{R}_1\mathbb{R}_3 + \mathbb{R}_2^2$. A negative Δ yields a single root at $\rho = 0$, indicating an attractive focus as illustrated in Fig. 19.3b, (iii). In contrast, a positive Δ results in three roots: one at $\rho = 0$ and two at $\rho \neq 0$ with the former corresponding to the attractive focus, and the latter delineating both an unstable and a stable cycle in the (z, z') plane (Fig. 19.3b, v). The intermediary scenario of $\Delta = 0$ (depicted in Fig. 19.3b, iv) portrays the fusion of both cycles. It is worth mentioning that as $\mathbb{R}_1$ turns to zero, the unstable cycle merges with the attractive focus, consequently forming an unstable focus as depicted by (ii) in Fig. 19.3a.

However, the case of $\mathbb{R}_3 < 0$ anticipates solutions with escalating $|z(\tau)|$, thus falling outside the bounds of the approximation applied in this methodology, and is consequently excluded from this analysis.

The condition $\mathbb{R}_1 = 0$ yields the equation

$$\alpha = \frac{k_1}{L^{2/3}} + \frac{\mathfrak{s}\beta}{\kappa S_0} - \frac{3\kappa^2 V_0^2}{4g} \tag{19.34}$$

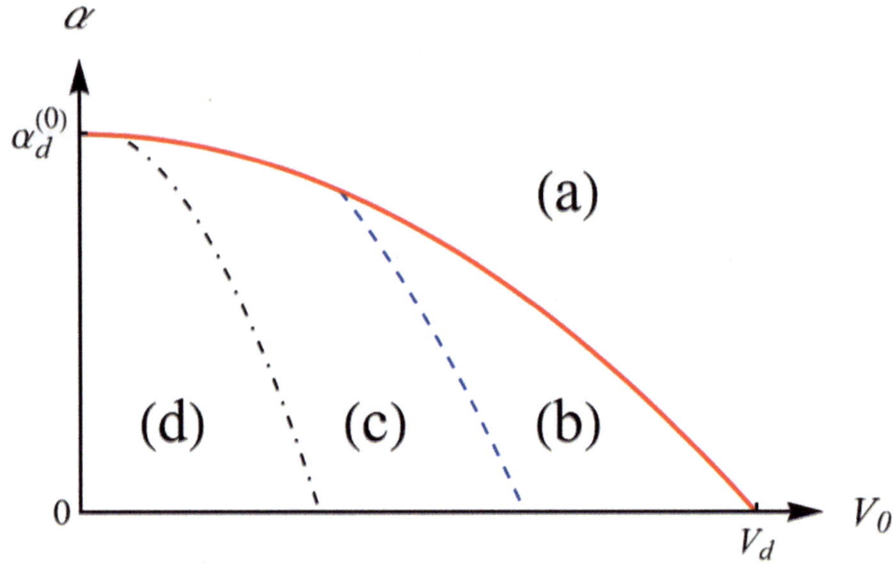

Fig. 19.4 Bifurcation diagram. The (V_0, α) plane is subdivided into regions (a)–(d), wherein (19.13) manifests distinct phase portraits. Here, $\alpha_{\mathrm{d}}^{(0)}$ is α_{d} taken at $V_0 = 0$

describing a line in the (V_0, α) plane. Concurrently, the condition $\Delta = 0$ is satisfied at

$$\alpha = \frac{k_1}{L^{2/3}} + \frac{\mathfrak{s}\beta}{\kappa S_0} - \frac{3\kappa^2 V_0^2}{4g} - \frac{4 \times 10^{-4} k_1^2 g m}{k_1 g m L^{2/3} - 0.94 q \mathfrak{s}^2 L^{4/3}} \tag{19.35}$$

Figure 19.4 elucidates the bifurcation diagram of (19.13).

This diagram divides the (V_0, α)-plane into four sectors (a)–(d), where, within the van der Pol-Andronov approximation, the (19.13) showcases diverging behavioral patterns, depicted in the phase portraits in Fig. 19.5.

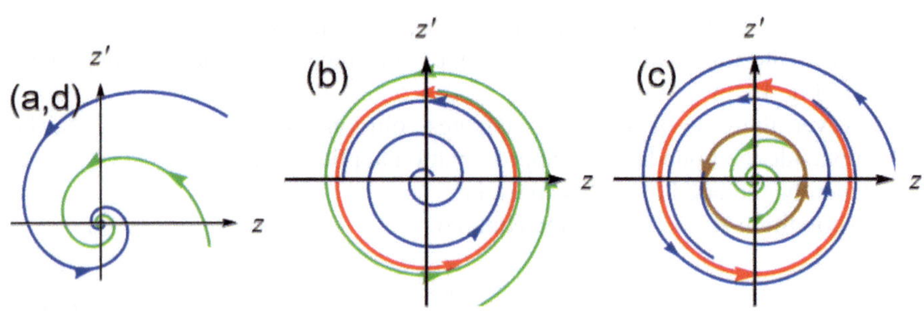

Fig. 19.5 Phase portraits. A showcase of three distinct phase portraits in the (z, z') plane that characterize the behavior of (19.13) within the confines of the van der Pol-Andronov approximation

In thc bifurcation diagram depicted in Fig. 19.4 we observe that the dynamics in domains (a) and (d) are characterized by a simple attractive focus, as detailed in panels (a) and (d) of Fig. 19.5.

In a more complex representation, panel (b) of Fig. 19.5 illustrates the presence of an attractive limit cycle, delineated by a red line. This cycle encompasses a repulsive focus represented by a blue trajectory that weaves around the cycle from an inward direction. Moreover, an external green trajectory symbolizes paths that encircle the cycle from the outside. This particular dynamic behavior is manifested in domain (b) of the bifurcation diagram shown in Fig. 19.4.

Delving deeper, panel (c) of Fig. 19.5 reveals a scenario with dual limit cycles taking place in the domain (c) of the bifurcation diagram Fig. 19.4. The internal cycle, highlighted in brown in panel (c) of Fig. 19.5, is unstable, harboring an attractive focus at its center, towards which the green trajectories are directed. Intriguingly, this unstable cycle is encapsulated within a stable cycle, which is marked in red. Notably, all trajectories situated between the inner (brown) and outer (red) cycles, in addition to those beyond the confines of the outer cycle, circulate around this stable cycle. These trajectories are visually represented in blue in the figure.

***

In conclusion, *applying the van der Pol–Andronov technique allowed us to partition the* (V_0, α) *plane into four regions, each representing a qualitatively distinct wedging regime. Among these, two regions correspond to different types of stick-slip crack propagation, influenced by the contributions of the crack tip process zone.*

19.5 Crack Propagation Regimes: Simulation

The van der Pol–Andronov technique simplifies the equations of motion by averaging their right-hand sides. To explore behavioral details of the original equations that this approximation does not capture, we simulate the original equation for specific parameter values. This approach enables us to visualize various types of crack tip motion, including behaviors not predicted by the van der Pol–Andronov approximation.

***

Before delving deeper, let us note the expression for α, given by $\alpha = \alpha_0(T - T_c)$. Consequently, the bifurcation diagram depicted in Fig. 19.4 isomorphically maps onto the plane (V_0, T). As illustrated in Fig. 19.4, when the temperature is sufficiently elevated (corresponding to region (a) on the bifurcation diagram), the phase portrait manifests as an attractive focus, as depicted in Fig. 19.5a. This behavior is mirrored in domain (d), where the trajectories converge at the point $z = z' = 0$. Consequently, regimes (a) and (d) represent scenarios where wedging occurs at a constant crack length, and any deviations from it vanish.

To gain a comprehensive understanding of the crack length, $l = l(t)$, and its velocity, $dl(t)/dt$, we numerically solved (19.4) as shown below. Equation of motion (19.4) is convenient to rewrite as follows:

$$mx'(t) = \frac{\mathfrak{q}}{l(t)} - 2\Sigma - \mathfrak{s}\,[V_0 + x(t)] -$$

$$-\frac{\kappa}{\beta}\,[V_0 + x(t)]\,\mathcal{M}\left[\frac{k_1}{l^{2/3}(t)} - \frac{\kappa^2}{4g}\,(V_0 + x(t))^2 - \alpha\right] \tag{19.36}$$

and

$$l'(t) = x(t) \tag{19.37}$$

where $y(t) \equiv dl/dt$.

Given the multitude of parameters defining both the wedging process and the phase transition, determining precise values corresponding to the stick-slip regime presents a challenge. For illustrative purposes, we selected the following arbitrary parameter values: $m = 1, k_1 = 1, \kappa = 2.5, \mathfrak{s} = 1.5, \beta = 0.7, g = 0.5, \mathfrak{q} = 1$ and $\Sigma = 2$. Additionally, we chose $S_0 \approx 4.9$ aligning with its average value (see Fig. 18.1). These selections yield the following reformulated equation:

$$x'(t) = \frac{1}{l(t)} - 2 - 1.5\,[V_0 + x(t)] -$$

$$-\,3.57\,[V_0 + x(t)]\,\mathcal{M}\left[\frac{1}{l^{2/3}(t)} - 3.125\,(V_0 + x(t))^2 - \alpha\right] \tag{19.38}$$

Equation (19.37) remains unchanged. By manipulating the two remaining parameters, α and V_0, we derived various phase portraits.

Utilizing Mathematica's standard NDSolve routine, we constructed the phase portraits of the system of (19.38) and (19.37) in the $(l, dl/dt)$ plane, in addition to the $l = l(t)$ dependence.

Case Studies.

Case 1: $\alpha = 1.3$ and $V_0 = 0.7$.. The phase portrait characterized by these parameters is illustrated in Fig. 19.6a.

This pattern corresponds to the one depicted in Fig. 19.5a and d. As observed in Fig. 19.6b, the crack length approaches a horizontal asymptote, signifying a constant crack length during the cleaving process within this regime.

Case 2: $\alpha = 0.5$ *and* $V_0 = 0.7$. The outcome of this scenario is displayed in Fig. 19.7a.

In this case, a repulsive focus is encased by an attractive cycle, with trajectories winding around the cycle–a phenomenon clearly illustrated by the magenta trajectory. This periodic variations in crack length represent self-oscillatory behavior, a characteristic feature in domain (b) of the bifurcation diagram in Fig. 19.4.

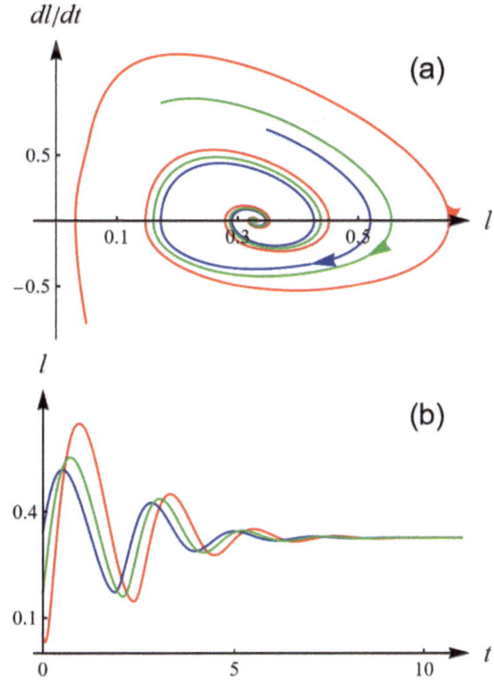

Fig. 19.6 Phase portrait from simulation showing a single focus and $l(t)$ behavior **a** The phase portrait of the system (19.38) and (19.37) in the plane $(dl/dt, l)$ for $\alpha = 1.3$ and $V_0 = 0.7$ highlights a singular attracting focus. **b** The crack length, l, stabilizes to a constant value

Case 3: $\alpha = 0.5$ *and* $V_0 = 1.2$. The results of this setup are showcased in Fig. 19.8.

This case aligns with the behavior documented in Fig. 19.5c. The phase portrait in Fig. 19.8a presents a central attractive focus encapsulated by an attractive cycle, which is highlighted by the black and red trajectories. This setup is feasible only if a repulsive cycle separates the focus and the attractive cycle. Although the repulsive cycle isn't visible in Fig. 19.8a its position can be accurately inferred, acting as a boundary between the blue trajectory around the focus and the red one circling over the attractive cycle. This complex dynamic results in two distinct stable behaviors concerning crack length as depicted in Fig. 19.8b. The trajectories whose initial conditions lie within the attraction basin of the focus converge to a constant crack value. The ones with the initial conditions far enough from the focus, progress to the self-oscillations.

Additional Observations: It is noteworthy that the system of (19.36) and (19.37) showcases a more intricate behavior compared to predictions made by equation (19.33). Upon scrutinizing system (19.36) and (19.37), we have identified distinct behavioral patterns that deviate qualitatively from those portrayed in Fig. 19.5. As an illustration, consider the behavior of the system at $\alpha = 0.5$ and $V_0 = 0.732$, displayed in Fig. 19.9.

At the core of the phase portrait, three fixed points are discernible. Standard methods for analyzing dynamic systems reveal these points to be: a repulsive knot at $l \approx 0.148$, $dl/dt = 0$, a saddle point at $l \approx 0.25$, $dl/dt = 0$ and an attractive focus

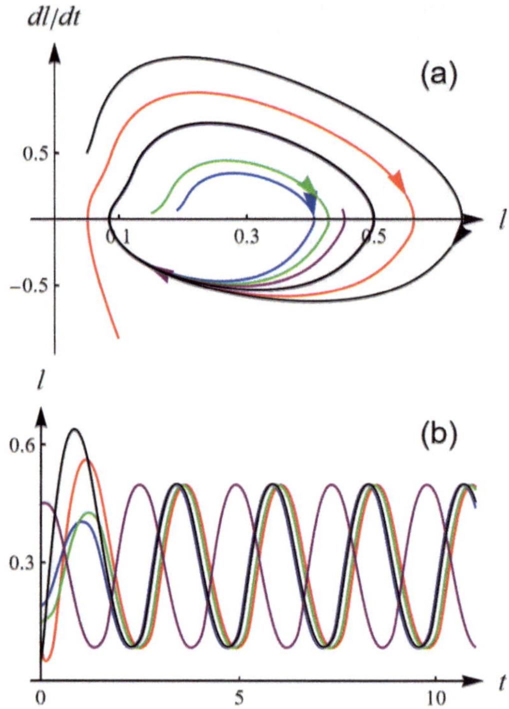

Fig. 19.7 Attractive cycle in the phase portrait and associated $l(t)$ dynamics **a** The phase portrait of the system (19.38) and (19.37) in the plane $(dl/dt, l)$ for $\alpha = 0.5$ and $V_0 = 0.7$ reveals a singular repulsive focus surrounded by an attractive cycle. Trajectories encircle a cycle, with the black trajectory closely winding around the cycle. This helps to visualize its position. **b** The crack length, l, exhibits periodic oscillations around some average value

at $l \approx 0.323, dl/dt = 0$. These points, marked in red, blue, and green respectively in Fig. 19.9a, are enveloped by an attractive cycle, visible due to a group of trajectories winding around it.

***

In summary, *the approximations obtained through the van der Pol-Andronov averaging method simplifies the original system significantly. Indeed, (*19.33*) can only depict a single fixed point, which can be either attractive or repulsive in nature. However, this approximation cannot adequately illustrate phase portraits with multiple fixed points.*

Despite this, the approximation effectively predicts the transition from steady states to self-oscillating regimes.

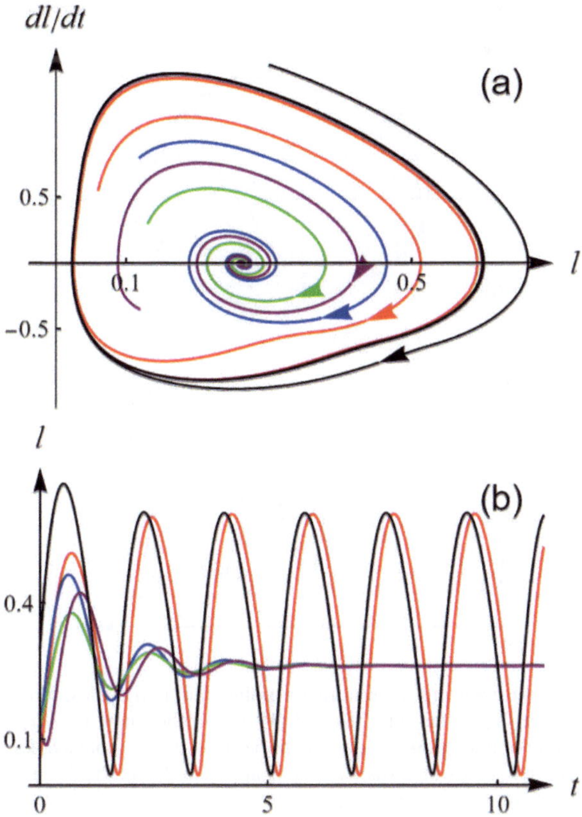

Fig. 19.8 Phase portrait illustrating both an attractive focus and limit cycles, along with $l(t)$ behavior **a** The phase portrait of the system (19.38) and (19.37) in the plane $(dl/dt, l)$ with $\alpha = 0.5$ and $V_0 = 1.2$ features a prominent attracting focus circled by an attractive cycle, discernible from the black trajectory. A repulsive cycle exists between them, albeit not visible. **b** Depending on their starting points, trajectories exhibit either a stabilization at a constant crack length or progress to self-oscillations

19.6 Chapter Summary

In conclusion, *in solids with non-primitive crystal cells, stress-induced local phase transitions can occur at the crack tip, playing a vital role in crack dynamics due to their contributions to velocity-dependent resistance to fracture. Our detailed analysis of the viscous friction force arising from the zone demonstrates that it engenders a self-oscillating behavior of the crack tip motion in certain regions of its dynamic phase diagram.*

***

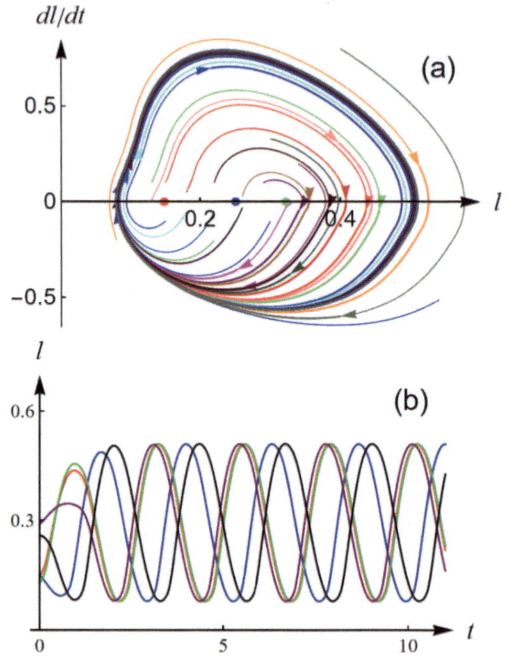

Fig. 19.9 Phase portrait featuring three attractor points and two cycles, alongside $l(t)$ behavior **a** The phase portrait of the system (19.38) and (19.37) in the plane $(dl/dt, l)$ for $\alpha = 0.5$ and $V_0 = 0.732$ exhibits three fixed points along the horizontal axis: a repulsive knot, a saddle and an attractive focus encompassed by an attractive cycle, apparent due to the dense bundle of black and blue trajectories. **b** $l(t)$ showcases a periodic pattern

References

1. D. Magis, J. Mater. Sci. **20**, 3041 (1985)
2. T.W. Webb, E.C. Aifantis, Int. J. Solids Struct. **32**, 2725 (1995)
3. J.S. Langer, Phys. Rev. E **62**, 1351 (2000)
4. E. Sharon et al., Nature **410**, 68 (2001); J.F. Boudet, S. Ciliberto, Physica **D142**, 317 (2000); R.D. Deegan et al., Phys. Rev. Lett. **88**, 014304 (2001)
5. J. Fineberg, M. Marder, Phys. Rep. **313**, 1 (1999); A. Yuse, M. Sano, Physica **D108**, 365 (1997); R. Deegan et al., Phys. Rev. E **67**, 066209 (2003); B. Yang, K. Ravi-Chandar,J. Mech. Phys. Solids **49**, 91 (2001); J.A. Hauch, M.P. Marder, Int. J. Fract. **90**, 133 (1998); T. Cramer et al., Phys. Rev. Lett. **85**, 788 (2000)
6. I.S. Aranson et al., Phys. Rev. Lett. **118**, (2000); L.O. Eastgate et al., Phys. Rev. **E65**, 036117 (2002); A. Karma et al., Phys. Rev. Lett. **87**, 045501 (2001)
7. Bouchbinder et al., Phys. Rev. Lett. **101**, 264302 (2008)
8. A. Livne et al., Phys. Rev. Lett. **101**, 264301 (2008)
9. G.I. Barenblatt, G.P. Cherepanov, J. Appl. Math. Mech. **24**, 993 (1960)
10. G.I. Barenblatt, R.L. Salganik, J. Appl. Math. Mech. **27**, 656 (1963)
11. G.I. Barenblatt, R.L. Salganik, J. Appl. Math. Mech. **27**, 1644 (1963)
12. B. Steverding, S.H. Lehnigk, Int. J. Fract. Mech. **6**, 223 (1970)
13. M.F. Kanninen, Int. J. Fract. **10**, 415 (1974)

14. L.B. Freund, *Dynamic Fracture Mechanics* (Cambridge University Press, Cambridge, 1998)
15. J.F. Malluck, W.W. King, Int. J. Fract. **13**, 655 (1977); G.T. Hahn, et al., in *2nd International Conference on Pressure Vessel Technology* (ASME, New York, NY, USA, San Antonio, TX, USA, 1973), p. 981; G.T. Hahn, et al., Eng. Fract. Mech. **7**, 583 (1975); Z.J. Bilek, S.J. Burns, J. Mech. Phys. Solids **22**, 85 (1974), L.B. Freund, J. Mech. Phys. Solids **25**, 69 (1977), K. Hellan, Int. J. Fract. **14**, 91 (1978); ibid, idem **14**, 173 (1978); ibid, idem **17**, 311 (1981)
16. A.A. Andronov, A.A. Vitt, S.E. Khaikin, *Theory of Oscillations* (Dover Publications, 1987), p. 583
17. N.B. Butenin, Y.I. Neimark, N.A. Fufaev, *Introduction into the Theory ofNonlinear Oscillations* (Moscow, Nauka, 1976). In Russian
18. A. Boulbitch, A.L. Korzhenevskii, Phys. Rev. Lett. **107**, 085505 (2011)

Part IV
Process Zones at Dislocations

Chapter 20
General Remarks on the Dislocations Process Zone

Abstract In this chapter, we examine the conditions under which the single dislocation model is applicable. Using simple arguments, we demonstrate how the sign of the binodal slope determines the relative positions of the daughter- and mother-phase zones at the dislocation. We then adapt the general rescaling relations to the dislocation framework and derive the corresponding equations of state and motion in terms of dimensionless parameters. Finally, we present dimensionless expressions for the zone's free energy and for the friction force acting on the propagating dislocation.

20.1 Introduction

The study of phase transitions in crystals and the mechanics of plasticity has historically developed along separate paths. However, a significant leap in the theoretical understanding of these phenomena occurred almost simultaneously. In 1937, Landau formulated his renowned theory of phase transitions, introducing the fundamental concept of the order parameter. Similarly, in 1934, Taylor, Orowan, and Polyani independently introduced the idea of crystalline defects—dislocations—whose collective movement they linked to macroscopically observed plastic deformation.

The theoretical significance of Landau theory was immediately recognized. The introduction of the order parameter was based on purely symmetry-related, and thus very general, arguments. Crystalline dislocations were introduced as topological linear objects, representing a general theoretical approach. Moreover, the ability of dislocations to move under the influence of small mechanical loads was geometrically evident. This was soon confirmed by Peierls's model calculation in 1940 and later by Nabarro (1947) and Eshelby (1949) with more realistic models.

The experimental development of Landau's and Taylor-Orowan-Polyani's ideas over the next twenty years took different trajectories. A multitude of complementary experimental methods for studying phase transitions were developed relatively quickly. In contrast, direct observation of individual dislocations remained virtually impossible until the invention of the electron microscope in the 1950s. Once this technology became available, the dislocation approach to describing plastic phenomena was rapidly accepted and widely utilized, especially in the study of metals, the main

© The Author(s), under exclusive license to Springer Nature Switzerland AG 2026
A. Boulbitch and A. Korzhenevskii, *Stress-Induced Local Phase Transformations Near Cracks and Dislocations*, Springer Series in Materials Science 356,
https://doi.org/10.1007/978-3-032-00945-6_20

plastic crystals. Dislocation theory led to significant progress in understanding the mechanical, electrical, and magnetic properties of these materials.

Many alloys—such as steel, and those based on titanium and cobalt, to name a few—experience phase transitions, often defining their utilization in manufacturing products from these materials. Thus, understanding the influence of dislocations on their desirable (or undesirable) properties became important. While semiconductors and dielectrics are generally less ductile than metal alloys, they also contain dislocations (usually of low mobility) and often exhibit phase transitions. In their low-symmetry phases, domains—regions with different values of the order parameter—exist. The restructuring of domain configurations under external mechanical loads or electric and magnetic fields occurs through the movement of domain boundaries interacting with dislocations. Therefore, even for macroscopically low-plasticity materials, it is necessary to describe the interaction of dislocations with the spatially distributed field of the order parameter.

So far, we have discussed plastic or low-plastic materials, considering their macroscopic mechanical properties. Similarly, phase transitions were understood as occurring uniformly throughout the entire bulk of the sample.

In this context, the influence of dislocations on the thermodynamic properties of the crystal, such as heat capacity and macroscopic susceptibility far from transition points, should not be particularly large. This is because the contribution of dislocations is proportional to their density, which is relatively small in samples that are not excessively plastically deformed.

However, the situation changes if we formulate the question differently: "*How does the phase transition influence the plasticity of the material by affecting the properties of the dislocation ensemble?*" Dislocations can significantly impact the structure-sensitive properties of the material, particularly its dynamic properties.

It is natural to commence the analysis with a problem involving a dislocation situated far from others, thereby interacting with them only negligibly. Henceforth, such a dislocation is referred to as *single* or *isolated*. Mathematically, this formulation eliminates the "small parameter"—the total (dimensionless) density of the dislocation ensemble—from the problem. The latter suggests that the expected effect of a phase transition can be strong, at least locally (i.e., in the vicinity of a dislocation, where the material's properties can change significantly, and its local configuration can also change). Whether this effect will be genuinely strong depends on the type and shape of the dislocation, as well as the properties and state of the surrounding crystalline matrix, particularly its proximity to the macroscopic phase transition point.

Assuming we have solved several specific problems and determined the conditions under which the effect is strong for a single dislocation (e.g., a restructuring of the dislocation shape occurs), we can then ask what to expect for the plastic properties of the sample as a whole. Even if the interaction of dislocations is weak, a substantial effect for one dislocation implies restructuring the entire dislocation ensemble. This means a change in the macroscopic properties of the sample due to a phase transition. Since dislocations interact with other defects, such as domain (twin) boundaries, the ensemble of the latter will also be rearranged, leading to effects such as ferroelastic, electrical, and magnetic repolarization. In addition to observing plastic flow, changes in the inhomogeneous internal structure of the sample should also manifest in experiments on "anomalous" scattering, primarily the scattering of neutrons, X-rays, and light.

The Part V of the book is devoted to addressing these issues.

20.2 Dislocation-Associated Zones on Opposite Sides of the Binodal

In this section, we identify the location on the phase diagram where the process zone associated with edge dislocations manifests (Boulbitch and Korzhenevskii, first published here).

***

This Part analyzes the process zone created by an isolated edge dislocation. First, it is necessary to discuss the conditions in which studying the zone of the single dislocation is appropriate. The latter is true when the average distance between dislocations within a crystal, denoted as $\mathbb{L}_{\text{disl}}$, significantly exceeds the zone's size, $\mathbb{R}_{\text{d}}$. Throughout this Part, we assume $\mathbb{R}_{\text{d}} \ll \mathbb{L}_{\text{disl}}$ as the default condition. Friedel argued that a "good crystal" contains $\rho_{\text{disl}} \sim 10^4$ to 10^6 dislocations per square centimeter [1] which gives $\mathbb{L}_{\text{disl}} \sim 10$ to $100\,\mu m$. This assumption leads to $\mathbb{R}_{\text{d}} \ll 1\mu m$ to $10\mu m$. In a solid subjected to plastic deformation, the zone size $\mathbb{R}_{\text{d}}$, allowing for the single dislocation approximation, is even smaller. According to this assumption, we neglect the elastic fields generated by dislocations and other defects in the environment of the dislocation of interest. This simplification yields the potential $U_0(\mathbf{r})$ for the case of a single edge dislocation as follows:

$$U_0(\mathbf{r}) = -B_{\text{d}} \frac{\sin(\theta)}{r} \tag{20.1}$$

where the constant, B_{d}, takes the following form:

$$B_{\text{d}} = \frac{(1 - 2\nu) A \mathfrak{b}}{\pi (1 - \nu)} \tag{20.2}$$

according to (1.13). The subscript "d" of the parameter B stands for "dislocation."

The sign of B_{d} corresponds to that of A. It is worth recalling that the necessary condition for the formation of the daughter-phase zone embedded within the bulk mother phase is expressed as:

$$U_0(\mathbf{r}) < 0 \tag{20.3}$$

Conversely, the inequality:

$$U_0(\mathbf{r}) > 0 \tag{20.4}$$

indicates the necessary condition for the emergence of the mother-phase zone within the daughter-phase bulk matrix.

When $A > 0$, the necessary condition (20.3) holds within the range of $0 < \theta < \pi$. The lattice stretches in this interval of angles, and consequently, the daughter-phase zone can form.

Continuing with the assumption of a positive striction constant $A > 0$, let us examine the condition (20.4) for the emergence of the mother-phase zone within the daughter-phase matrix. This condition holds at $\pi < \theta < 2\pi$, where the solid is compressed.

The sign of the binodal typically coincides with that of the striction constant. Hence, the necessary condition for zone formation dictates that:

Case 20.2.1 If the striction constant is positive, $A > 0$ implying $k_0 > 0$, the daughter-phase zone can form at the edge dislocation at $T > T_0$ within the angle range of $0 < \theta < \pi$. Conversely, at $T < T_0$, the mother-phase zone can emerge within the angle range of $\pi < \theta < 2\pi$.

Figure 20.1a schematically illustrates this scenario, showing dislocations with and without zones on the backdrop of the schematic phase diagram with a positive binodal slope, $k_0 > 0$. The orange-shaded areas represent domains occupied by the daughter phase, while the white areas indicate the mother phase.

For the case of a negative binodal slope, similar arguments demonstrate that:

Case 20.2.2 With $k_0 < 0$ $(A < 0)$, above the binodal at $T > T_0$, the daughter-phase zone emerges within the angles $\pi < \theta < 2\pi$, corresponding to a compressed lattice. Below the binodal at $T < T_0$, the mother-phase zone appears within the angles $0 < \theta < \pi$, signifying lattice stretching.

This scenario is depicted in Fig. 20.1b.

The above arguments imply that at any sign of A, the necessary conditions of the zone formation allow the process zone to occur on the phase diagram at both sides of the binodal.

To summarize, *the daughter-phase process zones consistently occur above, while the mother-phase zone resides below the binodal. The binodal inclination solely determines the side of the dislocation where the zone manifests.*

20.3 Rescaling Equations for the Dislocation

This section applies the general rescaling relations (8.2) and (8.3) *for the second- and first-order zones to the dislocation case, presenting a comprehensive list for reference. It also formulates the equations of motion and state in terms of the rescaled variables, as well as the zone's free energy and the friction force acting on the dislocation due to energy dissipation within the zone.*

***

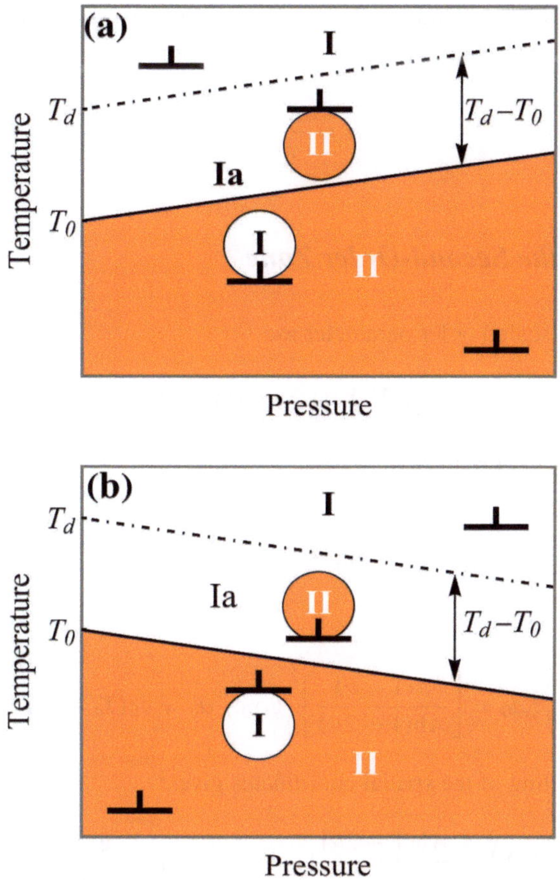

Fig. 20.1 Schematic demonstration of how the transition line slope affects the process zone location at the edge dislocation. The bulk mother phase I (white) and the daughter phase II (shaded) are separated by the binodal line, $T_0(p)$ (the solid line). The thumbnails of dislocations with and without a zone are shown as insets. **a** The case of the positive slope of the binodal $dT_0/dp = k_0 > 0$. The region of existence of the daughter-phase process zone, Ia, is situated at $T_0 \leq T \leq T_d$. Here, the daughter-phase process zone II (shaded) dwells at the dislocation. It is located within the region of the stretched lattice. Below the line $T_0(p)$, the daughter phase occupies the bulk of the crystal. The mother-phase zone dwells at the other side of the dislocation, where the latter compresses the crystal. **b** In the case of the negative slope, $k_0 < 0$, the daughter-phase process zone existence region, Ia, is occupied by the bulk mother phase I. It lies above the binodal, $T_0(p)$. In this domain, the daughter-phase zone II is embedded into the matrix of the mother-phase I. In the region II, the mother-phase zone I is embedded into the bulk daughter phase II localized at the side, where the dislocation compresses the lattice. The insets display dislocations with—and without zones in their vicinity. The dot-dashed line indicated by T_d shows the boundary at which the daughter-phase zone emerges

Using the general rescaling relations from 8.1 with $z = 1$, $f = \sin(\theta)$. For B defined by (5.3) we obtain the following expression:

$$B = \frac{A(1-2\nu)}{\pi(1-\nu)} \tag{20.5}$$

The Case of the Second-Order Zone

We define the rescaled order parameter as:

$$\eta(X, Y, t) = \eta_0 u(x, y, t),$$

which yields the expressions:

$$a = \left[\frac{\pi(1-\nu)}{A\mathfrak{b}(1-2\nu)}\right]^2 g\alpha; \qquad \eta_0 = \frac{A\mathfrak{b}(1-2\nu)}{\pi(1-\nu)\sqrt{g\beta}} \tag{20.6}$$

From these, we derive:

$$a_0 = \left[\frac{\pi(1-\nu)}{A\mathfrak{b}(1-2\nu)}\right]^2 g\alpha_0; \;\; a_\mathrm{d} = a_0(T_\mathrm{d} - T_\mathrm{c}) \tag{20.7}$$

Further rescaling of the spatial coordinates gives:

$$\{X, Y\} = \frac{A\mathfrak{b}(1-2\nu)}{\pi g(1-\nu)}\{x, y\}; \qquad \mathbb{R}_0 = \frac{g\pi(1-\nu)}{A\mathfrak{b}(1-2\nu)} \tag{20.8}$$

Substituting $z = 1$ and B (20.5) into the free energy expressions (5.13) and (5.14) and the dissipation function (5.15), we obtain the rescaled forms:

$$\mathcal{F} = \frac{1}{2\beta}\left[\frac{A\mathfrak{b}(1-2\nu)}{\pi(1-\nu)}\right]^2 \int_\Omega \left\{\frac{1}{2}(\nabla u)^2 + \frac{1}{2}\left(a + \frac{y}{x^2+y^2}\right)u^2 + \frac{1}{4}u^4\right\} dxdy \tag{20.9}$$

$$\mathcal{D} = \frac{\kappa}{2g\beta}\left[\frac{A\mathfrak{b}(1-2\nu)}{\pi(1-\nu)}\right]^2 \int_\Omega \left(\frac{\partial u}{\partial t}\right)^2 dxdy \tag{20.10}$$

Passing to co-moving coordinates, $u(x, y, t) = u(x - vt, y)$, and applying the general formula (3.58), we obtain the expression for the drag force exerted on the dislocation from the zone:

$$f = \frac{\kappa}{g\beta}\left[\frac{A\mathfrak{b}(1-2\nu)}{\pi(1-\nu)}\right]^2 v\int\limits_{\Omega}\left(\frac{\partial u}{\partial x}\right)^2 dxdy \tag{20.11}$$

According to (6.34), this becomes:

$$f = \frac{\kappa}{g\beta}\left[\frac{A\mathfrak{b}(1-2\nu)}{\pi(1-\nu)}\right]^2 v\xi^2\int\limits_{\Omega}\left(\frac{\partial\Psi_{\mathrm{d}}}{\partial x}\right)^2 dxdy$$

Using (6.36) we finally arrive at:

$$f = \frac{\kappa a_{\mathrm{d}}}{g\beta}\left[\frac{A\mathfrak{b}(1-2\nu)}{\pi(1-\nu)}\right]^2\left(1-\frac{a}{a_{\mathrm{d}}}-v^2\right)v\frac{J_2}{J_4}\int\limits_{\Omega}\left(\frac{\partial\Psi_{\mathrm{d}}}{\partial x}\right)^2 dxdy \tag{20.12}$$

valid in the domain $a/a_{\mathrm{d}} \le 1 - v^2 \wedge v \le 1$, where (6.36) is nonzero. Otherwise $f = 0$. The integral in (20.12) corresponds to I_2 (6.45). Thus, the last terms in (20.12) represent the combination $S(V) = J_2 I_2/J_4$ (18.5) which also appears in crack theory. In the case of a crack, $S(V)$ shows weak dependence on velocity v, except near $v = 1$ (see Fig. 18.1). A similar behavior can reasonably be expected for dislocations.

In contrast, the factor $\left(1 - a/a_{\mathrm{d}} - v^2\right) v$ shows strong nonmonotonous velocity dependence: at small v, the friction force is linear, $f \sim (1 - a/a_{\mathrm{d}})\, v$, reaches a maximum, and vanishes at $v = (1 - a/a_{\mathrm{d}})^{1/2}$. This predicts the existence of an anomalous descending branch of the $f = f(v)$ dependence, where an increase in velocity results in a decrease in friction force. Such a branch typically leads to instability, as discussed in Chap. 19 in the context of wedging.

20.3.1 The Case of the First-Order Zone

The constants a and b are straightforwardly obtained:

$$a = \left[\frac{\pi(1-\nu)}{A\mathfrak{b}(1-2\nu)}\right]^2 g\alpha; \quad b = -\sqrt{\frac{g}{\gamma}}\frac{\pi(1-\nu)}{A\mathfrak{b}(1-2\nu)}\beta \tag{20.13}$$

From the above relation one makes sure that the relations (20.7) are also valid in the first-order case. By using $a = a_0(T - T_{\mathrm{c}})$ one defines:

$$a - \frac{3b^2}{16} = a_0(T - T_0) \tag{20.14}$$

From the same general expressions, one obtains the relations of the dimensionless coordinates (x, y) to the original coordinates (X, Y), the dimensionless order parameter $u(x, y)$ to the original one $\eta(X, Y)$, and the expression for critical velocity V_d:

$$(x, y) = \frac{1}{\mathbb{R}_0}(X, Y); \quad \eta(X, Y, t) = \frac{1}{(g\gamma)^{1/4}}\left[\frac{A\mathfrak{b}(1-2\nu)}{\pi(1-\nu)}\right]^{1/2} u(x, y, \tau) \tag{20.15}$$

where the relation (20.8) works in this case as well. Finally,

$$V_d = \frac{2\sqrt{a_d}A\mathfrak{b}(1-2\nu)}{\pi\kappa(1-\nu)}; \quad v = \frac{V}{V_d}; \tag{20.16}$$

The latter relation is valid both for the second and first-order zone. Relations for the rescaled Gordon's velocity remain the same as (12.8) and (12.9) in the case of the crack.

After the rescaling, the equation of motion in the automodel regime (8.6) takes the form:

$$\frac{\partial^2 u}{\partial x^2} + \frac{\partial^2 u}{\partial y^2} + 2\sqrt{a_d}v\frac{\partial u}{\partial x} - \left(a - \frac{y}{x^2 + y^2 + \varepsilon}\right)u + bu^3 - u^5 = 0 \tag{20.17}$$

The (20.17) is central to our simulations and is governed by three dimensionless parameters: a, b, and v defined by (20.13)–(20.16). The potential: $U(x, y) = -y/(x^2 + y^2)$ is equivalent to the following one:

$$U(x, y) = -\frac{\sin(\theta)}{r} \tag{20.18}$$

and we introduce a cutoff parameter $\varepsilon = 10^{-6}$ to prevent singularities during the numerical process. As usual in this book, a_d represents the value of the dimensionless parameter a at which the soft bifurcation of the stationary problem ($v = 0$) occurs. We will derive the numerical value of a_d in the next chapter.

Equation (20.17) is structured optimally for simulations, and it provides the flexibility to explore both the second-order zone at $b < 0$ and the first-order zone at $b > 0$.

One derives equation of state from (20.17) by setting in it $v = 0$:

$$\frac{\partial^2 u}{\partial x^2} + \frac{\partial^2 u}{\partial y^2} = \left(a - \frac{y}{x^2 + y^2 + \varepsilon}\right)u - bu^3 + u^5 = 0 \tag{20.19}$$

To study the daughter-phase process zone embedded into the mother-phase matrix, we used the zero Dirichlet boundary conditions:

$$u_{\partial\Omega} = 0 \tag{20.20}$$

It is clear that (20.17) and (20.19) and the boundary condition (20.20) describe the first-order zone. To characterize the second-order zone one only has to omit the term u^5 and put $b = 1$.

In the case of the mother-phase zone within the bulk daughter phase and the first-order zone, $u_{\partial\Omega}$ is equal to the value (3.28) in the homogeneous daughter phase:

$$u_{\partial\Omega} = \sqrt{\frac{b + \sqrt{b^2 - 4a}}{2}} \tag{20.21}$$

By substituting (20.13)–(20.16) into the free energy expression (4.14), we can express the zone's free energy, $\mathcal{F}$ in terms of the dimensionless variables:

$$\mathcal{F} = \frac{A\mathfrak{b}(1-2\nu)\mathbb{L}}{\pi(1-\nu)}\sqrt{\frac{g}{\gamma}} \times F \tag{20.22}$$

where $\mathbb{L}$ is the length of the dislocation or its segment, and F in (20.22) is the *dimensionless free energy factor* expressed as:

$$F = \int \left\{ \frac{1}{2} (\nabla u)^2 + \frac{1}{2}\left(a - \frac{y}{x^2 + y^2}\right) u^2 - \frac{1}{4} b u^4 + \frac{1}{6} u^6 \right\} dxdy \tag{20.23}$$

We also provide the rescaled expressions for the friction force, f. Using the replacements (20.13)–(20.16) and the definition (3.58) one finds:

$$f = \frac{\kappa A\mathfrak{b}(1-2\nu)\mathbb{L}}{\pi(1-\nu)\sqrt{\gamma g}} V \int \left(\frac{\partial u}{\partial x}\right)^2 dxdy \tag{20.24}$$

Here, f is the dimensional friction force with the dimension [dyn], and $\mathcal{F}$ has the dimension of energy, [erg], in the CGS system.

***

In summary, *once the solution u to the equation of motion or state is determined, the corresponding solution η, zone's free energy $\mathcal{F}$, dissipation function $\mathcal{D}$, and friction force f can be obtained using the relations outlined in the previous section.*

Reference

1. J. Friedel, *Dislocations* (Pergamon Press, Oxford, 1964)

Chapter 21
In Search for the Bifurcation Point

Abstract In this chapter, we extend our general findings formulated previously to the case of an edge dislocation. We present a refined analytical approximation for the bifurcation point of the second-order zone at the edge dislocation, which closely aligns with previous estimates and simulation results obtained in the present section. We also conduct simulations for the second-order zone, enabling us to numerically determine the bifurcation point (Boulbitch and Korzhenevskii, first published here).

21.1 The Auxiliary Equation

In this section, we formulate the auxiliary equation that will be used to identify the bifurcation point in the subsequent analysis.

***

We use a universal approach outlined in Sects. 6.1 and 6.4, which relies on solving the auxiliary equation (6.3, 6.4). In the case of the dislocation, the potential, $U(\mathbf{r})$, is given by:

$$U(\mathbf{r}) = -\frac{\sin(\theta)}{r} \tag{21.1}$$

and the auxiliary equation takes the following form:

$$\begin{cases} \Delta\psi_n - \left(a_n - \frac{\sin\theta}{r}\right)\psi_n = 0 \\ \psi_n(\infty) = 0;\ |\psi_n| < \infty \end{cases} \tag{21.2}$$

Here Δ represents the 2D Laplace operator in the rescaled coordinates:

$$\Delta = \frac{\partial^2}{\partial r^2} + \frac{1}{r}\frac{\partial}{\partial r} + \frac{1}{r^2}\frac{\partial^2}{\partial \theta^2}$$

The auxiliary equation (21.2) takes on the form of the Schrödinger equation, as we have previously discussed in Sect. 6.1.2. Despite our efforts, an exact solution for this equation eludes us.

© The Author(s), under exclusive license to Springer Nature Switzerland AG 2026
A. Boulbitch and A. Korzhenevskii, *Stress-Induced Local Phase Transformations Near Cracks and Dislocations*, Springer Series in Materials Science 356,
https://doi.org/10.1007/978-3-032-00945-6_21

Nabutovskiĭ and Shapiro, pioneers in studying the formation of a superconductive process zone at the dislocation using this formalism, reported a value of $a_\mathrm{d} \approx 0.089$ derived from a variational solution [1]. We propose an alternative variational approach that yields a slightly better estimate for a_d.

21.2 The "Old Quantum Theory" Approximation

This section explores the 'Old Quantum Theory' approach to approximate the bifurcation point.

***

Let us now estimate a_d using the "Old Quantum Theory" approach (see Sect. 6.2).

If $A > 0$, the potential $U(\mathbf{r})$ in reference (21.1) is negative at $0 < \theta < \pi$. It is here that the zone may occur. Thus, averaging the potential over the angle interval $0 < \theta < 2\pi$ can be proceeded as:

$$\overline{U} = -\frac{1}{2\pi}\int_0^{2\pi} \frac{\sin(\theta)}{r}\Theta(\pi - \theta)d\theta,$$

where $\Theta(\pi - \theta)$ is the Heaviside theta-function taking care of integrating only over the angles where the potential $U(\mathbf{r})$ is negative. One finds:

$$\overline{U} = -\frac{1}{\pi r} \tag{21.3}$$

Applying (6.15), the estimate a_d becomes:

$$a_\mathrm{d} \approx 0.101 \tag{21.4}$$

21.3 Variational Solution for the Auxiliary Equation

In this section, we address the auxiliary equation using the variational method.

***

The rescaled auxiliary equation (21.2) resembles the Schrödinger equation featuring a unique potential (21.1). Although an exact solution is unavailable, Nabutovskiĭ and Shapiro reported a value of $a_\mathrm{d} \approx 0.089$, derived from a variational approach [1]. Here, we propose an alternative variational formulation that provides a slightly improved estimate for a_d.

In quantum mechanics, the ground-state eigenfunction minimizes the energy functional.[1] Following Sect. 6.1.2, the auxiliary equation (6.3) transforms into a Schrödinger equation through the substitution $a \to -E$. Consequently, we maximize a rather than minimizing the energy, leading to the definition of an a-functional:

$$a_\mathrm{d} = -\frac{\int \left\{[\nabla\psi(\mathbf{r})]^2 + U(\mathbf{r})\psi^2(\mathbf{r})\right\} r dr d\theta}{\int \psi^2(\mathbf{r}) r dr d\theta} \tag{21.5}$$

The maximum value of a, derived from this functional, represents the ground eigenvalue of the auxiliary equation (21.2). A range of probe functions can be tested, and the one yielding the highest a provides the best approximation. However, these functions must adhere to the correct asymptotic behavior as $r \to \infty$, where the potential $U(r)$ diminishes, reducing the equation to:

$$\Delta\psi - a_\mathrm{d}\psi = 0$$

The yields the asymptotic solution $\psi \sim \exp(-r\sqrt{a_\mathrm{d}})$. To satisfy this behavior, we select probe functions of the form $\psi = f(r,\theta)\exp(-r\sqrt{a_\mathrm{d}})$, where $f(r,\theta)$ represents a new probe function. Specifically, we use:

$$f(r,\theta) = 1 + k_1 r^{k_2}\sin(\theta) \tag{21.6}$$

where k_1 and k_2 are the variational parameters. Substituting $f(r,\theta)$ into the functional (21.5), integrating and maximizing the result with respect to k_2 yields the implicit equation:

$$F(k_1, a_\mathrm{d}) = (4a_\mathrm{d})^{k_2} + k_1^2\left(1 + k_2 + k_2^2\right)\Gamma(2k_2) - \frac{k_1\Gamma(1+k_2)}{2^{k_2} a_\mathrm{d}^{(k_2+1)/2}} = 0$$

where $\Gamma(z)$ is the Gamma-function. Calculating the implicit derivative with respect to k_1, one finds the condition of maximum: $da_\mathrm{d}/dk_1 = 0$, which reduces to $\partial F/k_1 = 0$, simplifying to:

$$k_1\left(1 + k_2 + k_2^2\right)\Gamma(2k_2) = \frac{\Gamma(1+k_2)}{(2\sqrt{a_d})^{k_2+1}}$$

Solving these two equations together yields:

$$k_1 = \frac{(2\sqrt{a_\mathrm{d}})^{k_2-1}\Gamma(1+k_2)}{\left(1 + k_2 + k_2^2\right)\Gamma(2k_2)}; \qquad a_\mathrm{d} = \frac{\Gamma^2(1+k_2)}{4\left(1 + k_2 + k_2^2\right)\Gamma(2k_2)}$$

---

[1] Proof and details one finds in §20 of the book of L. D. Landau and E. M. Lifshitz "Quantum Mechanics" [2].

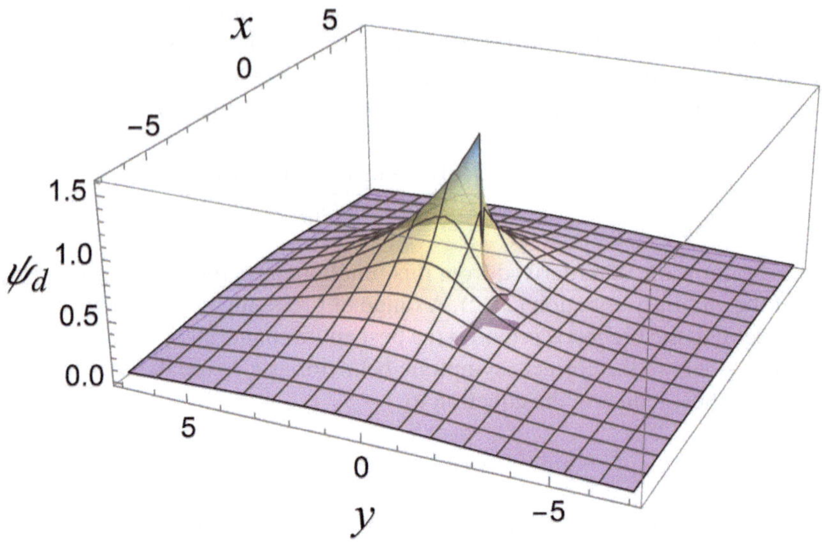

Fig. 21.1 Ground-state eigenfunction (21.8) from the variational approach

Maximizing a_d with respect to k_2 gives $k_2 \approx 0.51933$ corresponding to

$$a_\mathrm{d} \approx 0.11229 \tag{21.7}$$

The corresponding value of k_1 is approximately 0.62. Compared to the earlier estimate by Nabutovskiĭ and Shapiro [1], this value is closer to the true ground-state eigenvalue, although the difference remains small.

With $\sqrt{a_\mathrm{d}} \approx 0.34$, the ground-state eigenfunction is approximated as:

$$\psi_\mathrm{d} \approx \left[1 + 0.62\sqrt{r}\sin(\theta)\right] e^{-0.34r} \tag{21.8}$$

Figure 21.1 displays the spatial distribution of the function (21.8).

While the variational method provides a robust estimate for the eigenvalue, it does not guarantee an accurate eigenfunction [2]. Consequently, we do not calculate process zone properties here and instead transition to simulations for further insights.

References

1. V.M. Nabutovskiĭ, V.Y. Shapiro, Sov. Phys. JETP **48**(3), 480 (1978)
2. L.D. Landau, E.M. Lifshitz, *Quantum Mechanics*, 3 edn., §97 (Butterworth-Heinemann, Burlington, MA, 2003)

Chapter 22
Motionless Dislocation Studied by Simulations: The Phase Diagram

Abstract This chapter addresses the process zone at a stationary edge dislocation, which we investigated through simulations. For the dislocation case, we applied the same methodology as previously described for the crack. We begin with an analysis of the daughter-phase zone embedded in the mother-phase matrix and obtain its dimensionless free energy from simulations. This enables us to establish the local binodal and upper spinodal, which govern the emergence of the first-order process zone. We then turn to the mother-phase zone embedded in the daughter-phase matrix and conclude by constructing a phase diagram that delineates the domain of existence for both process zones against the backdrop of the bulk phase diagram. In line with predictions for the quasi-soft case, our results confirm that the process zone is confined to a specific finite region of the phase diagram (Boulbitch and Korzhenevskii, first published here).

22.1 The Daughter-Phase Zone Embedded into the Bulk Mother Phase

We numerically examined the daughter-phase process zone at a motionless dislocation by simulating (20.19), leveraging the approach previously described for the daughter-phase zone at the crack tip. The smaller spatial extent of the dislocation-induced zone requires fewer computational studies[1] *than those needed for crack tips. Equation (20.19) allows for investigating the second-order zone at* $b < 0$ *and the first-order zone at* $b > 0$*. For* $b < 0$*, we determine the bifurcation point through simulations, comparing results with previous findings. At* $b > 0$*, our study numerically derives the free energy of the zone and identifies the conditions for hard bifurcation.*

***

[1] The term "studies" used in the COMSOL software we explain in detail in Annex E.2.6. Five to seven studies typically suffice for dislocation zones, compared to the approximately twenty studies needed for crack tip zones.

© The Author(s), under exclusive license to Springer Nature Switzerland AG 2026

A. Boulbitch and A. Korzhenevskii, *Stress-Induced Local Phase Transformations Near Cracks and Dislocations*, Springer Series in Materials Science 356,
https://doi.org/10.1007/978-3-032-00945-6_22

22.1.1 *Qualitative Results of the Numerical Analysis of the Daughter-Phase Zone*

In the (b, a) plane, we observed that for large positive values of a, the numerical solution converges to $u = 0$, as in the case of the crack tip zone. As expected, the simulations return nontrivial solutions for specific region of the (b, a) plane. Figure 22.1 showcases such solutions for four values of $b = -0.1$, 10^{-3}, 0.5, and 1.2.

We captured all the solutions shown in Fig. 22.1 close to the point of zone emergence. The solution with $b = -0.1$ corresponds to the second-order zone, while the one with $b = 10^{-3}$ represents the case of the quasi-second-order one. In appearance, they are very similar. Noticeably, the more prominent first-order solutions, achieved with $b = 0.5$ and $b = 1.2$, exhibit a different shape with a broader peak.

Figure 22.2a illustrates the dependency of the Hilbert norm $||u|| = ||u(a)||$ (22.1) for $b = -0.1$ and $b = 0.001$. The former exhibits the typical behavior of a soft bifurcation.

The triangles in Fig. 22.2a represent the variation of $||u||$ at $b = 10^{-3}$. The positive value of b signifies the first-order zone. However, the relatively small value of its norm and especially of its jump in the transition point, δu, allows us to categorize it as a quasi-second-order one. The dependence of $||u||$ on a qualitatively resembles the behavior of the order parameter during a bulk phase transition in a homogeneous solid, as depicted in Fig. 3.2.

In contrast, Fig. 22.2b reveals a distinct behavior. In this case, the discontinuity δu surpasses unity, indicating a hard bifurcation. Additionally, as a decreases, the

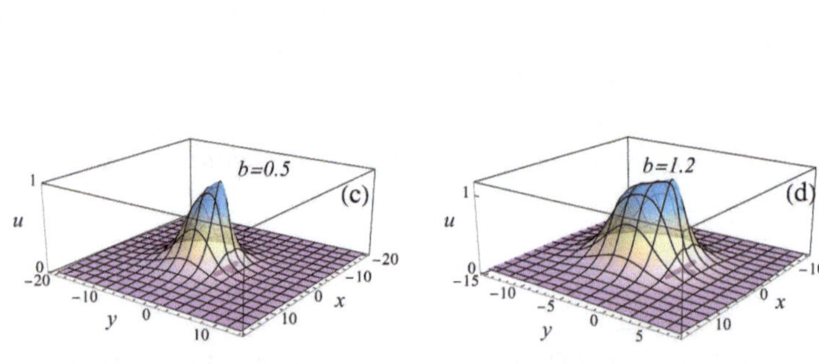

Fig. 22.1 Zone at the edge dislocation obtained from simulations. The results were obtained by simulations with **a** $b = -0.1, a = 0.13$, **b** $b = 10^{-3}, a = 0.13$, **c** $b = 0.5, a = 0.188$ and **d** $b = 1.2$, $a = 0.33$. The position of the dislocation is visible through the transparent surface $u(x, y)$

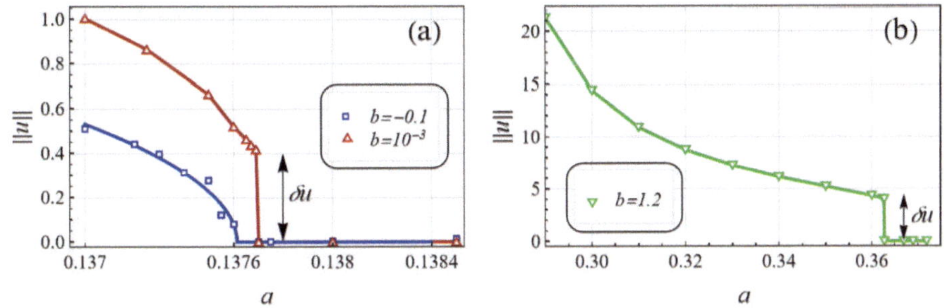

Fig. 22.2 Behavior of the Hilbert norm of the solution (22.1). **a** The run of $||u||$ at $b = -0.1$ (squares) and $b = 0.001$ (triangles). In the former case, the solid line represents the fit with the function (22.2). In the latter one, the solid line serves as a visual guide. **b** The behavior of the norm at $b = 1.2$. The solid line guides the eye. In both cases, $b = 0.001$ and $b = 1.2$, we observed the discontinuity of the norm, δu, in the bifurcation point

order parameter norm increases faster than is typical for bulk phase transformations. One attributes this accelerated increase to the expansion of the zone's area rather than changes in the order parameter amplitude.

22.1.2 *Finding the Bifurcation Point by Simulations*

Let us address finding the bifurcation point by simulations. We simulate the equation of state (20.19). This equation contains both the terms $-bu^3$ and u^5. From the theory of phase transitions (see Sect. 3.3.2), one knows that the soft bifurcation only takes place if $b < 0$. We performed the simulation with $b = -0.1$ and the boundary conditions (20.20). The reader finds details of our approach in Annex II.

Like in the case of the crack, the simulations revealed that at large positive a values (20.19) has only a trivial solution. However, below a certain value $a = a_{\rm d}$, a nontrivial solution emerges.

This section aims to find the value $a_{\rm d}$. A convenient way for its determination is using Hilbert's norm of the solution:

$$||u|| = \left[\int u^2(x, y) dx dy \right]^{1/2} \tag{22.1}$$

We calculated the norm (22.1) during each simulation. Figure 22.3 demonstrated the norm (22.1) obtained by the simulations in the very close vicinity of the bifurcation point shown by blue squares.

By fitting this data to the function:

$$||u|| = \begin{cases} 0, & a > a_{\rm d} \\ k_1, \sqrt{a_{\rm d} - a}\ , & a \le a_{\rm d} \end{cases} \tag{22.2}$$

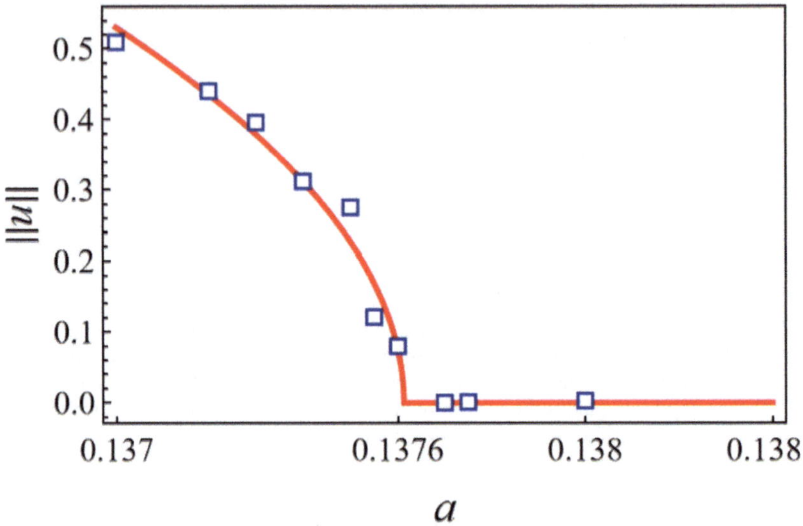

Fig. 22.3 Hilbert norm of the solution. We show here the solution obtained with $b = -0.1$ at various a. Squares show the simulation results and the solid line exhibits their fitting with (22.2)

and using k_1 and a_d as the fitting parameters we obtain the amplitude, $k_1 \approx 21.47$, and

$$a_{\mathrm{d}} \approx 0.138 \tag{22.3}$$

The solid line in Fig. 22.3 showcases the run of the function (22.2) with the fitting parameters given above.

Recalling that $a_{\mathrm{d}} = a_0(T_{\mathrm{d}} - T_{\mathrm{c}})$ and passing to the dimensional parameters with the relation (20.13) one finds the bifurcation temperature:

$$T_{\mathrm{d}} - T_{\mathrm{c}} \approx \frac{a_{\mathrm{d}}}{\alpha_0 g}\left[\frac{A\mathfrak{b}(1-2\nu)}{\pi(1-\nu)}\right]^2 \tag{22.4}$$

Let us juxtapose our estimated value of $a_{\mathrm{d1}} \approx 0.101$ given by the "old quantum theory" approach (21.4) and the value $a_{\mathrm{d2}} \approx 0.089$ found by Nabutovskiĭ and Shapiro [1] with a more precise result of the numeric solution, $a_{\mathrm{d3}} \approx 0.138$. The discrepancy between the former value and the result (22.3) occurs only in the second decimal place, differing by

$$\frac{a_{\mathrm{d3}} - a_{\mathrm{d1}}}{a_{\mathrm{d3}}} \approx 0.3$$

The value a_{d} mentioned above represents the critical point of the soft bifurcation. The value (22.3) is the best of the three above results.

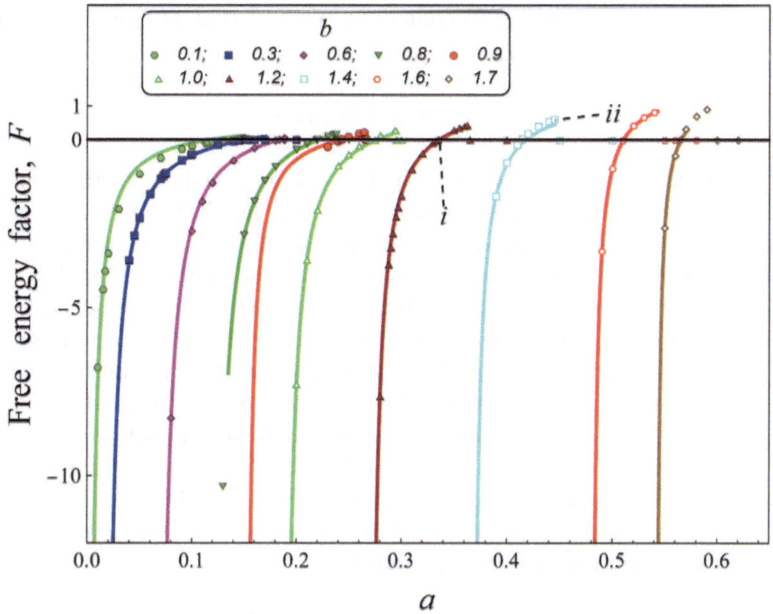

Fig. 22.4 Dependence of the free energy factor F (20.23), on a and b. (i) indicates a point where $F(a, b)$ turns into zero, while (ii) shows the tooth extremity point corresponding to the a_{sp} value of the parameter a. Solid lines show fitting of the corresponding data according to (22.7)

22.1.3 The Zone's Free Energy

The simulation enables one to extract the dependence of the zone dimensionless free energy factor, F (20.23), on parameters a and b. Our findings are showcased in Fig. 22.4 for a comprehensive understanding of these dependencies.

In Fig. 22.4, we present the dimensionless free energy factor, $F(a)$, at various positive values of b, all corresponding to the first-order zone. We refer to the dimensionless free energy factor as *free energy* for shortness throughout this discussion. It is noteworthy that for each b, one finds specific intervals of a values resulting in *negative* free energy.

Note 22.1.1 The latter manifests a fundamental principle: *the equilibrium zone can only exist if it reduces the crystal's free energy.*

As we continue to increase a, the free energy experiences a gradual increase until it reaches the point at which $F = 0$. One such point is indicated by (i) in Fig. 22.4. Beyond this point, the equilibrium zone ceases to exist. In the homogeneous case, one refers to the points in the (b, a) plane at which the free energy equals zero as the binodal (3.32). In the zone case, we indicate this line as the *local binodal*. Collecting the points of the local binodal, we fitted them using the function $a_d(b) \approx a_{d0} + k_1 b^2$, where $a_{d\,0} = 0.138$ is the bifurcation point (22.3) at $b = 0$, and k_1 serves as the fitting parameter yielding the parabola:

$$a_d(b) \approx 0.138 + 0.144 \times b^2 \tag{22.5}$$

With a further increment in a, the zone free energy becomes positive, indicating a metastable state,[2] the free energy curve forming the shape of a tooth with the extremity at $a = a_{sp}$. It is pointed out as (ii) in Fig. 22.4. As soon as a exceeds a_{sp}, the free energy reaches zero value jumpwise. This critical point coincides with the disappearance of the local free energy minimum associated with the metastable state, rendering the zone unsustainable in both stable and metastable states. Beyond this threshold, even the metastable zone becomes impossible. As previously, by analogy with the homogeneous case, we refer to the line in the (b, a)-plane as the *local upper spinodal*. This behavior holds for all b values greater than zero, with the most visible examples occurring within the b range of 0.8–1.7. It is important to note that this behavior is consistent across all cases illustrated in Fig. 22.4. However, in some cases, the tooth size is too small to be discernible at the plot's scale. We have successfully fitted the local upper spinodal using the function:

$$a_{sp}(b) \approx a_{d0} + 0.157 \times b^2 \tag{22.6}$$

For future analysis, we also need an analytical expression for perturbation of the free energy factor F generated by the daughter-phase process zone. Having not found it from theoretical consideration, we turned to fit the simulation data shown by dots in Fig. 22.4. We subjected the choice of the fitting function to the following conditions. First, the function must change its sign as a crosses the bifurcation point, a_d (22.5). Second, it should turn to infinity as a approaches the binodal. That is, it should be inversely proportional to some degree of $a - 3b^2/16$. Finally, it should be simple. By trial and error, we established a function satisfying these conditions and best fitting all the simulation data:

$$F = k\frac{a - a_d(b)}{a - 3b^2/16} \tag{22.7}$$

where k is the fitting parameter.

The fitting by (22.7) is shown by solid lines in Fig. 22.4. Comparing the simulation results shown in Fig. 22.4 by dots with the fitting drawn by solid lines, one can see that the accuracy of the formula appears to be higher with increasing both b and the difference $a - a_d$.

In principle, $k = k(b)$. However, it exhibits some scattering around unity plausibly related to errors in the simulation data. Its mean value, $\langle k \rangle$, and standard deviation, δk, are:

$$\langle k \rangle \approx 1.13; \quad \delta k \approx 0.34$$

---

[2] Our simulations, devoid of thermal noise, enable us to identify and analyze these metastable states.

Generally, the fit by the formula (22.7) is reasonably accurate and appropriate to use for estimates.

Representing a, a_{d}, and $a - 3b^2/16$ according to (20.14) and using the free energy expression (20.22) one transitions from the dimensionless free energy factor (22.7) to the daughter-phase zone's free energy in terms of the physical parameters:

$$\frac{\Delta\mathcal{F}_{\mathrm{d}}}{\mathbb{L}} \approx -1.13\frac{A\mathfrak{b}(1-2\nu)}{\pi(1-\nu)}\sqrt{\frac{g}{\gamma}\frac{T_{\mathrm{d}}-T}{T-T_0}} \tag{22.8}$$

where the temperature belongs to the interval $T_0 \leq T \leq T_{\mathrm{d}}$. Let us note that $\Delta\mathcal{F}_{\mathrm{d}}$ diverges as $T \rightarrow T_0$, as one can guess, inspecting the behavior of the simulation results in Fig. 22.4.

***

In summary, *simulations for the second-order process zone ($b < 0$) yield a bifurcation point of $a_{\mathrm{d}} \approx 0.138$, closely aligning with the approximation from the 'Old Quantum Theory' method. Nabutovskiĭ and Shapiro's result is slightly less precise, but all three values are in close agreement. The importance of this numerical value (22.3) lies in its relevance across numerous subsequent relationships.*

For the first-order zone ($b > 0$), the simulations provide the free energy of the zones across various parameter values, allowing for an accurate fit to an approximate analytical function (22.8). Additionally, the simulations determine the local upper spinodal, denoted $a_{\mathrm{sp}}(b)$, above which ($a > a_{\mathrm{sp}}$) the daughter-phase zone ceases to exist, and the local spinodal $a_{\mathrm{d}} = a_{\mathrm{d}}(b)$, below which ($a < a_{\mathrm{d}}$) the daughter-phase zone represents an equilibrium state with negative free energy.

22.2 The Edge Dislocation in the Bulk Daughter Phase

We conduct a numerical investigation of the order parameter distribution around a stationary edge dislocation within the bulk daughter phase of a crystal, focusing on the first-order zone. The simulations reveal the formation of a mother-phase zone encapsulated inside the bulk daughter phase, a phenomenon that becomes increasingly pronounced for small, positive values of b. This section also presents an approximate analytical solution for the order parameter far from the dislocation. This solution facilitates deriving a simple expression for the free energy attributed to the order parameter field generated by the dislocation.[3]

***

[3] These results are presented here for the first time.

22.2.1 The Order Parameter Distribution

To study the mother-phase zone embedded into the daughter-phase matrix in the vicinity of the motionless dislocation, we simulated (20.19) using the boundary condition (20.21). The qualitative outcomes from the simulations elucidate the order parameter distribution in proximity to the edge dislocation. They are shown in Fig. 22.5.

The numerical results align with those observed for the mother-phase zone at a crack tip (see Chap. 16), indicating a similar behavior pattern across different forms of defects within the crystal. Far from the binodal, the order parameter's distribution forms a shallow dip configuration with a shape resembling a beak. It transitions to a dip achieving the level $u = 0$ as it approaches the binodal (as depicted in panels c and d). After the dip has achieved $u = 0$, further approach to the binodal (panels e and f) results in its lateral growth.

Differing from the crack tip zone configuration, the distribution at the edge dislocation features a distinctive peak on one side and a dip on the other. This occurs due to two domains at the dislocation: at $0 < \varphi < \pi$, the potential $U(\mathbf{r}) = \sin(\varphi)/r$ is positive, leading to the formation of a dip,[4] while at $\pi < \varphi < 2\pi$, it becomes negative, resulting in the emergence of a peak.

Note 22.2.1 Like the mother-phase zone at the crack tip, the order parameter distribution at the dislocation remains inhomogeneous across all values of the parameters a and b.

Let us recall that in the bulk mother phase, the local binodal line (22.5) separates the phase diagram domain where the daughter phase zone takes place from that where the dislocation is bare. It, thus, separates two dislocation states that are qualitatively different from one another, the line (22.5) representing the bifurcation boundary.

Unlike the daughter-phase zone case, the property described in the Note 22.2.1 makes it impossible to define the bifurcation boundary separating qualitatively different states. Indeed, some order parameter inhomogeneity persists everywhere below the binodal of the bulk phase diagram.

In this situation, we introduce conditional boundaries. Let us call a *developed mother phase zone* the state where the minimal value of the order parameter, u_{min}, is below a few percent of its value in infinity, $u_{\partial\Omega} \equiv u_{\text{d}}$, or smaller: $|u_{\text{min}}|/u_{\text{d}} \lesssim 10^{-2}$. It corresponds to the distributions shown in Fig. 22.5c–f. The simulation shows that this state lies below the bulk binodal. It is bounded from below by the line fitted by the parabola:

$$a \approx a_{\text{m}} + 0.23b^2 \tag{22.9}$$

with $a_{\text{m}} \approx -0.037$.

---

[4] Recall that the condition for the formation of the mother-phase zone embedded in the daughter-phase matrix is $U(\mathbf{r}) > 0$.

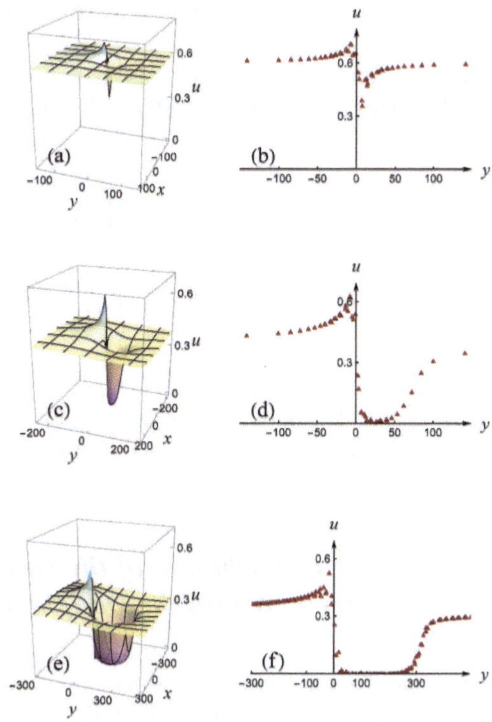

Fig. 22.5 Dimensionless order parameter u: lateral profile near the edge dislocation below the binodal. Panels **a, c, e** depict 3D distributions while **b, d, f** show the cross-sections of these distributions along the plane $x = 0$. Successive images are captured at $b = 0.1$ close to the binodal. At this b value, the binodal is at $a \approx 0.0019$. (**a, b**) Show the distribution at $a = -0.1$, (**c, d**) at $a = -0.01$, and (**e, f**) at $a = -0.001$

The value $a_{\rm m} = -0.037$ yields the minimal temperature, $T_{\rm m}$, below which the developed mother-phase zone gives place to the less developed one at $b = 0$. Passing to the dimensional variables, one finds:

$$T_{\rm m} = T_{\rm c} - 10^{-4}\frac{1}{\alpha_0 g}\left[\frac{6A\mathfrak{b}(1-2\nu)}{1-\nu}\right]^2 \tag{22.10}$$

22.2.2 *The Daughter Phase at Large Distances from the Dislocation*

In the bulk daughter phase, the equation of state in the presence of the dislocation takes the form (20.19).[5] For $r \gg 1$, where the potential $U(\mathbf{r}) = \sin(\varphi)/r$ is small, the solution can be sought by expanding it in terms of $U(\mathbf{r})$. Considering the expansion up to $U^2(\mathbf{r})$, we obtain:

[5] Here, we focus on the first-order transition; the simpler second-order case can be treated analogously.

$$u(\mathbf{r}) \approx k_1 U(\mathbf{r}) + k_2 U^2(\mathbf{r}) + \cdots,$$

where k_1 and k_2 are coefficients determined by substituting this expansion into (20.19). Through this process, the solution for $u(\mathbf{r})$ approximates as follows:

$$u(\mathbf{r}) \approx \sqrt{\frac{b+\sqrt{b^2-4a}}{2}} \left\{1 - \left[b\left(b+\sqrt{b^2-4a}\right) + 4a\right] \frac{\sin(\varphi)}{4a(b^2-4a)r} + \right.$$

$$\left. + \left[2a(5b^2-12a) - b^5(b^2-8a)\sqrt{b^2-4a}\right] \left[\frac{\sin(\varphi)}{4a(b^2-4a)r}\right]^2 \right\} + O(r^{-3}) \tag{22.11}$$

Note 22.2.2 The derived solution (22.11) exhibits a long-range character, extending throughout the entire crystal or, at minimum, across the interdislocation distance $\mathbb{L}_{\mathrm{disl}}$.

22.2.3 The Contribution of the Order Parameter to the Free Energy Within the Daughter Phase

The *density of the dimensionless free energy factor* (20.23)—hereafter referred to as the *factor density* for shortness—in the bulk daughter phase, in the presence of the dislocation, takes the form:

$$\Phi = \frac{1}{2}(\nabla u)^2 + \frac{1}{2}\left[a - \frac{\sin(\varphi)}{r}\right]u^2 - \frac{b}{4}u^4 + \frac{1}{6}u^6 \tag{22.12}$$

By inserting here the expansion (22.11), we obtain:

$$\Phi \approx \Phi_0 + \Delta\Phi \tag{22.13}$$

where Φ_0, representing the homogeneous daughter phase's factor density, is given by:

$$\Phi_0 = -\frac{(b^2-4a)^{3/2} + b^3 - 6ab}{24} \tag{22.14}$$

We focus on the perturbation $\Delta\Phi$, arising from the dislocation's influence. By substituting (22.11) into (22.12) after some simplifications one finds:

$$\Delta\Phi \approx \frac{1}{4}\left(\sqrt{b^2-4a} + b\right)\frac{\sin(\varphi)}{r} - \frac{1}{4\sqrt{b^2-4a}}\frac{\sin^2(\varphi)}{r^2} + O(r^{-3}) \tag{22.15}$$

Let us find the free energy factor (20.23), $F = F_0 + \Delta F_{\mathrm{m}}$, where F_0 describes the free energy factor of the homogeneous crystal in the daughter phase equal to Φ_0

multiplied by the crystal dimensionless volume, and ΔF_{m} is its far-field perturbation by the mother-phase zone. We calculate ΔF_{m} by integrating $\Delta\Phi$ over the entire crystal, expressed in terms of the rescaled coordinates:

$$\Delta F_{\mathrm{m}} = \int_0^{2\pi} d\varphi \int_{\mathfrak{b}}^{\mathbb{R}} \Delta\Phi \times r dr$$

Integration over φ eliminates the first term in (22.15). The integration of the second term over r reveals a logarithmic divergence as $r \to 0$ and $r \to \infty$. To address this, cutoffs are introduced at $r = \mathfrak{b}$, corresponding to the Burgers vector, and at $r = \mathbb{R}$, where $\mathbb{R}$ is either on the order of $\mathbb{R}_{\mathrm{cr}}$, representing the crystal's size, or $\mathbb{L}_{\mathrm{disl}}$, which estimates the interdislocation distance:

$$\Delta F_{\mathrm{m}} \approx -\frac{\pi}{4\sqrt{b^2 - 4a}} \ln\left(\frac{\mathbb{R}}{\mathfrak{b}}\right) \tag{22.16}$$

Note 22.2.3 During the derivation of the dimensionless free energy factor (22.16), we assumed the contribution of the configuration of the order parameter close to the dislocation to be small and disregarded it. We justified this assumption numerically in Annex III. It demonstrates that, indeed, the primary influence on the free energy perturbation stems from (22.16).

Transitioning as usual to the physical coordinates, one finds the perturbation of the free energy of the dislocation, $\Delta\mathcal{F}_{\mathrm{m}}$, given rise by the mother-phase zone:

$$\frac{\Delta\mathcal{F}_{\mathrm{m}}}{\mathbb{L}} \approx -\frac{1}{\pi\sqrt{\beta^2 - 4\alpha\gamma}} \left[\frac{A\mathfrak{b}(1-2\nu)}{2(1-\nu)}\right]^2 \ln\left(\frac{\mathbb{R}}{\mathfrak{b}}\right) \tag{22.17}$$

The maximum value of $\Delta\mathcal{F}_{\mathrm{m}}$, reached at the binodal, is:

$$\frac{\Delta\mathcal{F}_{\mathrm{m}}^{(\max)}}{\mathbb{L}} \approx -\frac{1}{2\pi|\beta|} \left[\frac{A\mathfrak{b}(1-2\nu)}{1-\nu}\right]^2 \ln\left(\frac{\mathbb{R}}{\mathfrak{b}}\right) \tag{22.18}$$

Although this expression approaches infinity at the upper spinodal ($\alpha = \beta^2/4\gamma$), its valid domain of application is strictly below the binodal ($\alpha < 3\beta^2/16\gamma$), making the upper spinodal unattainable. This maximum value diverges as β^{-1} when $\beta \to 0$, indicating significant implications for the free energy as the parameter β decreases.

***

In conclusion, *the numerical solution allows us to determine the formation conditions of the mother-phase zone, identifying the temperature T_{m} below which the developed mother-phase zone ceases to exist. The order parameter distribution within the bulk daughter phase significantly differs from that in the bulk mother phase, extending throughout the entire crystal rather than being confined to the vicinity of* *the dislocation. The contribution of the far-field solution to the free energy is substantially greater than that associated with the near-field solution describing the mother-phase zone.*

22.3 Phase Diagram of the Process Zone at the Edge Dislocation

The relations obtained during the analysis of the daughter- and mother-phase zones allow us to construct the phase diagram. It delineates the domains where the process zone manifests, layered upon the bulk phase diagram.

***

Figure 22.6 presents the phase diagram, delineating both daughter-phase and mother-phase zones within the (b, a) plane. Only bulk and local binodals are depicted in Fig. 22.6, omitting spinodals from the image to maintain clarity. Simulation-derived phase boundary points are marked with dots.

In our previous discussion (see Sect. 3.3.2), the bulk phase diagram is generally divided into two major domains, with the bulk mother phase residing above the bulk binodal, and the bulk daughter phase below. The relation:

$$a = \begin{cases} 0, & b < 0 \\ \frac{3b^2}{16}, & b \geq 0 \end{cases} \tag{22.19}$$

defines the bulk binodal. Equation (22.19) corresponds to the solid blue line in Fig. 22.6.

Regions I and II lie above the bulk binodal line (22.19), indicating the bulk mother phase. Below the line (22.19), the bulk daughter phase belongs to Regions III and IV.

In Region I, the dislocation remains zone-free within the bulk mother phase. The local binodal, marked by brown squares and dot-dashed brown line, signifies the threshold at which a daughter-phase zone emerges within the mother-phase matrix in the vicinity of the dislocation. The parabolic equation (22.5) approximates this line. It intersects the bulk binodal (22.19) at the point with $b = b_*^{(1)}$:

$$b_*^{(1)} \approx 1.698 \tag{22.20}$$

The point of intersection is denoted by a black asterisk in Fig. 22.6. The Region II is highlighted by hatching. Here, the daughter-phase zone embedded into the mother-phase matrix is stable.

Region III, depicted in light red, points out the bulk daughter phase, in which the dislocation generates the well-developed mother-phase zone in its vicinity. Red triangles obtained from the simulations show the boundary III–IV.

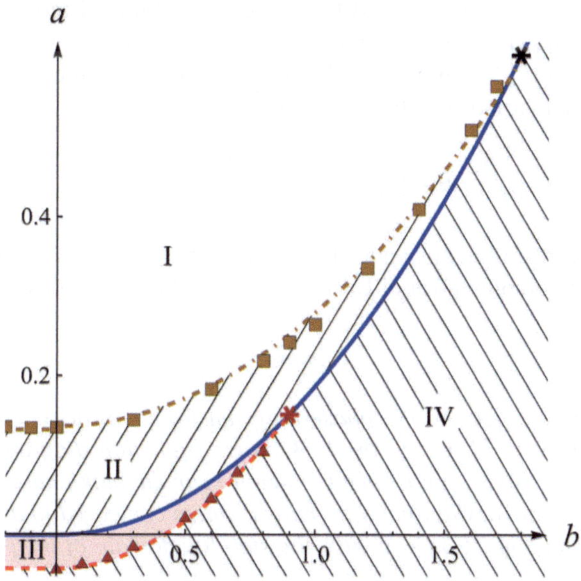

Fig. 22.6 Existence regions of mother- and daughter-phase zones on the background of the bulk phase diagram. In Region I, no zone forms at the dislocation. The emergence of daughter-phase zone is delineated by the dot-dashed line (squares). This zone occupies Region II. The bulk binodal is depicted as a solid line. The line of the daughter-phase zone emergence intersects the bulk binodal at a point indicated by the asterisk. Region III features a prominent mother-phase zone, where the order of magnitude of $|u|/u_d$ does not exceed 10^{-2}. Region IV is characterized by the beak-shaped order parameter distribution, suggesting a nascent mother-phase zone

Region IV's boundary with Region III, shown by the red triangles, is approximated by (22.9). The line (22.9) crosses the bulk binodal at

$$b = b_*^{(2)} \approx 0.86 \tag{22.21}$$

indicated by a red asterisk.

In region IV (hatched), the order parameter distribution exhibits the shape of a beak (as shown in Fig. 22.5a, b). We categorize this region as the one with a *nucleating mother-phase zone*.

Our analytical theory and simulations indicate that the daughter-phase and mother-phase process zones are confined to finite domains within the phase diagram at positive b. Specifically, the daughter-phase process zone only manifests at $b \leq b_*^{(1)}$. At $b > b_*^{(1)}$, the bulk daughter phase arises within the whole crystal before the daughter-phase zone can form. Analogously and for the same reason, the mother-phase process zone only shows up at $b \leq b_*^{(2)}$.

Let us observe that both $b_*^{(1)}$ and $b_*^{(2)}$ are close to unity. The bifurcation can be regarded as a manifested hard one in this range of values. Correspondingly, the physical (dimensional) $\beta_*^{(1,2)}$ can occur rather large in their absolute values. Converting to the dimensional variables with (20.13) one obtains:

$$\beta_*^{(1,2)} \approx -\{0.5, 0.3\} \times \frac{6A\mathfrak{b}(1-2\nu)}{1-\nu}\sqrt{\frac{g}{\gamma}}; \tag{22.22}$$

Note 22.3.1 It remains an open question whether these findings would persist when applying a more complex form of Landau's potential than that described in (3.8).

***

In summary, *we derived the phase diagram for the process zone at an edge dislocation and overlaid it with the bulk phase transition diagram. As expected, the process zone exists within a specific region near the bulk phase transition line, with the local binodals exhibiting different steepness compared to the bulk binodal, leading to intersections. Consequently, the stability domain of the process zones is bounded at* $b > 0$.

Reference

1. V.M. Nabutovskiĭ, V.Y. Shapiro, Sov. Phys. JETP **48**(3), 480 (1978)

Chapter 23
Manifested First-Order Zone at the Dislocation

Abstract The study of process zones of a manifested first order holds significant importance since manifested first-order transitions are more prevalent in nature when viewed objectively. The order parameter undergoes a substantial jump at the binodal in such transitions, often reaching values close to the saturation point. Consequently, atomic displacements within this context can become comparable to the elementary lattice cell's size. The latter determines specific properties of such transitions. This exploration sheds light on the manifested first-order process zone's unique characteristics and implications (Boulbitch and Korzhenevskii, first presented here).

In nature, first-order transitions are much more common than second-order ones, or the so-called first-order transitions close to second-order [1].

A manifested first-order transition is characterized by a substantial discontinuity in the transition point. In our approach, it occurs when b is about one or more prominent. In such cases, the process zone exhibits distinct first-order characteristics.

23.1 The Zone at $b \gtrsim 1$

Based on the simulation results, we qualitatively characterize the manifested process zone at an edge dislocation, focusing on the details of the order parameter distribution.

***

23.1.1 *Qualitative Study*

When b is about 1 or larger, the system exhibits a manifested first-order process zone, characterized by unique behavior. Let us immerse ourselves in the specifics of this phenomenon.

© The Author(s), under exclusive license to Springer Nature Switzerland AG 2026

A. Boulbitch and A. Korzhenevskii, *Stress-Induced Local Phase Transformations Near Cracks and Dislocations*, Springer Series in Materials Science 356,
https://doi.org/10.1007/978-3-032-00945-6_23

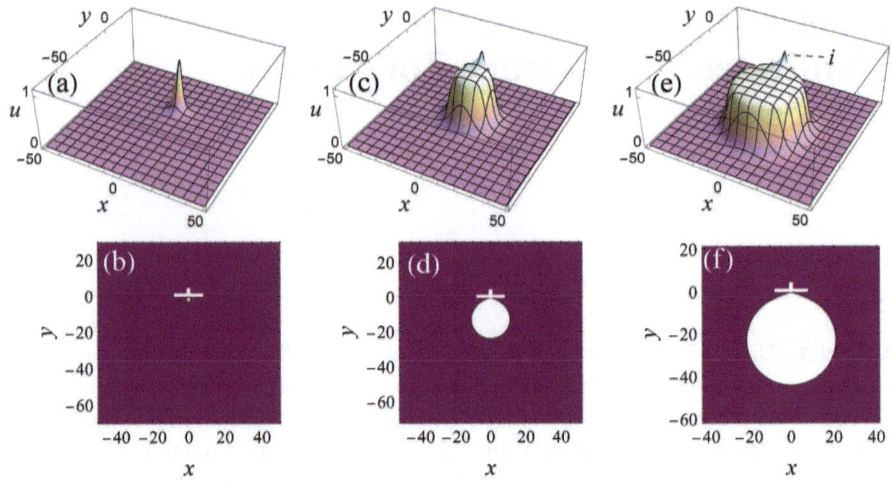

Fig. 23.1 Numerical solutions of (20.19). **a**, **c** and **e** Show the 3D images of the solutions, while **b**, **d**, and **f** display the contours of the phase interface. The solutions are obtained by simulation with $b = 1$ and $a = 0.2945$ (**a, b**), $a = 0.21$ (**c, d**) and $a = 0.2$ (**e, f**). The dislocation position at $x = y = 0$ is invisible in (**a**), (**c**), and (**e**) but shown in white in (**b**), (**d**), and (**f**). In images (**b**), (**d**), and (**f**), the yellow color indicates the region within the zone (the daughter phase), while the purple one shows that outside (the mother phase). At the panel (**e**), the topknot is indicated by i

Figure 23.1 showcases three solutions of equation of state (20.19) obtained at $v = 0$ and $b = 1$, with the dislocation positioned at $x = y = 0$. The dislocation's presence is concealed in Fig. 23.1a, c, e but is highlighted in white in Fig. 23.1b, d, f. Figure 23.1a, c, e display a 3D representation, while Fig. 23.1b, d, f show the zone interface contour.

23.1.2 Details of the Image

The solutions shown in Fig. 23.1a, b are obtained very close to the upper spinodal (at $a = 0.2945$, $b = 1$). Notably, at $a = 0.2945$, the zone's lateral size is relatively small.

As a decreases, we observe a slight increase in the solution amplitude and a significant enlargement of its lateral dimensions. Figure 23.1e displays the distribution $u(x, y)$ at $a = 0.2$, which is close to the bulk crystal binodal. The latter has the value $a_{\text{bin}} = 3/16 \approx 0.1875$ at $b = 1$. Figure 23.1f shows the corresponding zone boundary. For comparison, all images are presented at the same spatial scale. The phase boundary is approximately circular, with its dimensionless radius, R, increasing approximately twentyfold from $R \approx 1$ at $a = 0.2945$ to $R \approx 20$ at $a = 0.2$. Therefore, decreasing a results in a significant lateral expansion of the zone within the manifested first-order zone regime.

Figure 23.2 provides another insight into the evolution of the process zone as a varies. This figure presents the cross-section of two-dimensional solutions along y at $x = 0$, obtained at $b = 1$ with different a values, ranging from $a = 0.2945$ to $a = 0.2$. The red dislocation symbol marks its location at $x = y = 0$.

23.1.3 Topknot

In this cross-section, one can see a prominent, narrow peak near the origin, referred to as the *topknot*. It is indicated by i in Figs. 23.1e and 23.2a.

The Topknot Origin.[1] The topknot's formation is attributed to the potential $U(r)$ behavior. Indeed, as r approaches the dislocation, $U(r)$ tends toward negative infinity, causing an increase in the order parameter u in this region. However, at $r < 1$, the first term $\Delta u \sim u/r^2$ of (20.17) exceeds its third term, $\sim u/r$. Therefore, the term Δu dominates in this region, limiting the order parameter growth. Substituting the estimate $\Delta u \sim u/r^2$, $\sin(\varphi)u/r \sim u/r$, and $r \sim 1$ into (20.19), one finds the following estimate for the topknot height:

$$u_{\text{topknot}}^2 \approx \frac{1}{2}\left(b + \sqrt{b^2 - 4a + 8}\right)$$

Assuming $a = 0.2$ and $b = 1$ corresponding to the data shown in Fig. 23.2a in red, one gets $u_{\text{topknot}} \approx 1.4$. This estimate agrees with the result shown in Fig. 23.2a.

There are arguments suggesting that the topknot might originate from limitations inherent in Landau's theory.

On the one hand, the relation $a = a_0(T - T_c)$ represents the first term of a series expansion of the function $a = a(T)$ that is valid for small values of $T - T_c$. Substituting $a + U(r)$ (20.1) in place of a can be interpreted as introducing an effective Curie temperature, $T_c^{(\text{eff})}$, such that $T_c^{(\text{eff})} = T_c - U(r)/a_0$. In the vicinity of the dislocation, $T_c^{(\text{eff})} \to \infty$, making the difference $T - T_c^{(\text{eff})}$ significant. As a result, Landau's theory may no longer be valid in this regime. On the other hand, as demonstrated in the earlier estimate in this section, the surface tension $\sim\Delta u$ effectively reduces the difference $T - T_c^{(\text{eff})}$, thereby restoring the problem to the domain where Landau's theory is still applicable. This restoration is further justified by the agreement between the estimates and numerical results.

Further, consider for a moment the case of displacive-type transitions. Here, η corresponds to an atomic displacement, which cannot exceed the size of the crystal lattice. Classical Landau's theory includes no mechanism to enforce such a limitation, relying instead on the assumption that η remains small. However, in certain cases, η values within the topknot may violate this assumption. Similarly, in the case of order-disorder transitions, η is associated with a probability that cannot exceed unity.

[1] This section may be skipped during a first reading.

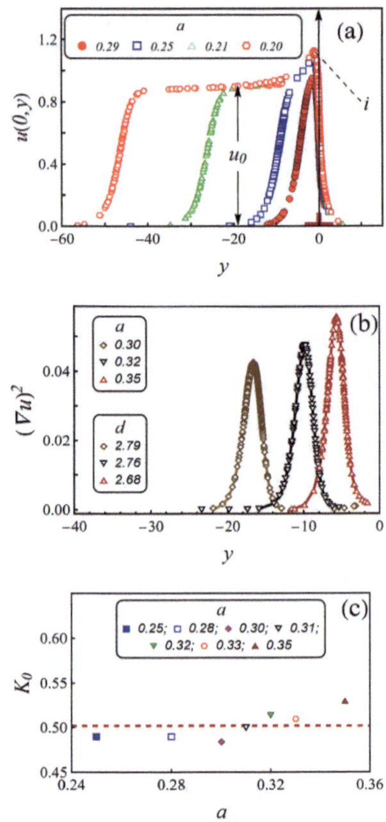

Fig. 23.2 Cross-sectional analysis of numerical solutions. **a** The cross-section of the 3D solutions, $u(x, y)$ along y at $x = 0$ obtained with $b = 1$ at various values of a. The topknot of one of the distributions is indicated by i. **b** The cross-sections of $(\nabla u)^2$ along the same line show the structure of the phase interface. The dots show the simulation results, while the solid lines represent their interpolations. The insets show the values of parameter a and wall width, d. The latter is determined as the half-width of each peak. **c** The values of the integrals K_0 (23.1) for the peaks shown in (**b**)

It is also worth noting that direct comparisons have shown the topknot significantly contributes to the free energy of the dislocation. Throughout this part, all the dislocation's free energy calculations are based on simulations that include the topknot, reflecting its influence on the solutions.

***

In conclusion, *there remains some uncertainty regarding the existence of the topknot. Nevertheless, we are inclined to believe that it exists in many cases.*

23.1.4 Plateau and $(\nabla u)^2$

A plateau forms beyond the topknot when a becomes somewhat smaller than its local spinodal value. In this book, we refer to its top surface as the *roof*. The roof is shallow, with a slight gradient ($\partial u(0, y)/\partial y \sim 10^{-3}$), and its height u_0 is close to unity and remains almost independent of a. We have summarized u_0 values in Table 23.1.

Table 23.1 u_0: roof height in the middle of the process zone

a	0.20	0.21
u_0	0.906	0.912
u_d	0.85	0.84

In Table 23.1, we list the heights of the roof, u_0, for two a values, 0.20 and 0.21. These values are compared to the order parameter value, u_d, in the bulk daughter phase of the homogeneous crystal at $b = 1$. Despite their slight disparity, u_0 and u_d are closely aligned and close to unity.

Note 23.1.1 For numerical estimates of our predictions concerning the manifested first-order zone, one can safely use $u_0 \approx 1$.

This observation bears similarity to the process zone behavior near a crack tip, specifically, the roof structures seen in both cases (Figs. 13.1 and 23.2).

Figure 23.2b demonstrates the dependence of the square of the order parameter gradient, $(\nabla u)^2$, on y at $x = 0$. These data points correspond to $b = 1.2$ for various a values. $(\nabla u)^2$ is significantly different from zero only at the zone's interface, where the order parameter experiences a rapid change. Multiple peaks visible in Fig. 23.2b correspond to phase fronts in the vicinity of the zone's boundary point, diametrically opposite to the dislocation. They differ by the values of a. We interpolated the data obtained from the simulations. Analyzing these interpolation curves allows us to determine the phase interface's dimensionless thickness d. One can see that $d \approx 3$ (see the inset in Fig. 23.2b). Figure 23.2c shows the values of the expression:

$$K_0 = \int_{-\infty}^{\infty} \left(\frac{\partial u}{\partial n}\right)^2 dn \tag{23.1}$$

obtained by way of the integration of the interpolation curves. Here, $\partial u/\partial n$ is the derivative along the normal to the boundary, and the integration runs along the same normal. The mean value of all these integrals is

$$\langle K_0 \rangle \approx 0.502 \tag{23.2}$$

We will use it in what follows.

***

In conclusion, *the manifested first-order process zone exhibits intricate behaviors, particularly when b is near or exceeds* unity. *The zone size varies with changes in the parameter a. The formation of a topknot near the dislocation and the subsequent development of a shallow roof are noteworthy characteristics. This observation bears similarities to the behavior of process zones near crack tips. The zones interface thickness practically does not vary with the variation of the parameter a.*

23.2 Estimating the Size of the Motionless Process Zone

This section analytically estimates the dimensionless size, R_d, of the manifested first-order process zone associated with the stationary edge dislocation.

***

Let us proceed with estimating the parameter R_d. Assuming that $R_d \gg 1$, we can neglect the surface tension term in (8.37). In dimensionless terms, this equation for the first-order case simplifies to

$$a - \sin(\theta)/R = 3b^2/16$$

Solving this equation yields the following result:

$$R(\theta) = 2R_d \sin\theta; \quad R_d \approx \frac{8}{16a - 3b^2} \tag{23.3}$$

Figure 23.3 illustrates the contour described by the first equation (23.3). It represents a regular circle with the radius R_d. Applying the same transformation as in Sect. 14.2.1, we get a simple expression for the zone radius in terms of the temperature as follows:

$$R_d \approx \frac{1}{2a_0\,(T - T_0)} \tag{23.4}$$

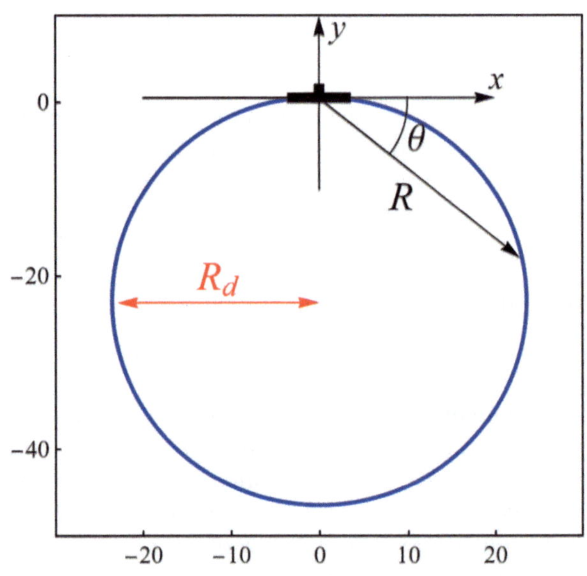

Fig. 23.3 Boundary near a stationary edge dislocation for a manifested first-order zone. The zone contour can be accurately approximated by the expression (23.3), where R_d is the dimensionless zone radius. It corresponds to Fig. 23.1f, obtained from a simulation with $a = 0.2$, $b = 1$, and $v = 0$

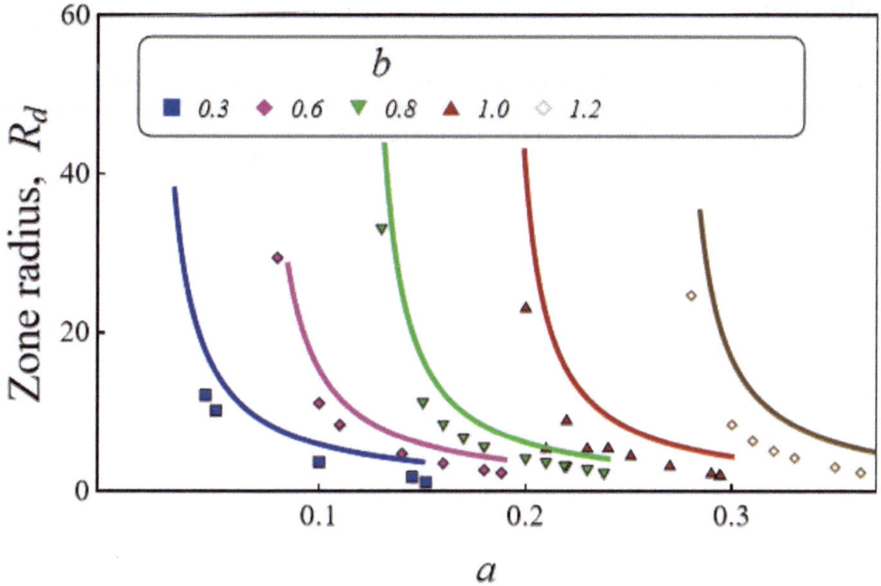

Fig. 23.4 Zone radius. The zone radius, R_d, as the function of a at various b values. Dots show the results of the simulations, while the solid lines demonstrate the run of the curves (23.3)

According to the rescaling formulas (20.8) and (23.3) one obtains the physical, dimensional zone size $\mathbb{R}_\text{d} = \mathbb{R}_0 \times R_\text{d}$:

$$\mathbb{R}_\text{d} = \frac{A\mathfrak{b}(1 - 2\nu)}{2\pi\alpha_0(T - T_0)(1 - \nu)} \tag{23.5}$$

Figure 23.4 presents R_d as the function of a at various b values. The dots represent the results obtained from simulations, while the solid lines depict the curves defined by (23.3). A comparison between the two sets of data demonstrates that the formulas provided by (23.3) offer an acceptable approximation of the corresponding values.

These estimates provide us with accurate analytical expressions for the zone's size in the case of the manifested first-order zone.

***

In summary, *for a large process zone ($R_\text{d} \gg 1$), the gradientless approximation holds. This approximation describes a circular zone, with its shape and size aligning well with simulation results.*

Reference

1. E.Y. Tonkov, *High Pressure Phase Transformations: A Handbook*, vols. 1–3 (Gordon and Breach, Amsterdam, 1992); E.Y. Tonkov, E.G. Ponyatovsky, *Phase Transformations of Elements Under High Pressure*, vol. 4 (CRC Press, Boca Raton, FL, 2005)

Chapter 24
Process Zone at the Propagating Dislocation

Abstract This chapter examines a dislocation moving at constant velocity while carrying an associated manifested process zone. We focus on simulations to investigate changes in the zone's configuration, followed by an analysis of the evolving structure of the order parameter distribution. Next, we analytically estimate key geometrical parameters of the zone, which serve as a basis for calculating the friction force exerted on the dislocation (Boulbitch and Korzhenevskii, first presented here).

24.1 Configuration of the Propagating Process Zone

This section examines the configuration of the propagating process zone as a function of its velocity.

***

24.1.1 Simulation Results

Let us address the dislocation-zone complex, the dislocation propagating along its sliding plane coinciding with the x-axis. Its dimensionless velocity is denoted as v. Figure 24.1 compares the process zones of a stationary dislocation (a) to those of dislocations propagating at velocities $v = 0.05$ (b), 0.2 (c), 0.4 (d), 0.6 (e), and 0.642 (f).

The process zone vanishes at a critical velocity, denoted as v_{cr}, approximately equal to 0.6425, as we revealed from the simulations.

Note 24.1.1 Notably, the upper portion of the distribution, which we refer to as the 'roof,' forms a plateau-like structure, as indicated by arrow (*i*) in Fig. 24.1.

Figure 24.2 illustrates the phase boundaries of these zones. We observe that while the stationary zone exhibits a *lopsided melon-like* shape (Fig. 24.2a), the configuration becomes deformed, resembling a *stretched-out melon* shape, as the dislocation

© The Author(s), under exclusive license to Springer Nature Switzerland AG 2026

A. Boulbitch and A. Korzhenevskii, *Stress-Induced Local Phase Transformations Near Cracks and Dislocations*, Springer Series in Materials Science 356,
https://doi.org/10.1007/978-3-032-00945-6_24

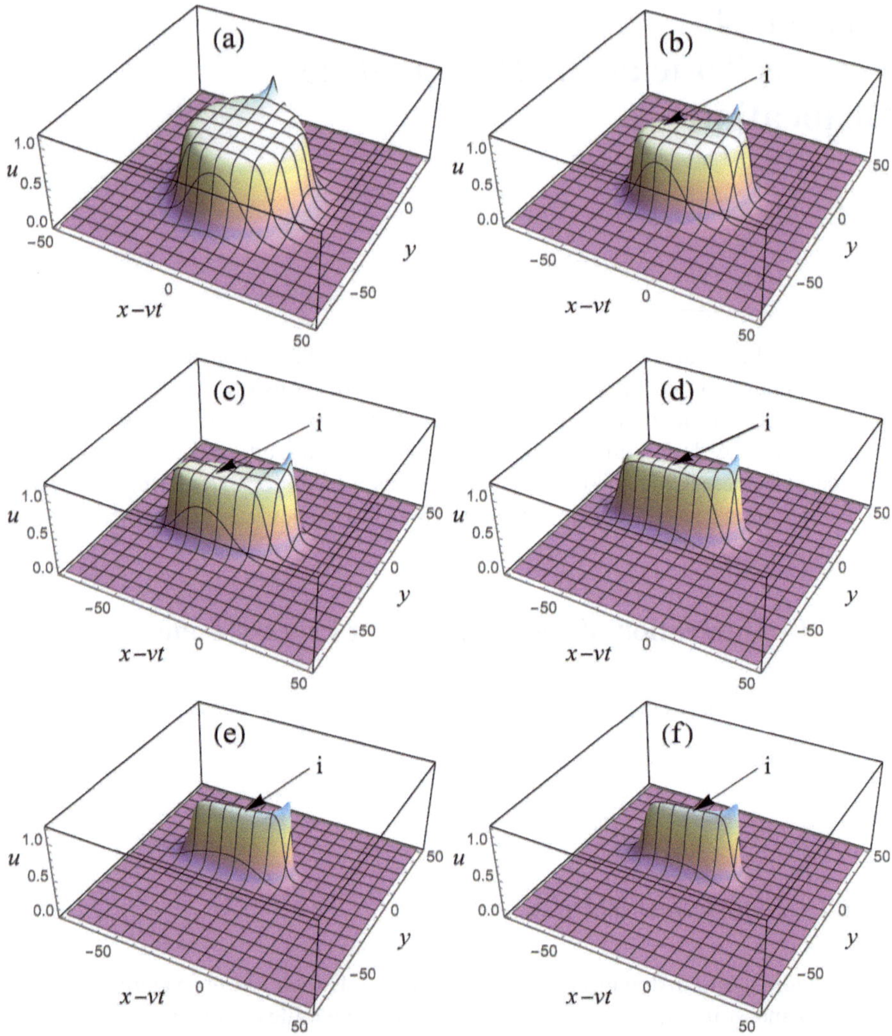

Fig. 24.1 Order parameter distributions within the process zone at a propagating dislocation. Simulations were conducted with $a = 0.2$ and $b = 1$. Panel **a** shows the zone at a stationary dislocation ($v = 0$) for reference. Panels **b–f** depict the zones at dislocation velocities: (**b**) $v = 0.05$, (**c**) $v = 0.2$, (**d**) $v = 0.4$, (**e**) $v = 0.6$, and (**f**) $v = 0.642$. The zone disappears at $v \approx 0.6425$ (not shown). Arrow *i* indicates the plateau of the order parameter distribution

propagates at slow velocities (Fig. 24.2b, c). With increasing velocity, the zone's contour transforms into an irregular, *cucumber-like* shape (Fig. 24.2d–f).

Let us also note that the cucumber's long axis length is only 2–3 fold larger than the motionless zone's radius, R_d. In other words, they exhibit the same order of magnitude.

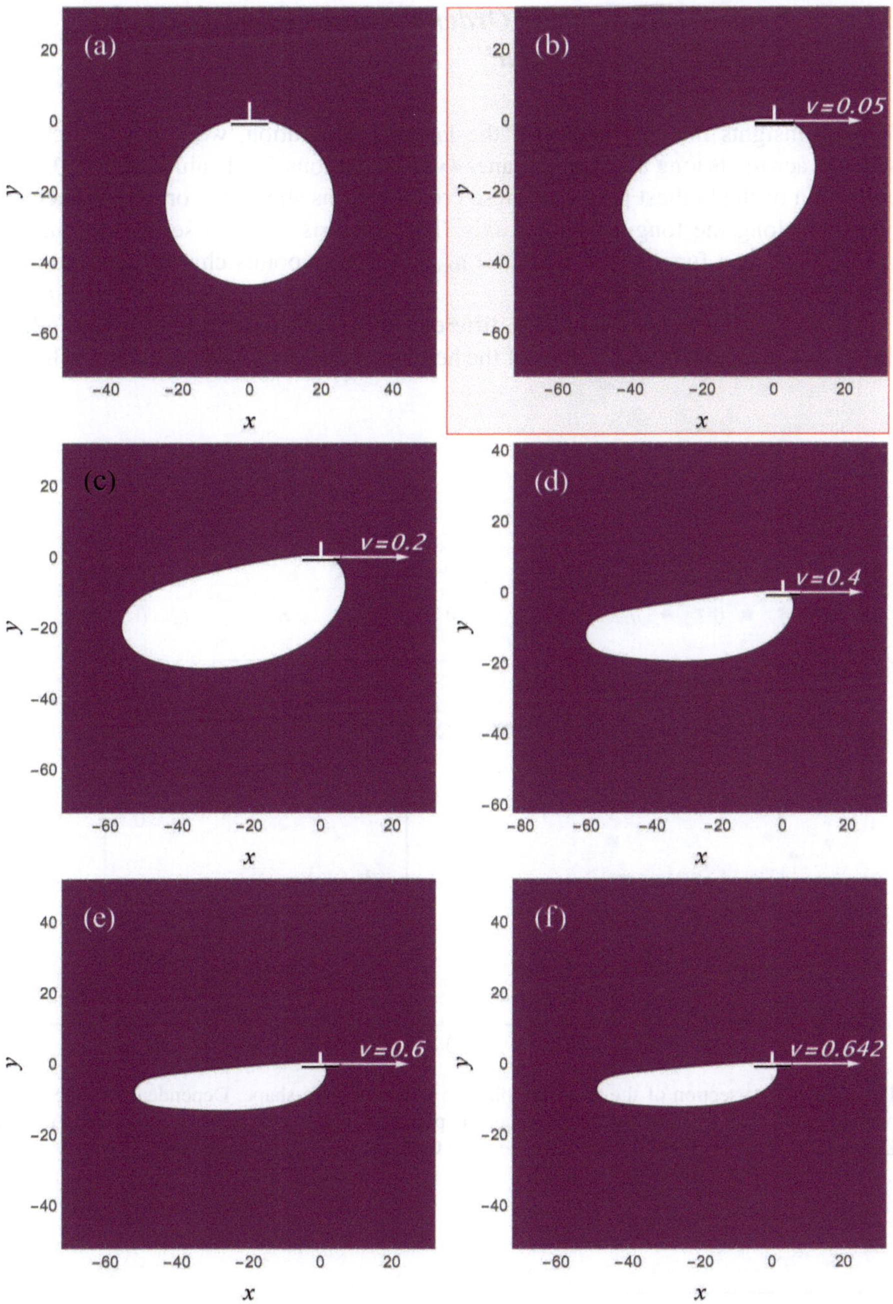

Fig. 24.2 Process zone contours shown in Fig. 24.1. The dislocation symbol is shown in white; where it appears against a white background, it is distinguished by a thin black line drawn just below it. The motion direction with velocity v is indicated by the arrow

24.1.2 Cross-Section of the Order Parameter Distribution Along the "Cucumber"

To gain insights into the nature of the latter configuration, we conducted cross-sections across its long axis using planes (y, u) at various fixed values of $x < 0$. The collection of the highest points of these cross-sections shows the order parameter's behavior along the long cucumber axis. This analysis yields a set of top plateau points, u_{top}, as a function of x: $u_{\text{top}} = u_{\text{top}}(x)$. These points characterize the roof behavior illustrated in Fig. 24.3.

The roofs of the zones obtained at different velocities closely resemble each other. Table 24.1 presents the u_{top} values of the heights of these roofs for various v values.[1]

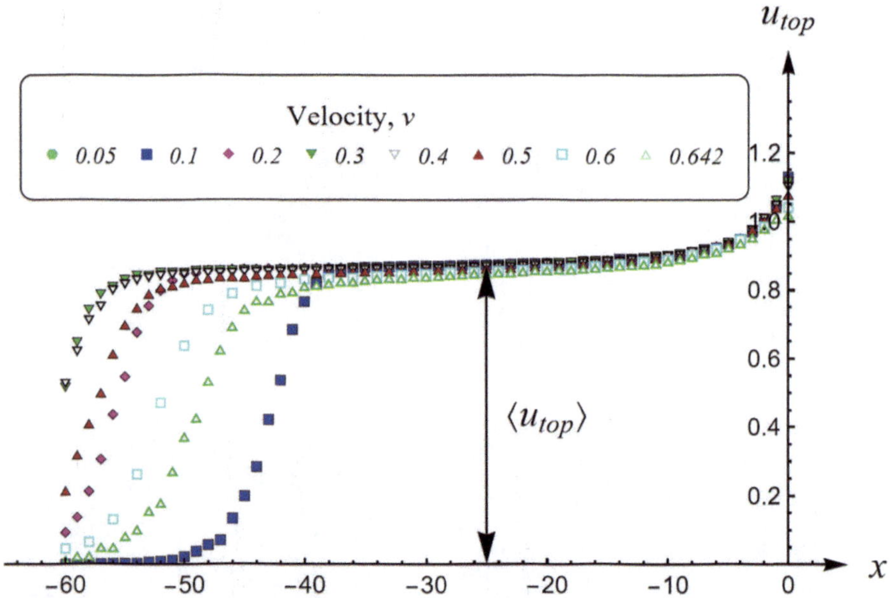

Fig. 24.3 Cross-section of the zone exhibiting a cucumber-like shape. Dependence of the order parameter, u_{top}, on x at the highest points of the plateaux cross-sections by various planes $x < 0$. Dependencies are obtained for the velocities $v = 0.05, 0.1, 0.2, \ldots, 0.6$ and 0.642

[1] For each v, we specifically consider u_{top} values within the interval $-30 < x < -15$, situated in the middle of the roof. This selection avoids abrupt variations in u visible at the trailing edge of the $u(x)$ distribution or the topknot close to $x = 0$. To account for its gentle variations with x, we calculate the average u_{top} over the interval $-30 < x < -15$. The averaged value we denoted as $\langle u_{\text{top}} \rangle$.

Table 24.1 Average roof height at $a = 0.2, b = 1$

v	0.05	0.1	0.2	0.3	0.4	0.5	0.6	0.642
$\langle u_{\text{top}} \rangle$	0.874	0.874	0.873	0.871	0.867	0.862	0.854	0.848

The data reveals that the roof height experiences only slight decreases with increasing velocity, primarily varying in the third to the second decimal digit.

***

In summary, *as the dislocation velocity increases, the shape of the zone evolves from a purely circular form to a lopsided, melon-like shape, then to an elongated, melon-like configuration, and finally to an irregular, cucumber-like form. The "roof" height of the order parameter distribution behind the dislocation remains independent of velocity.*

24.2 Geometrical Parameters of the Propagating Dislocation's Zone

For the propagating dislocation, we were able to find approximations for the curvature radius of the trailing end and the wake length. This section presents these findings.

***

24.2.1 Analyzing Curvature Radius of the Zone's Trailing Edge

Figure 24.4 presents dependence of the curvature radius, ρ, of its trailing edge, as a function of the dislocation velocity, provided the wake exhibits a cucumber shape.

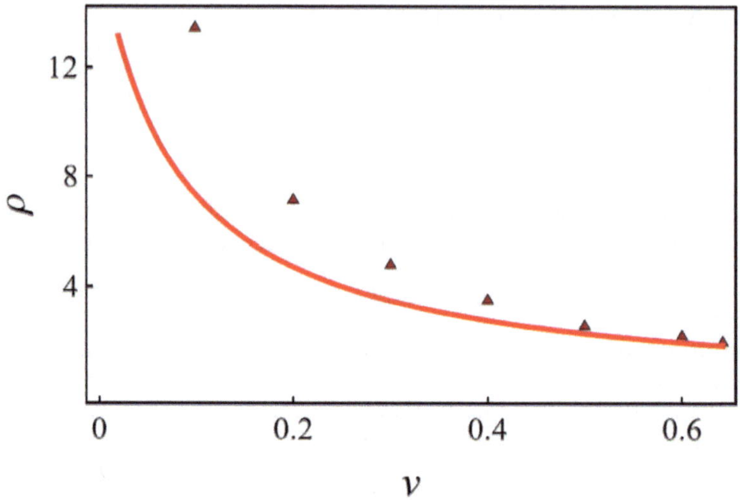

Fig. 24.4 Curvature radius of the zone's trailing edge. The radius was obtained from simulations and compared with the analytical expression. Here, we show the simulations performed with $a = 0.2$ and $b = 1$ (dots). The solid line shows the behavior of the function (24.2)

Utilizing the relation for the trailing edge curvature radius (3.74) we easily find:

$$\rho = \frac{1}{2\sqrt{a_{\mathrm{d}}}(v - v_{\mathrm{G}})} \tag{24.1}$$

and using Gordon's velocity, v_{G} (3.69) and $a_{\mathrm{d}} \approx 0.14$, we can derive the following approximate expression:

$$\rho \approx 1.34 \left(v - 0.55 \frac{8a - 3b^2 + 3b\sqrt{b^2 - 4a}}{\sqrt{b^2 - 2a - b\sqrt{b^2 - 4a}}} \right)^{-1} \tag{24.2}$$

When $a = 0.2$ and $b = 1$, this expression is depicted by the solid line in Fig. 24.4. One can see that at small velocities, the solid curve demonstrates a deviation from the simulation (24.2) points. One expects it since, at small velocities, it is the potential $U(\mathbf{r})$ rather than the motion that determines the lopsided melon-like shape of the zone. We observe it in Fig. 24.2. In contrast, the more significant the velocity, the closer the zone shape to the cucumber, the better the analytical result (24.2) describes the simulations. The zone adapts its configuration so that the interface surface tension contributes to the driving force exerted on its boundary. The latter enables the zone to keep pace with the dislocation. This adaptation entails a decrease in the curvature radius.

24.2.2 The Wake's Length

To estimate the geometry of the wake at $v \gg v_G$, let us roughly regard its configuration as follows. The upper zone's boundaries in Fig. 24.2d–f are visible from the dislocation at a small angle with respect to Ox axis. For the present estimate, we neglect this angle and regard the zone as a region enclosed between two parallel horizontal plane boundaries bounded from left and right by two semi-circles with the radii ρ, as shown in Fig. 24.5.

The length of the boundaries we denote as l_{wake}. Since the upper boundary is (almost) horizontal with $\varphi \approx \pi$, the action of the potential on this boundary vanishes. One concludes that it propagates as a stress-free phase front. As we have discussed in Sect. 3.5, this can only occur with Gordon's velocity, v_G. The dislocation propagates with the velocity v from left to right. The trailing edge of the wake must keep pace with the dislocation. In this regime, we use the expression (24.1). The trailing edge runs over the distance, l_{wake}, with the velocity v. During the same time, the upper boundary moves over the distance ρ with the velocity v_G. This yields the relation:

$$\frac{l_{wake}}{v} = \frac{\rho}{v_G} \tag{24.3}$$

Combining this relation with (24.1) one finds the expression for the length l_{wake}:

$$l_{wake} \approx \frac{k_g v}{2\sqrt{a_d} v_G (v - v_G)} \tag{24.4}$$

where we introduced the dimensionless geometric factor, $k_g \sim 1$, to account for a difference between the simulated configuration (Fig. 24.2d–f) and that displayed in Fig. 24.5 grounding the expression (24.4). From the simulations at large velocities: $v = 0.4, 0.6$, and 0.642 at $a = 0.2$ and $b = 1$ (as shown in Fig. 24.2d–f), we get the values $l_{wake} = 60, 51$, and 48 dimensionless units. Comparison with those obtained according to the formula (24.4) at the same conditions yields:

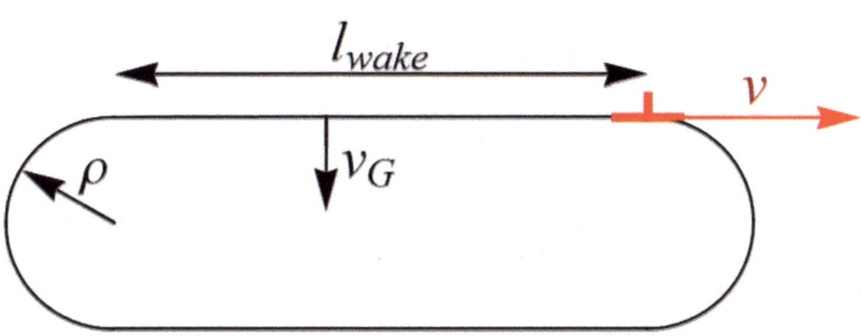

Fig. 24.5 Sketch of the zone's wake. The wake propagates together with the dislocation. See the description in the text

$$k_g \approx 2.7 \tag{24.5}$$

making sure that the approach, indeed, yields a reasonable estimate.

Passing to the dimensional coordinates, if the temperature is close to the binodal, one finds:

$$\mathbb{L}_{\text{wake}} \approx \frac{2k_g(T_0 - T_c)^{3/2}}{3(T - T_0)^2}\sqrt{\frac{g}{\alpha_0}} \tag{24.6}$$

***

In summary, *the standard expression for the curvature radius of the zone's trailing end closely matches simulation results, allowing for a rough estimation of the wake length. The latter diverges as the temperature approaches the binodal, following a* $(T - T_0)^{-2}$ *law.*

24.3 Friction Force Acting on the Moving Dislocation

Using the approximate wake representation sketched in Fig. 24.5, *we derive expressions for the drag force exerted on the dislocation due to energy dissipation within the zone. We analyze two distinct cases: slow motion* ($V < V_G$) *and fast motion* ($V > V_G$).

***

Expression (20.24) gives a general expression for the friction force, f, arising due to the energy dissipation within the process zone. In the dislocation literature, it is conventional to define an equivalent stress, $\sigma = f/\mathfrak{b}\mathbb{L}$, necessary to generate the corresponding force. Using (20.24) we obtain:

$$\sigma = \frac{\kappa A(1 - 2\nu)}{4\sqrt{\gamma g}(1 - \nu)} V K \tag{24.7}$$

Here, K we define as a dimensionless integral[2]:

$$K = \int \left(\frac{\partial u}{\partial x}\right)^2 dxdy \tag{24.8}$$

expressed in terms of the dimensionless order parameter, u, and dimensionless coordinates, x and y. The other parameters of the expression (24.7) we have already introduced previously.

[2] This integral is similar to, but not identical to, I_2 introduced previously (6.45).

To derive the stress σ, one only needs to estimate the value of the integral K. We do the latter separately in the case of the slow and fast regimes.

Like in the case of the propagating crack, we define the regime of the slow and fast propagation by comparison of the dislocation's speed, v, with Gordon's velocity, v_G.

24.3.1 Regime of the Slow Propagation

This section focuses on the slow regime, $v < v_G$. Our simulations have demonstrated that the zone's configuration is melon-like in this regime. In other words, its boundary can be satisfactorily regarded as approximately a circle with the radius R_d (23.4).

Figure 24.6 showcases the cross-section of the function $(\partial u/\partial x)^2$ along Ox, representing two symmetric bell-shaped peaks. The integration over x runs from left to right yielding twice the same result. Therefore, we will duplicate the integration over the leading zone's boundary. By recasting in terms of the polar coordinates in (24.8), one finds:

$$K = \int \left(\frac{\partial u}{\partial r} \cos(\varphi) - \frac{1}{r} \frac{\partial u}{\partial \varphi} \sin(\varphi) \right)^2 r dr d\varphi$$

As we know from the simulations, $\partial u/\partial \varphi$ is small, except for a close vicinity of the dislocation. Therefore, we omit the second term within the parentheses. This brings us to the integral:

$$K \approx 2 \int_0^{\infty} \left(\frac{\partial u}{\partial r} \right)^2 r dr \times \int_{-\pi/2}^{\pi/2} \cos^2(\varphi) d\varphi \equiv \pi \int_0^{\infty} \left(\frac{\partial u}{\partial r} \right)^2 r dr$$

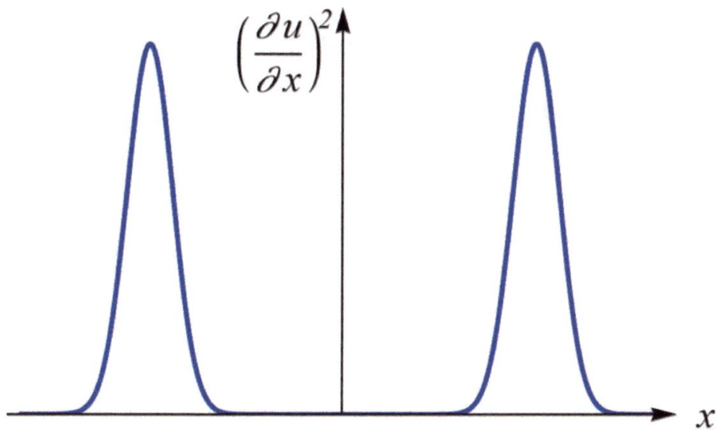

Fig. 24.6 Cross-section of the integrand in K (see 24.8)

Factor 2 in front of the first expression considers that the zone represents two semicircles making equal contributions. The last expression integrates the narrow bell-shaped function. Therefore, one replaces r under the integral by the radius of the zone's boundary, R_{d}. One finds:

$$K \approx \pi \int_0^\infty \left(\frac{\partial u}{\partial r}\right)^2 r dr \approx \pi R_{\mathrm{d}} \int_0^\infty \left(\frac{\partial u}{\partial r}\right)^2 dr \tag{24.9}$$

The latter integral coincides with $K_0 \approx 1/2$, as defined in (23.1). Using its average value (23.2), one obtains the expression for the dimensionless integral, $K \approx \pi/4a_0(T - T_0)$. Applying the rescaling relations (20.13) we get in physical parameters:

$$K \approx \frac{1}{4\pi g \alpha_0 (T - T_0)} \left[\frac{A\mathfrak{b}(1 - 2\nu)}{1 - \nu}\right]^2 \tag{24.10}$$

24.3.1.1 Viscous Friction Regime During the Slow Motion

By substituting (24.10) in the expression (24.7), we finally arrive at the stress corresponding to the friction force due to the energy dissipation within the zone in the course of the slow propagation:

$$\sigma_{\mathrm{slow}} = \frac{\kappa \mathfrak{b}^2}{4\pi^2 \alpha_0 (T - T_0)\sqrt{\gamma}} \left[\frac{A(1 - 2\nu)}{\sqrt{g}(1 - \nu)}\right]^3 V \tag{24.11}$$

Let us highlight that σ_{slow} is proportional to the dislocation velocity. One can regard it as a viscous-like contribution. Further, $\sigma_f^{(\mathrm{slow})} \sim (T - T_0)^{-1}$. It diverges at $T \to T_0$.

One can determine the velocity, V_{PN}, at which the stress σ_{slow} becomes comparable to the Peierls-Nabarro stress, σ_{PN} [1]:

$$\sigma_{\mathrm{PN}} \sim 10^{-3}\mu \tag{24.12}$$

V_{PN} takes the following form:

$$V_{\mathrm{PN}} = \frac{4a_{\mathrm{d}}\sigma_{\mathrm{PN}}\sqrt{g\gamma}}{A\kappa} \frac{1 - \nu}{1 - 2\nu} \tag{24.13}$$

At $V > V_{PN}$ the friction due to the process zone dominates the resistance to the dislocation nonconcervative motion,[3] while as soon as V becomes smaller than V_{PN}, it is Peierls-Nabarro barrier that forms the resistance, while the zone-generated friction can be neglected.

24.3.1.2 Comparison of the Zone-Generated Viscous Friction with That Engendered by the Phonons Radiation and Scattering

There are several sources of the retarding force acting on the dislocation. The common opinion is that the main contribution is due to (i) the irradiation of phonons in the course of its motion and (ii) scattering of phonons on the dislocation's stress field. Its estimate has been obtained by Leibfried [3]. Expressed in terms of the corresponding shear stress, σ_r, it reads:

$$\sigma_r \sim 10^{-3}\frac{\mu V}{c_t}, \tag{24.14}$$

however, yielding an underestimated friction value.[4] It is informative to estimate the ratio σ_{slow}/σ_r:

$$\frac{\sigma_{slow}}{\sigma_r} \sim 10^2 \times \frac{\kappa c_t \mathfrak{b}^2(1+\nu)}{E\alpha_0(T-T_0)\sqrt{\gamma}}\left[\frac{A(1-2\nu)}{\sqrt{g}(1-\nu)}\right]^3 \tag{24.15}$$

24.3.2 *Regime of the Fast Motion*

We regard the zone as moving fast if $V \gg V_G$. In this case, the zone configuration is roughly shown in Fig. 24.5, its leading and trailing boundaries representing semi-circles. Contributions to K (24.8) from these boundary portions reduce themselves to (24.9), where one replaces R_d by ρ (24.1). In contrast, the other parts of the boundary are almost parallel to the Ox axis, as revealed by the simulations (Fig. 24.2), and their contribution is negligible. Therefore, one finds:

$$\sigma_{fast} = \frac{\mathfrak{b}}{2\pi\sqrt{g\gamma}}\left[\frac{A(1-2\nu)}{1-\nu}\right]^2\frac{V}{V-V_G}. \tag{24.16}$$

Close to the binodal, according to (3.72) Gordon's velocity is given by:

$$V_G \approx \frac{2(\alpha_0 g)^{1/2}(T-T_0)}{\kappa(T_0-T_c)} \tag{24.17}$$

---

[3] The analogous resistance is also known for cracks travelling through the crystal lattice [2].

[4] See the footnotes in Sect. 7.8 of the Russian version of the Hirth and Lothe book [4] (Moscow, Atomizdat, 1972).

where the velocity is small. In this regime ($V \gg V_G$), σ_{fast} takes the following form:

$$\sigma_{\text{fast}} = \frac{\mathfrak{b}}{\pi\sqrt{g\gamma}}\left[\frac{A(1-2\nu)}{1-\nu}\right]^2. \tag{24.18}$$

It becomes independent of both the dislocation velocity and κ exhibiting, thus, the similarity to the dry friction. This is related to decreasing the curvature radius, ρ, of the trailing boundary with increasing velocity.

Notably, we derived the results (24.16) and (24.18) based on the geometry shown in Fig. 24.5 which differs from those displayed in Fig. 24.2. Therefore, one should understand these results as the order of magnitude estimates.

***

In summary, *we derived analytical expressions for the stresses associated with the drag force for both slow and fast dislocation motion. For slow motion, the drag force is proportional to the dislocation velocity, exhibiting a viscous-like behavior, and diverging as* $(T - T_0)^{-1}$ *near the binodal. In contrast, for fast motion, the drag force is independent of both velocity and temperature, resembling dry friction in mechanics. These results are based on the geometry depicted in Fig.* 24.5, *which slightly differs from that given by simulations (Fig.* 24.2), *and should therefore be considered as estimates.*

24.4 Comparative Analysis of the Dislocation and Crack Wakes

This section compares the wake structures of dislocations and cracks, explaining the physical origins of their differing geometries.[5]

***

The remarkable similarity between $\langle u_{\text{top}}\rangle$ and u_{d} hints at the plateau section (indicated by arrow i in Fig. 24.1) serves as a representation of the metastable wake. This conceptually mirrors the wake left behind by a propagating crack tip zone, a topic explored in Sect. 13.4. However, a striking distinction emerges when we scrutinize the shapes of these wakes in the context of cracks and dislocations.

24.4.1 Shapes During a Slow Motion

First and foremost, the region at the crack tip undergoes a notable morphological transformation, as detailed in Chap. 13. Below Gordon's speed, v_G, the crack tip

[5] This section may be skipped during a first reading.

zone assumes a concave shape, while at velocities above v_G, it adopts a convex form. This transformation unfolds due to the distinctive angular dependency of the crack potential $U \sim \cos(\theta/2)$ (10.7), illustrated in Fig. 10.1. This potential results in a concave configuration when θ is close to $\pm\pi$ for the stationary zone. As the crack advances with a modest velocity ($0 < v < v_G$), the boundary invagination persists at the zone's trailing edge. Consequently, the zone elongates in the direction opposite to its velocity, with the degree of elongation increasing in tandem with the crack's speed. This elongation causes a gradual reduction in the depth of invagination until it ultimately vanishes at $v = v_G$. The vanishing signifies the transition from concave to convex, a phenomenon termed a morphological transformation.

Conversely, in the case of the dislocation, the scenario takes a different turn. The potential $U \sim \sin(\theta)$ (21.1) does not give rise to any invagination at $v = 0$, and hence, there is no rationale for invagination to manifest during the dislocation motion.[6]

24.4.2 Droplet Versus Cucumber Shape During the Fast Motion

Secondly, when $v > v_G$, the crack tip zone-wake combination assumes a droplet-like shape. One can roughly liken its wake section to a triangular configuration (refer to Figs. 14.1c, 13.7 and 13.9). This unique configuration arises once more from the angular dependency of the potential $U \sim \cos(\theta/2)$, which diminishes as θ approaches $\pm\pi$. Consequently, the wake at these angles turns out to be an overheated phase unaffected by any potential. In this case, the phase front takes on an almost planar form and advances at Gordon's velocity, v_G. Under these specific conditions, the potential remains nonzero but small. As a result, the front generates the triangular configuration, as elaborated upon in Sect. 14.4.2.

In stark contrast, the zone at a propagating dislocation assumes an irregular cucumber-like shape. This shape evolves due to the potential (21.1) configuration, causing it to slenderize with increasing velocity (cf. the evolution from (d) to (f) in Fig. 24.2). In the case of dislocations, the whole zone is exposed to the nonvanishing potential, thereby precluding the adoption of droplet-like or triangular forms.

***

In conclusion, *the wake configuration of a dislocation differs qualitatively from that of a crack. The dislocation wake evolves from a circular shape to melon-like and then cucumber-like, without undergoing a concave-to-convex transformation. By contrast, the crack wake starts with a concave shape at low crack velocities, progressing to a convex, droplet-like shape when* $V > V_G$*. This difference arises from*

---

[6] At specific parameter values ($a = 0.2$, $b = 1$, and $a_d \approx 0.14$), Gordon's velocity (3.69) attains $v_G \approx 0.081$. The images illustrate both the zones at velocities $v = 0$ and 0.05 (Fig. 24.2a, b) below v_G, as well as $v = 0.1$–0.642 (Fig. 24.2c–f) above v_G. In both cases, we exclusively observe a zone characterized by a convex configuration, which agrees with the above arguments.

the angular dependence of the potential $U = U(\theta)$ in each case: in the crack case, $U(\theta)$ vanishes as θ approaches $\pm\pi$, creating a concave zone shape at low velocities, while the dislocation case lacks this behavior.

This analysis primarily applies to a one-component order parameter. If the order parameter becomes multicomponent, with an interaction term involving several components $\varepsilon_{\alpha\beta}$, a wider variety of potential configurations may arise for process zones at both cracks and dislocations. For example, if a dislocation's configuration includes invagination at the sliding plane, this may lead to a morphological transition and the formation of a triangular wake or similar structure, akin to what is observed in the crack case.

References

1. J. Friedel, *Dislocations* (Pergamon Press, Oxford, 1964)
2. R. Thomson, Physics of fracture, in *Atomistic of Fracture*, ed. by R.M. Latanision, J.R. Pickens (Plenum Press, New York, 1983), pp. 167–204
3. G.Z. Leibfried, Z. Phys. **135**, 23 (1953)
4. J.P. Hirth, J. Lothe, *Theory of Dislocations* (Krieger Publishing Company, Malabar, Florida, 1991)

Chapter 25
A Low-Angle Grain Boundary

Abstract A low-angle grain boundary is a dislocation wall characterized by a large spacing between dislocations, a specific type of defect that generates a process zone influencing the boundary's behavior. Near a phase transformation, the low-angle grain boundary forms a chain of highly localized process zones, each centered at a dislocation within the wall. These zones differ structurally from those at isolated dislocations. The order parameter's dependence on the parameter a (and thus on the temperature T) is pronounced near the bifurcation point and grows significantly beyond it. Unlike the behavior seen with isolated dislocations, this increase corresponds to a rapid growth in the order parameter's peak height with a negligible increase in its lateral size. The free energy of the zone chain exhibits a dependence of $F \sim (a_d - a)^2$ (Boulbitch and Korzhenevskii, first presented here).

25.1 Dislocation Wall: Introduction

A low-angle grain boundary is a dislocation wall characterized by a large spacing between dislocations, a specific type of defect that generates a process zone influencing the boundary's behavior. This section introduces the key properties of the low-angle grain boundary that are central to our analysis.

***

We examine the low-angle grain boundary aligned along the OY axis, as depicted in Fig. 25.1.

The parameter $\mathbb{D}_\text{w}$ denotes the distance between the dislocations within the wall. The subscript "w" stands for "wall." This parameter is expressed as:

$$\mathbb{D}_\text{w} = \frac{\mathfrak{b}}{\theta} \tag{25.1}$$

© The Author(s), under exclusive license to Springer Nature Switzerland AG 2026

A. Boulbitch and A. Korzhenevskii, *Stress-Induced Local Phase Transformations Near Cracks and Dislocations*, Springer Series in Materials Science 356,
https://doi.org/10.1007/978-3-032-00945-6_25

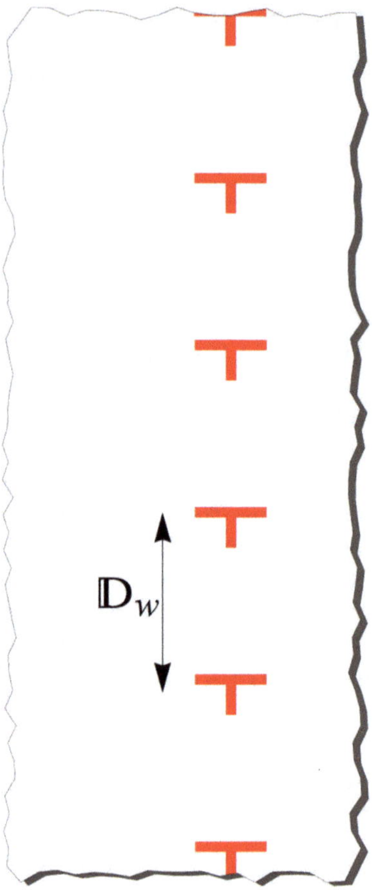

Fig. 25.1 Dislocation wall

where θ represents the grain boundary angle [1]. The exact solution for the stress tensor trace generated by this wall can be found in the book by Hirth and Lothe.[1] It is given by:

$$\sigma_{ii} = \frac{\mu \mathfrak{b} \sin(2\pi y)}{\mathbb{D}_\text{w}(1-\nu)\,[\cos(2\pi y) - \cosh(2\pi x)]} \tag{25.2}$$

where μ is the shear modulus, x, and y are dimensionless coordinates related to the original coordinates, X and Y with the dimension in centimeters, defined as:

$$x = \frac{X}{\mathbb{D}_\text{w}}; \quad y = \frac{Y}{\mathbb{D}_\text{w}}$$

Using the relation:

[1] It is given by (19.77) in the book [2].

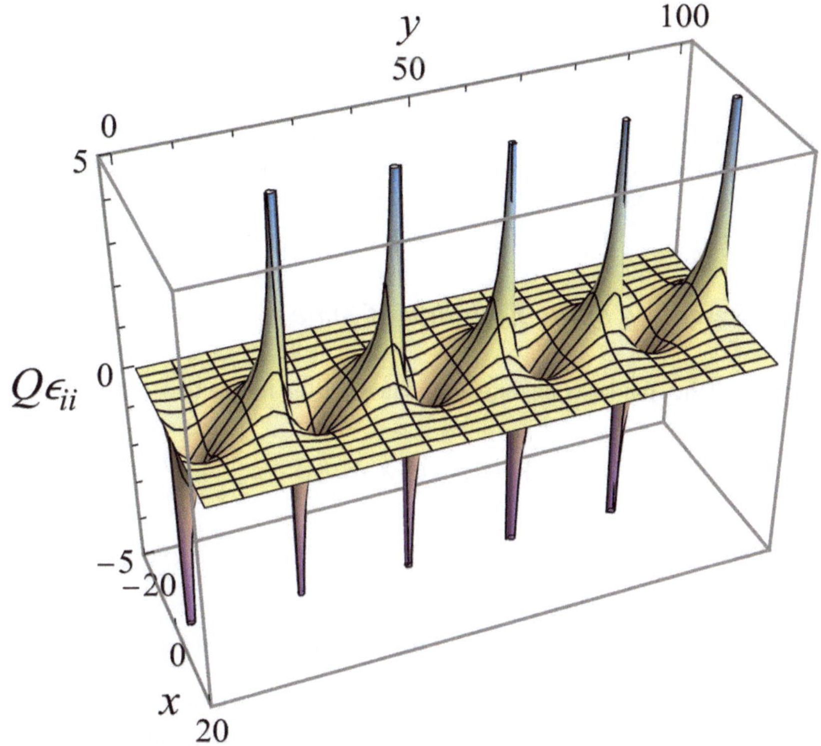

Fig. 25.2 Strain distribution across the dislocation wall. The factor Q is specified in the text. The image is clipped at ±5

$$\varepsilon_{ii} = \frac{1-2\nu}{E}\sigma_{ii}$$

one can obtain the strain tensor trace:

$$\varepsilon_{ii} = \frac{\mathfrak{b}(1-2\nu)\sin(2\pi y)}{2\mathbb{D}_{\mathrm{w}}(1-\nu^2)\left[\cos(2\pi y)-\cosh(2\pi x)\right]} \tag{25.3}$$

Figure 25.2 illustrates the potential $U(x, y)$ consisting of a chain of highly localized potential wells. Process zones are also expected to be concentrated within these wells. In this figure, we denote $Q = 2\mathbb{D}_{\mathrm{w}}(1-\nu^2)/\mathfrak{b}(1-2\nu)$.

The surface tension of the dislocation wall, $\tau^{(0)}_{\mathrm{wall}}$, arises purely from mechanical effects. It is defined as the wall energy W_{wall} per unit dislocation length $\mathbb{L}$ and per period $\mathbb{D}_{\mathrm{w}}$:

$$\tau^{(0)}_{\mathrm{wall}} = \frac{W_{\mathrm{wall}}}{\mathbb{L}\mathbb{D}_{\mathrm{w}}}$$

It is given by[2]:

$$\tau_{\text{wall}}^{(0)} = \frac{\mathfrak{b}\theta\mu}{4\pi(1-\nu)} \ln\left(\frac{e\alpha_{\text{w}}}{2\pi\theta}\right) \tag{25.4}$$

where α_{w} is a dimensionless parameter. For non-metals, α_{w} typically ranges around 4, whereas for metals, it is approximately between 0.5 and 2 [2], and e is the base of the natural logarithms.

25.1.1 Rescaling of the Equation of State

We begin with the standard Ginzburg-Landau equation for the first-order zone, where $\beta < 0$.

$$g\left(\frac{\partial^2\eta}{\partial X^2} + \frac{\partial^2\eta}{\partial Y^2}\right) - [\alpha + U(X,Y)]\,\eta - \beta\eta^3 - \gamma\eta^5 = 0 \tag{25.5}$$

The potential:

$$U(X,Y) = \frac{A\mathfrak{b}(1-2\nu)}{\mathbb{D}_{\text{w}}(1-\nu^2)} \times \frac{\sin(2\pi Y/\mathbb{D}_{\text{w}})}{\cos(2\pi Y/\mathbb{D}_{\text{w}}) - \cosh(2\pi X/\mathbb{D}_{\text{w}})}$$

is more complex than those encountered in cases of a crack or isolated dislocation. Unlike those scenarios, this potential depends on an additional parameter, $\mathbb{D}_{\text{w}}$, which has the dimension of length. Therefore, one rescales it in a manner different from our usual method. We apply the following substitutions:

$$(x_1, y_1) = \left[\frac{A\theta(1-2\nu)}{10\pi g(1-\nu^2)}\right]^{1/2}(X,Y); \quad \eta(X,Y) = \left[\frac{A\theta(1-2\nu)}{10\pi\gamma(1-\nu^2)}\right]^{1/4} u(x_1, y_1) \tag{25.6}$$

resulting in the following rescaled equation:

$$\frac{\partial^2 u}{\partial x_1^2} + \frac{\partial^2 u}{\partial y_1^2} - \left[a + \frac{10\pi\sin(2\pi q y_1)}{\cos(2\pi q y_1) - \cosh(2\pi q x_1)}\right]u + bu^3 - u^5 = 0 \tag{25.7}$$

where

$$a = \frac{10\pi(1-\nu^2)}{A\theta(1-2\nu)}\alpha; \quad b = -\left[\frac{10\pi(1-\nu^2)}{A\gamma\theta(1-2\nu)}\right]^{1/2}\beta \tag{25.8}$$

Here, q is a dimensionless parameter defined by $q = \mathbb{D}_{\text{w}}/\mathbb{R}_0$, where $\mathbb{R}_0$ is the natural scale given by $\mathbb{R}_0 = x_1/X$. The values of q and $\mathbb{R}_0$ are:

---

[2] See (19.85) in the book of Hirth and Lothe [2].

$$q = \frac{1}{\mathfrak{b}}\left[\frac{10\pi g\theta(1-\nu^2)}{A(1-2\nu)}\right]^{1/2}; \quad \mathbb{R}_0 = \left[\frac{10\pi g(1-\nu^2)}{A\theta(1-2\nu)}\right]^{1/2} \tag{25.9}$$

The factor 10π was chosen in the rescaled potential to simplify the numerical procedure, leading to the corresponding factors in a, b, and q.

25.1.2 Free Energy Rescaling

By applying the rescaling relations to the free energy, it transforms into the following form:

$$\mathcal{F} = g\mathbb{L}\left[\frac{A\theta(1-2\nu)}{10\pi\gamma(1-\nu^2)}\right]^{1/2} F \tag{25.10}$$

where

$$F = \int\left\{\frac{1}{2}(\nabla u)^2 + \frac{1}{2}\left[a + \frac{10\pi\sin(2\pi q y_1)}{\cos(2\pi q y_1) - \cosh(2\pi q x_1)}\right]u^2 - \frac{1}{4}bu^4 + \frac{1}{6}u^6\right\}dx_1 dy_1 \tag{25.11}$$

Here, ∇ denotes the gradient operator with respect to x_1 and y_1: $\nabla = (\partial/\partial x_1, \partial/\partial y_1, 0)$, and $\mathbb{L}$ represents the dislocation length.

***

In summary, *the low-angle grain boundary generates a strain tensor trace that forms the potential $U(X,Y)$, which is periodic along the wall and decays exponentially across it. These characteristics of the potential give rise to the unique features of its process zone, as demonstrated in the following analysis. The physical behavior of the zone-free wall is influenced by its surface tension, $\tau^{(0)}_{\text{wall}}$. The potential $U(X,Y)$ is more complex compared to those of crack or dislocation, leading the rescaled equation to depend on the dimensionless parameter q in addition to a and b.*

25.2 Process Zones Within the Grain Boundary

This section examines the process zones of a low-angle grain boundary through simulations, initially focusing on qualitative aspects such as the chain of the process zones, the shape of individual chain elements, and the behavior of the order parameter norm. Subsequently, we analyze the free energy behavior of the process zone chain.

***

25.2.1 Qualitative Results

Equation (25.7) is now prepared for simulations. We conducted simulations at various small angles, including $\theta = 0.1°$, $0.5°$, $1°$ and $2°$. Our attempt to make simulations with ultra small angles $\theta \sim 10^{-4°}$ only gave trivial results $u = 0$. We concluded that at such angles, the bifurcation does not occur. The simulation technique employed for this problem is consistent with our previous methods, so we will not describe it in detail here. Figure 25.3 illustrates the order parameter distributions within one (a) and three (b) periods of the wall.

To further examine the order parameter distributions, we analyzed their cross-sections, as shown in Fig. 25.4.

We observe that the order parameter is localized along both the longitudinal and transverse directions within the narrow regions near the maxima of $u(x_1, y_1)$. As anticipated, the distribution is anisotropic, with different sizes and slopes along and across Oy_1.

To quantitatively assess the localization of the order parameter, we determined the zone size $\mathbb{R}_y$ in the OY direction. We extracted $\mathbb{R}_y$ as the width of the distribution

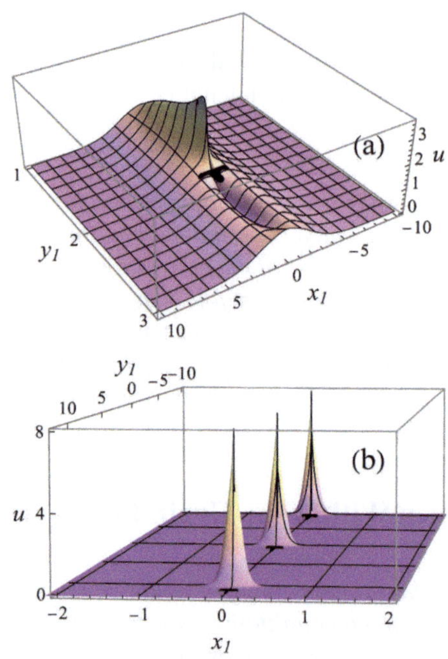

Fig. 25.3 3D view of the order parameter distribution within the low-angle boundary. Panel **a** provides a close-up view showing the order parameter, $u = u(x_1, y_1)$, distribution in the vicinity of one of the dislocations of the chain. Panel **b** offers a broader view, showing distributions from neighboring dislocations. Dislocation positions are marked in black

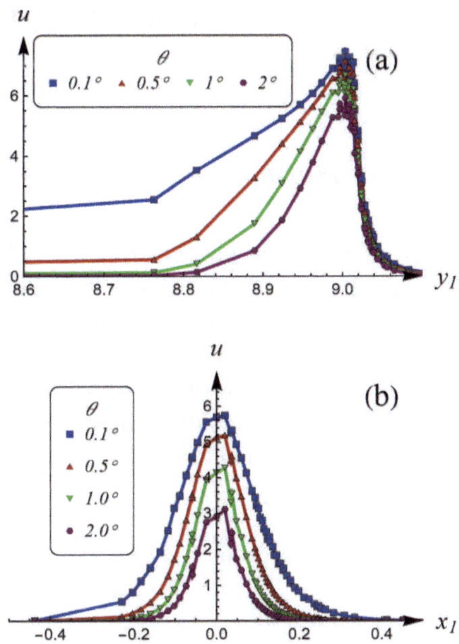

Fig. 25.4 Zone size in the OX and OY directions as a function of grain boundary angle θ

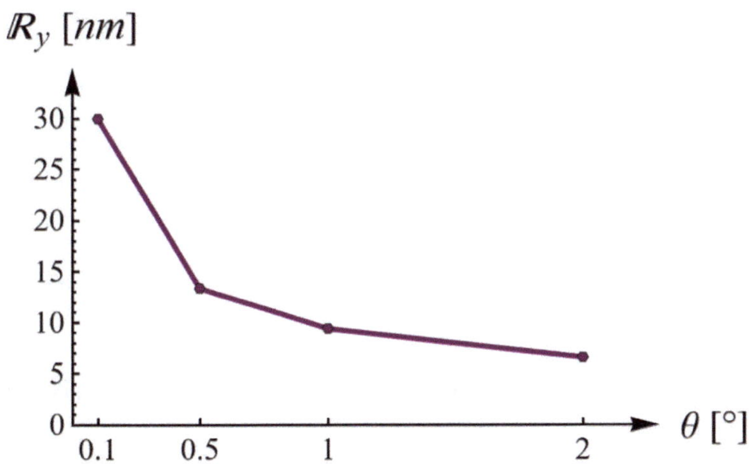

Fig. 25.5 Zone size in the OY direction as a function of grain boundary angle

at half of $u_{\max}$ along y_1 using the data from Fig. 25.4a and the rescaling relations (25.6). The result is presented in Fig. 25.5.

The size of the zone decreases with increasing grain boundary angle θ. The same trend is observed for the thickness of the order parameter in the OX direction.

Note 25.2.1 Thus, the order parameter distribution is highly localized in the vicinity of the dislocation.

25.2.2 Order Parameter Norm

To identify the emergence point of the process zone, it is helpful to monitor the Hilbert norm of the order parameter (22.1). Figure 25.6 illustrates the behavior of Hilbert's norm of the order parameter near the bifurcation point for the different angles θ under investigation.

Note that the parameter a, as given by the expression (25.8), exhibits unexpectedly large values not found in our previous results. These high values result from the different rescaling transformations applied to the dislocation wall.

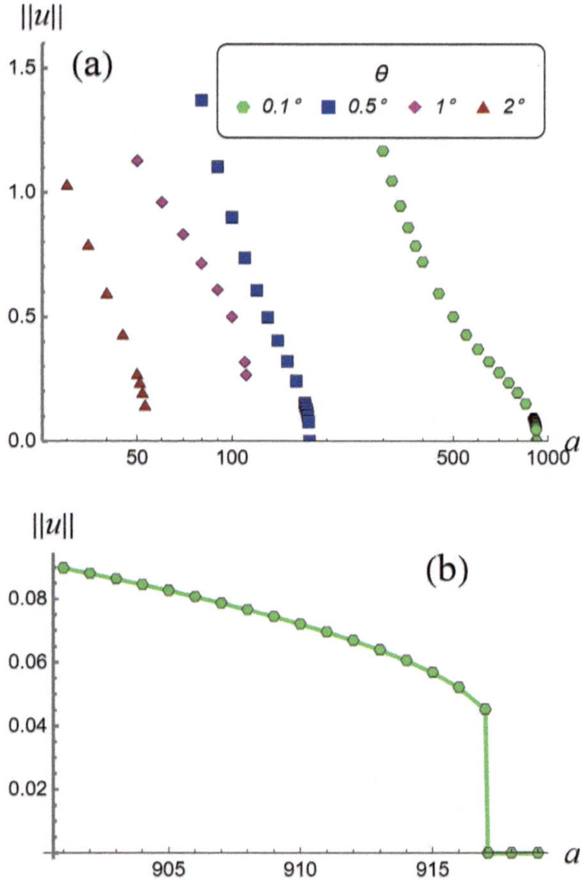

Fig. 25.6 Hilbert norm $\|u\|$ versus a for four values of θ. Panel **a** shows the log-linear behavior of the order parameter norm for four values of θ across a broad range of the parameter a. Panel **b** provides a magnified view of the norm for $\theta = 0.1°$ close to the bifurcation point, where the order parameter u exhibits typical behavior characteristic for the quasi-second-order zone

It is evident that the emergence point of the zone, a_d, is highly dependent on the grain boundary angle θ:

θ (°)	0.1	0.5	1.0	2.0
a_d	917	175	111	53

Near the bifurcation point, the order parameter displays typical behavior: as the parameter a decreases, the order parameter undergoes a jump, characteristic of first-order transitions. After this jump, with further decreases in a, there is a modest increase in the order parameter. This behavior resembles the patterns discussed in Sect. 3.3.2.3. It is depicted in Fig. 3.2b.

However, with a further reduction in a, we observe a transition to a more rapid increase in the order parameter norm, which is atypical for standard bulk phase transformations.

Note 25.2.2 Comparing with the content of Note 25.2.1, it can be concluded that this dramatic growth in $||u||$ is not solely related to an increase in the height u_{max} of the distribution $u(x_1, y_1)$, but rather to an expansion in the lateral size of the zone.

25.2.3 Contribution of the Process Zone to the Surface Energy of the Domain Wall

We derived the free energy factor, F, given by (25.11), from the simulations. Figure 25.7a illustrates its behavior across the range of angles under investigation.

It is important to note that F is negative. To plot it in double-logarithmic form, we display its absolute value. The dimensionless free energy behavior of the dislocation wall deviates significantly from that of single dislocations, reaching values in the range of 10^2 to 10^3 dimensionless units.

Figure 25.7b indicates that, near the bifurcation point where $a_d - a \leq 30$, all the curves can be roughly approximated by the analytical expression:

$$F \approx -5 \times 10^{-3}(a_d - a)^2 \tag{25.12}$$

as shown by the solid black line in Fig. 25.7b. Here, a varies from zero to a_d.

The surface tension of the wall consists of two components:

$$\tau_{wall} = \tau_{wall}^{(0)} - \Delta\tau_{wall}$$

where $\tau_{wall}^{(0)}$ represents the mechanical contribution to the dislocation wall's surface tension (25.4), and $-\Delta\tau_{wall}$ arises from the negative free energy of the process zone:

$$\Delta\tau_{wall} = \frac{|\mathcal{F}|}{\mathbb{LD}_w}$$

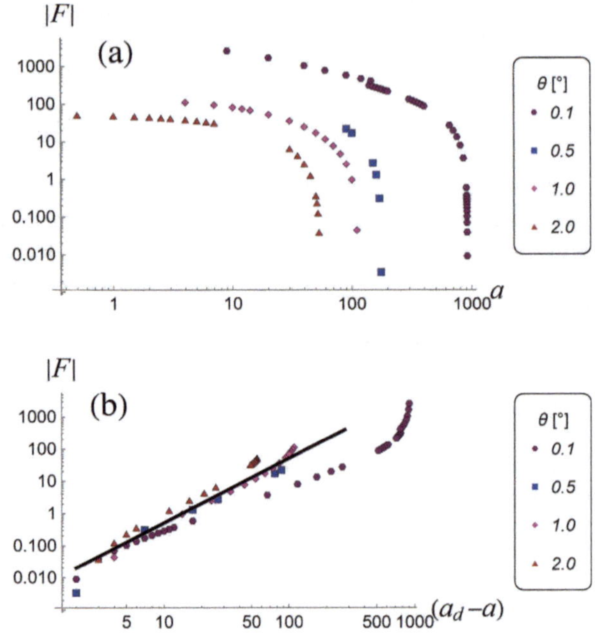

Fig. 25.7 Absolute value of the free energy factor $|F|$ of the dislocation wall. **a** The free energy as a function of parameter a. **b** Plotting $|F|$ against $a_{\mathrm{d}} - a$ aligns all curves to the same starting point. The curves collapse into a narrow sheaf around an average direction, which is shown in black

Using references (25.10–25.12), and the rescaling relations (25.8), we arrive at the following expression:

$$\Delta\tau_{\mathrm{wall}} \approx -\frac{g{\alpha_0}^2}{2\mathfrak{b}\sqrt{10\gamma\theta}}\left[\frac{\pi(1-\nu^2)}{A(1-2\nu)}\right]^{3/2}(T - T_{\mathrm{d}})^2 \tag{25.13}$$

This (25.13) allows us to determine the dimensionless plasticity factor for the wall:

$$k_{\mathrm{pl}} = \frac{|\Delta\tau_{\mathrm{wall}}|}{\tau_{\mathrm{wall}}^{(0)}}$$

which yields:

$$k_{\mathrm{pl}} = \sqrt{\frac{2}{5\gamma}}\frac{g\alpha_0^2\left[\pi(1-\nu^2)\right]^{5/2}}{\mathfrak{b}^2 E\left[A\theta(1-2\nu)\right]^{3/2}\ln(2e/\pi\theta)}(T - T_{\mathrm{c}})^2 \tag{25.14}$$

***

In summary, *near a phase transformation, the low-angle grain boundary forms a chain of highly localized process zones, each centered at a dislocation within the wall. These zones differ structurally from those at isolated dislocations. The order parameter's dependence on a (and thus on T) is pronounced near the bifurcation* *point and grows significantly beyond it. Unlike the behavior seen with isolated dislocations, this increase corresponds to a rapid growth in the order parameter's peak height with a negligible increase in its lateral size. The free energy of the zone chain exhibits a dependence of* $F \sim (a_{\mathrm{d}} - a)^2$.

References

1. C. Kittel, *Introduction to Solid State Physics*, 4th edn. (Wiley, New York, 2005)
2. J.P. Hirth, J. Lothe, *Theory of Dislocations* (Krieger Publishing Company, Malabar, Florida, 1991)

Chapter 26
Formation of Plastic Instability

Abstract Forming a process zone reduces the free energy of the dislocation-zone complex, effectively diminishing the dislocation's line tension. This reduction enhances the crystal's plastic properties. Plastic instability commences once the zone's negative energy surpasses that of the dislocation in absolute value. This chapter derives the temperature that dictates the onset of plastic instability (Boulbitch and Korzhenevskii, first presented here).

26.1 Plastic Instability Induced by Dislocation Dressing

The formation of a process zone reduces the free energy of the dislocation-zone complex, increasing its flexibility and rendering certain elements of the dislocation ensemble prone to instability. This chapter begins by examining the general conditions under which the process zone initiates plastic instability, analyzing the underlying mechanisms and thresholds. We then investigate the onset of instability for the daughter-phase zone-dislocation complex embedded in the mother-phase matrix. Subsequently, we define a corresponding interval for dislocations within the bulk daughter phase, below the bulk binodal, where plastic instability emerges due to the formation of a mother-phase zone embedded in the daughter-phase matrix at the dislocation.[1]

***

The plastic behavior hinges on the dislocation line tension, τ_{disl}, as outlined by Friedel [1], and Hirth and Lothe [2]. Traditionally, for an unaltered dislocation, its line tension $\tau_{\text{disl}}^{(0)} > 0$, is equal to its elastic energy, $W^{(0)}$ (1.14), per unit length, as given by (1.18):

$$\tau_{\text{disl}}^{(0)} = \frac{W^{(0)}}{\mathbb{L}} \tag{26.1}$$

[1] The results presented in this section are published here for the first time.

© The Author(s), under exclusive license to Springer Nature Switzerland AG 2026

A. Boulbitch and A. Korzhenevskii, *Stress-Induced Local Phase Transformations Near Cracks and Dislocations*, Springer Series in Materials Science 356,
https://doi.org/10.1007/978-3-032-00945-6_26

As soon as the process zone emerges at the dislocation, its free energy contributes to that of the dislocation. Its net energy, W, takes the following form:

$$W = W^{(0)} + \Delta\mathcal{F}$$

Let us denote the line tension perturbation as $\Delta\tau_{\text{disl}}$:

$$\Delta\tau_{\text{disl}} = \frac{\Delta\mathcal{F}}{\mathbb{L}} \tag{26.2}$$

The net line tension of the dislocation takes the form:

$$\tau_{\text{disl}} = \tau_{\text{disl}}^{(0)} + \Delta\tau_{\text{disl}} \tag{26.3}$$

The free energy perturbation is negative: $\Delta\mathcal{F} < 0$, a principle dictated by thermodynamics, as we already discussed in the Note 22.1.1. The perturbation of the line tension, $\Delta\tau_{\text{disl}}$, is also negative. The net line tension decreases below its value, $\tau_{\text{disl}}^{(0)}$. The latter increases the dislocation flexibility along with the plasticity of the crystal.

Let us introduce the parameter:

$$k_{\text{pl}} = \frac{|\Delta\tau_{\text{disl}}|}{\tau_{\text{disl}}^{(0)}} > 0, \tag{26.4}$$

to gauge the variation of the crystal plasticity.

A dislocation structure maintains stability provided that $\tau_{\text{disl}} > 0$. The transition to a negative τ_{disl} takes place as soon as

$$k_{\text{pl}} = 1 \tag{26.5}$$

1. Such a transition makes the formation of edge dislocations, edge components of dislocation loops, and bending of screw dislocation energetically preferable, encouraging their proliferation. Formation of the process zones at dislocations enhances, thus, the plasticity in the solid.
2. The growth of the edge components of the dislocations mentioned above, however, is curtailed by two main factors:
   (a) Interaction with other dislocations, reconfiguring the order parameter field, ultimately reducing $\Delta\tau_{\text{disl}}$ as per Le Chatelier's principle.
   (b) Adequate time is required to form plastic structures. However, dislocation motion faces hindrances by several mechanisms [1, 2], with additional friction forces (20.24) emerging upon process zone formation, decelerating dislocation movement. Factors (a) and (b) collectively contribute to decrease the plasticity of the crystal.

The specific expression of the inequality (26.5) and its implications differ significantly depending on whether it applies to a daughter-phase zone embedded in a bulk mother phase or to a mother-phase zone encompassed by a daughter-phase matrix. Each context yields a distinct form and outcome of the inequality. Subsequent sections explore these scenarios.

26.2 Plastic Instability Due to the Daughter-Phase Zone at the Dislocation

This section identifies a region within the phase diagram where edge dislocations in the bulk mother phase become unstable due to a daughter-phase process zone. We delineate the temperature range below the local binodal but above the bulk binodal lines where plastic instability manifests.

***

26.2.1 Derivation of the Plasticity Factor

Using the formulas from references (22.8), (26.1), (26.2), and the definition (26.4), we derive the plasticity factor, $k_{\mathrm{pl}}^{(\mathrm{d})}$, related to the edge dislocation with the daughter-phase process zone:

$$k_{\mathrm{pl}}^{(\mathrm{d})} \approx q_{\mathrm{d}} \times \frac{T_{\mathrm{d}} - T}{T - T_0} \tag{26.6}$$

where

$$q_{\mathrm{d}} = 9.04 \frac{A(1-2\nu)(1+\nu)}{\mathfrak{b} E(1-\nu)\ln(\mathbb{R}_{\mathrm{cr}}/\mathfrak{b})} \sqrt{\frac{g}{\gamma}} \tag{26.7}$$

Considering:

- the crystal size $\mathbb{R}_{\mathrm{cr}} \sim 10\,\mu\mathrm{m}$ and the elementary cell size $\mathfrak{b} \sim 3\,\mathrm{Å}$, or
- the dislocation spacing in a "good" crystal $\mathbb{L}_{\mathrm{d}} \sim 10\ \mu\mathrm{m}$, as described by Friedel [1]

we obtain $\ln(\mathbb{R}_{\mathrm{cr}}/\mathfrak{b}) \sim \ln(\mathbb{L}_{\mathrm{d}}/\mathfrak{b}) \sim 10$. Using the parameters from the Tables A.1 and A.2 listed in Annex I, we can estimate the typical value of the constant q_{d} as:

$$q_{\mathrm{d}} \sim 10^{-2}$$

26.2.2 Temperature of Instability Onset

Applying the criterion of plastic instability (26.5), the temperature, T_{inst}, below which the instability onset takes place is:

$$T_{\mathrm{inst}} = \frac{T_0 + q_{\mathrm{d}} T_{\mathrm{d}}}{1 + q_{\mathrm{d}}} \tag{26.8}$$

We aim to plot (26.8) on the phase diagram. This can be done by expressing the temperatures in terms of the parameters a and b using the relations:

$$a_{\mathrm{inst}} = a_0(T_{\mathrm{inst}} - T_{\mathrm{c}}); \quad a_{\mathrm{d}} = a_0(T_{\mathrm{d}} - T_{\mathrm{c}}); \quad a_0(T_0 - T_{\mathrm{c}}) = \frac{3b^2}{16} \tag{26.9}$$

Extracting the expressions for T_0 and T_{d} from these relations, substituting them into (26.8), and recalling the expression for a_{d} (22.5), we find:

$$a_{\mathrm{inst}} \approx \frac{1}{1 + k_{\mathrm{pl}}^{(\mathrm{d}0)}} \left[0.14 k_{\mathrm{pl}}^{(\mathrm{d}0)} + \left(\frac{3}{16} + 0.14 k_{\mathrm{pl}}^{(\mathrm{d}0)} \right) b^2 \right] \tag{26.10}$$

The instability domain lies below the local binodal, a_{d} (22.5), and above the bulk binodal, $a = 3b^2/16$.

26.3 Formation of Plastic Instability in the Daughter Phase

26.3.1 Parameter of Instability

As we shift our focus to the physical parameters, we follow the transformation rule indicated in (20.22), which defines the relation $\Delta\mathcal{F}$. Additionally, it is necessary to adjust the values of a and b as per the equations detailed in (20.13). Representing the perturbation of the dislocation linear tension, $\tau_{\mathrm{disl}}^{(0)}$ (see Sect. 1.4.1), by the order parameter contribution. By expressing this in terms of the binodal slope, k_0, the relationship can be summarized as:

$$\Delta\tau_{\mathrm{disl}} \approx -\frac{1}{\pi\sqrt{\beta^2 - 4\alpha\gamma}} \left[\frac{k_0 \mathfrak{b} \alpha_0 E}{12(1-\nu)} \right]^2 \ln\left(\frac{\mathbb{R}_{\mathrm{cr}}}{\mathfrak{b}} \right) \tag{26.11}$$

This formulation encapsulates the transition to utilizing physical parameters within the free energy expression.

One finds:

$$k_{\mathrm{pl}}^{(\mathrm{m})} \approx \frac{2A^2(1-2\nu)^2(1+\nu)}{E(1-\nu)\sqrt{\alpha_0\gamma(T_{\mathrm{up}} - T)}} \tag{26.12}$$

where $\Delta T_{up} = T_{up} - T_0$ is the temperature interval between the upper spinodal, T_{up}, and the binodal, T_0 (3.33). Since the mother-phase process zone only occupies the temperature interval $T < T_0$, the expression for k_{pl} never diverges and achieves its maximum value, $k_{pl}^{(max)}$, at the binodal ($T = T_0$):

$$k_{pl}^{(max)} \approx \frac{k_0^2 \alpha_0^{3/2} E(1+\nu)}{36(1-\nu)\sqrt{\gamma \Delta T_{up}}} \tag{26.13}$$

This effect reduces the line tension within the bulk daughter phase, similarly to how it occurs in the bulk mother phase. However, there are significant differences between these two scenarios:
In the Bulk Mother Phase:

1. The process zone's energy $\Delta \mathcal{F}$ does not depend on the crystal size $\mathbb{R}_{cr}$, as the zone is localized at the dislocation. In contrast, $k_{pl}^{(d)}$ depends on $[\ln (R_{cr}/\mathfrak{b})]^{-1}$. Although the logarithmic function changes slowly, for macroscopic crystals, this relationship reduces $k_{pl}^{(d)}$ by approximately an order of magnitude.
2. The influence of plastic instability is limited to areas of the phase diagram where the process zone is present, specifically in region II, as shown in Fig. 22.6.

In Contrast, in the Bulk Daughter Phase:

1. The domain of plastic instability extends beyond the mere presence of the mother-phase zone, covering broader regions than just regions III and IV in the phase diagram (Fig. 22.6).
2. The perturbation caused by the order parameter field is long-range, scaling with the crystal size as $\ln (\mathbb{R}_{cr}/\mathfrak{b})$, which means that $k_{pl}^{(m)}$ remains independent of $\mathbb{R}_{cr}$ and significantly amplifies its impact compared to that in the bulk mother phase.

Thus, the instability condition (26.5) is directly applicable. The derived formula for the plastic factor $k_{pl}^{(m)}$ (26.12) applies to the mother-phase zone embedded in the bulk daughter phase, defining the temperature range indicative of plastic instability in the daughter phase:

$$0 < T_{up} - T_{pl}^{(m)} < \frac{4}{\alpha_0 \gamma} \left[\frac{A^2(1-2\nu)^2(1+\nu)}{E(1-\nu)} \right]^2 \tag{26.14}$$

This is supplemented by the requirement that the mother-phase process zone occurs only below the binodal: $T_{pl} < T_0$.

In the next section, we will map the relation (26.14) onto the phase diagram in the (b, a) plane. To do this, we express the condition in terms of a and b as follows:

$$\frac{b^2}{4} - \frac{g}{\gamma} \left[\frac{\pi A(1-2\nu)(1+\nu)}{\mathfrak{b} E} \right]^2 < a_{inst}^{(m)} < \frac{3b^2}{16} \tag{26.15}$$

***

In summary, *the simple condition $k_{\text{pl}} \geq 1$governs the onset of plastic instability induced by the process zone at the dislocation. This condition translates into temperature intervals defining where plastic instability arises in the bulk mother and daughter phases, though these intervals differ.*

In the bulk mother phase, *the process zone remains localized near the dislocation, with its influence confined to the region where the daughter-phase zone exists.*

Conversely, in the bulk daughter phase, *the dislocation generates long-range perturbations in the order parameter field, contributing to the zone's free energy. This influence extends beyond the existence domain of the mother-phase zone.*

26.4 Mapping Plastic Instability Domains on the Phase Diagram

26.4.1 Overview of the Phase Diagram

Building upon our prior discussions, we now integrate the identified domains of plastic instability into the crystal's phase diagram, best visualized in the (b, a) plane. Figure 26.1 schematically illustrates these domains against the backdrop of the zone phase diagram, with notations for regions I to VII and color codes and hatching.

The diagram delineates seven distinct regions, with region I representing the crystal wholly in the mother phase, where the dislocation remains bare. Dislocation dressing occurs in regions II and VI, delineated by the local binodal, the dot-dashed line (a), (22.5) and the global binodal (the solid line line (c), (22.19)). Beneath the global binodal, the crystal transitions entirely to the daughter phase, showcasing fully-developed and less-developed mother-phase zones within the bulk daughter phase in regions III and IV, respectively. The demarcation between these regions is the dashed line (d) (22.9).

Above the binodal, plastic instability initiates within the hatched area of region VI, delineated by dashed line (b) (26.10) and terminating where it intersects the local binodal. The intersection point is shown by a disk. The domain VI extends to the critical value $b = b_{\text{cr}}^{(1)} \approx 1.698$ (22.20), beyond which plastic instability ceases.

The continuation of the plastic instability domain highlighted by hatching lies beneath the bulk binodal. It encompasses the entirety of regions III, IV and VII without necessitating separate shading. This domain is framed by the global binodal (the solid line c) and the lower onset line of plastic instability (the dotted line (e), (26.15)), yielding the relation:

$$a = -\Delta a_{\text{pl2}} + b^2/4; \quad \Delta a_{\text{pl2}} > 0 \tag{26.16}$$

Notably, the upper spinodal equation $a = b^2/4$ shifts downwards to form a line (26.16), intersecting with the global binodal, $a = 3b^2/16$, at the point $b_{\text{cr}}^{(2)}$ specified by (26.17), indicating the rightward limit of plastic instability beneath the binodal.

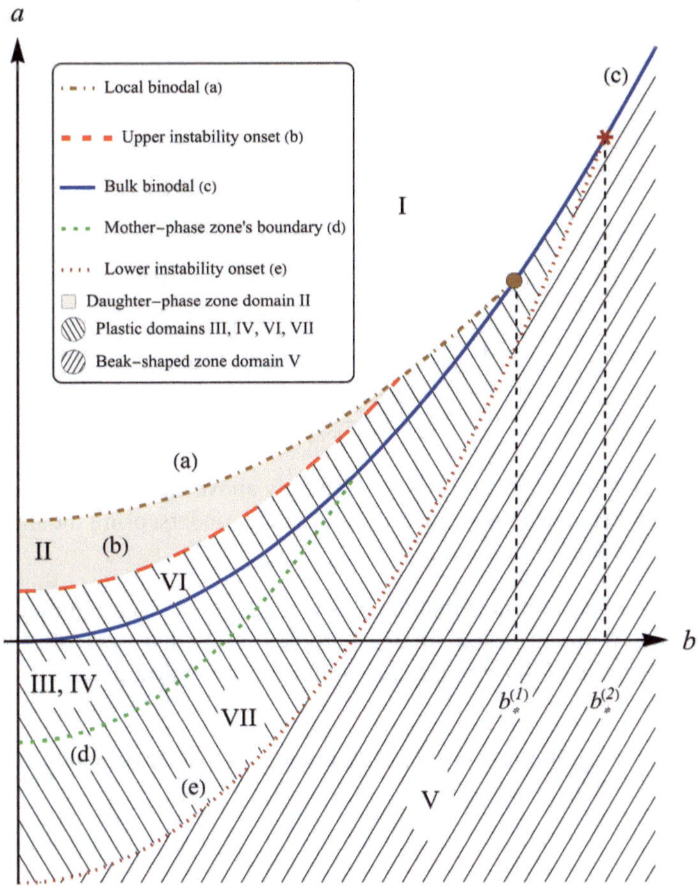

Fig. 26.1 Plastic instability domains overlaid on the zone phase diagram. Not in scale. Find the description in the text

$$b_{\mathrm{cr}}^{(2)} = 4(\Delta a_{\mathrm{pl2}})^{1/2} \tag{26.17}$$

Finally, the hatched area V, located below the dotted line (e) for $b < b_{\mathrm{cr}}^{(2)}$ and below the solid line (c) for $b \geq b_{\mathrm{cr}}^{(2)}$, contains the dislocation along with the beak-shaped mother-phase zone embedded within the bulk daughter phase.

26.4.2 Physical Interpretation and Temperature Intervals

The shifts Δa_{pl1} and Δa_{pl2} translate to temperature intervals $\Delta T_2^{(\mathrm{up})}$ and $\Delta T_2^{(\mathrm{low})}$, respectively, as described by (26.18) and (26.15). Additionally, $b_{\mathrm{cr}}^{(1)}$ and $b_{\mathrm{cr}}^{(2)}$ correlates to specific $\beta_{\rm cr}^{(1)}$ and $\beta_{\rm cr}^{(2)}$ value (26.19), providing a direct connection to physical parameters:

$$\Delta T_2^{(\rm up)} \approx 1.4\times 10^{-5}\frac{\alpha_0}{g}\left[\frac{k_0 \mathfrak{b} E}{(1-\nu)}\right]^2$$

$$\Delta T_2^{(\rm low)} = \frac{1}{\alpha_0\gamma}\left[\frac{A^2(1+\nu)(1-2\nu)}{E(1-\nu)}\right]^2 \quad (26.18)$$

$$\beta_{\rm cr}^{(1)} \approx -0.54\frac{A\mathfrak{b}(1-2\nu)}{1-\nu}\sqrt{\frac{\gamma}{g}} \quad (26.19)$$

$$\beta_{\rm cr}^{(2)} = -\frac{4A^2(1+\nu)(1-2\nu)}{E(1-\nu)} \quad (26.20)$$

Note 26.4.1 Distinct origins of plastic instability above and below the binodal lead to a significant difference between $\Delta a_{\rm pl1}$ and $\Delta a_{\rm pl2}$, underscoring the intricate relationship between the two domains.

Note 26.4.2 We demonstrated that the plastic domain occupies a limited part of the phase diagram. While the latter finding is rooted in Landau's potential framework (3.8), the persistence of these properties within more complex potentials remains an open question.

In the ensuing section, we further elaborate on the numerical parameters for this phase diagram and their implications.

26.4.3 Numerical Estimates of the Plastic Instability Domains

26.4.3.1 Temperature Interval Estimates in the Bulk Mother Phase

By estimating the expressions (26.8) and (26.14) using the data provided in Tables A.1 and A.2 (Annex 1), we obtained the following estimates for the ratio $T_{\rm pl}/T_0$:

$$T_{\rm pl}^{(\rm d)}/T_0 \approx 1.3 \;\text{ and }\; T_{\rm pl}^{(\rm m)}/T_0 \approx -1.4 \quad (26.21)$$

The first relation suggests that the plastic instability temperature $T_{\rm pl}^{(\rm d)}$ reaches up to approximately $1.3T_0$. For $BaTiO_3$, this corresponds to $T_{\rm pl}^{(\rm d)} \approx 496\,\rm K$. However, the second result appears non-physical, implying that the plastic domain extends to "negative" temperatures. It is also worth noting that according to the estimate listed in Table 27.1, $T_{\rm m} \sim T_c - 10^3\,\rm K$. The latter is also unphysical. The explanation of this seeming contradiction is that the latter estimate is subjected to one more necessary condition that we discussed only equivocally. Namely, the conditions (26.21) are valid as long as they apply to the bulk mother- and daughter phases. Beyond their area of existence the conditions do not apply.

***

In summary, *the key qualitative result of our study is the identification of an extensive region on the phase diagram where plastic instability occurs. Estimates indicate that this region spans a significant temperature interval.*

References

1. J. Friedel, *Dislocations* (Pergamon Press, Oxford, 1964)
2. J.P. Hirth, J. Lothe, *Theory of Dislocations* (Krieger Publishing Company, Malabar, Florida, 1991)

Chapter 27
Numerical Estimates of the Zone's Key Parameters

Abstract In this chapter, based on material parameters for $BaTiO_3$, we provide estimates for the process zone generated by a single dislocation and by a dislocation wall. As in the previous chapter, we estimate key geometric characteristics, including the natural scale of the zone, its size, the length of the wake, the temperature intervals of the zone's existence, and the critical velocity. In addition, we examine key energetic and mechanical characteristics, including the plastic factor, the stress generated by energy dissipation in the zone relative to the Peierls–Nabarro stress, and the velocity at which these stresses become equal.

27.1 Parameters of the Zone Generated by the Dislocation

In this section, we provide estimates for the process zone generated by a single dislocation, based on the material constants found in Sect. A.1 and A.2. Similar to the previous chapter, we estimate key geometric characteristics such as $\mathbb{R}_0$, $\mathbb{R}_\mathrm{d}$, *and* $\mathbb{L}_\mathrm{wake}$, *as well as temperature intervals* $T_\mathrm{d} - T_\mathrm{c}$ *and* $T_\mathrm{m} - T_\mathrm{c}$, *and the critical velocity* V_d. *In addition, we examine key characteristics, including the ratio of the zone's free energy to the energy of a bare dislocation* $\Delta\mathcal{F}_{\mathrm{d,m}}/W^{(0)}$, *the stress generated by energy dissipation in the zone relative to the Peierls-Nabarro stress* $\sigma_{\mathrm{slow,\,fast}}/\sigma_\mathrm{PN}$, *and the velocity* V_PN, *at which* $\sigma_\mathrm{slow}/\sigma_\mathrm{PN} \sim 1$.

***

27.1.1 Estimates for the Single Dislocation

The estimates in Table 27.1 are rounded to the order of magnitude.

***

In summary, *the dislocation's zone exhibits the following peculiarities:*

1. **Temperature Intervals and Natural Spatial Scale**. *The temperature intervals* $T_\mathrm{d} - T_\mathrm{c}$ *and* $|T_\mathrm{m} - T_\mathrm{c}|$ *can reach values* $\sim 10^3$ K, *while* $\mathbb{R}_0 \sim \mathfrak{a}$ *(where* $\mathfrak{a}$ *is the*

© The Author(s), under exclusive license to Springer Nature Switzerland AG 2026

A. Boulbitch and A. Korzhenevskii, *Stress-Induced Local Phase Transformations Near Cracks and Dislocations*, Springer Series in Materials Science 356,
https://doi.org/10.1007/978-3-032-00945-6_27

Table 27.1 Key parameters of a zone at the dislocation

Property	Estimate for displacive transitions	Estimate for order-disorder transitions	References
$T_d - T_c$ (K)	$\sim 10^3$	10 to 10^2	(22.4)
$T_m - T_c$ (K)	$\sim -10^{-3}$	-1 to -10	(22.10)
$\mathbb{R}_0$ (nm)	$\sim 10^{-1}$	$\sim 10^{-1}$	(20.8)
$\mathbb{R}_d$* (nm)	$\sim 10^3$	10 to 10^2	(23.5)
$\Delta\mathcal{F}_d / W^{(0)}$ (d/l)**	$\sim -10^{-2}$	$\sim -10^{-2}$	(22.8), (1.17)
$\Delta\mathcal{F}_m / W^{(0)}$ (d/l)	~ -10	~ -1	(22.18), (1.17)
V_d (m/s)	$\sim 10^2$	1	(20.16)
V_{PN} (cm/s)	~ 10	10^{-1}	(24.13)
$\mathbb{L}_{wake}$*** (μm)	~ 1	10^{-1}	(24.6)
$\sigma_{slow}/\sigma_{PN}$ (d/l)	$\sim 10^{-2}$	1	(24.11), (24.12)
$\sigma_{fast}/\sigma_{PN}$ (d/l)	$\sim 10^2$	$\sim 10^2$	(24.18), (24.12)
σ_{slow}/σ_r (d/l)	$\sim 10^5$	$\sim 10^4$ to 10^5	(24.15)

* $T = T_0 + 1\,\text{K}$
** d/l is dimensionless
*** $T - T_0 \approx T_d - T_c;\ T_0 - T_c \approx 10\,\text{K}$

lattice constant). Therefore, the points made in Clauses 2 and 4 of Sect. 16.3.1 are also relevant here. Unlike the crack case, where $T_d - T_c$, $|T_m - T_c|$, and $\mathbb{R}_0$ depend on K_I and can vary with the applied load, in the case of dislocations, these parameters are independent of any variable factors and remain constant.

2. **Second-Order Zones During Superconductive Transitions**. *Superconductive transitions exhibit phase transition line slopes on the phase diagram ranging from approximately* $\sim 10^{-4}$ *to* 10^{-2} K/kbar, *indicating that the corresponding striction constants are small, which results in small values for* $T_d - T_c$. *For instance, Nabutovskii and Shapiro* [1] *estimated* $T_d - T_c$ *to be around* 10^{-2} *and* 10^{-3} K *for* V_3Ge *and* Nb_3Sn, *respectively, illustrating the distinctive nature of superconductive transitions.*[1]
3. **Size of the Mature Zone**. *At* $T = T_0 + 1$ K, *the zone size is* $R_d \sim 1\,\mu\text{m}$. *As the temperature approaches the binodal, the zone size increases rapidly, potentially challenging the assumption of isolated dislocations, i.e.,* $\mathbb{R}_d \ll \mathbb{L}_{disl}$ *(see Sect.* 20.2*). In this regime, the isolated dislocation model may no longer apply.*
4. **Stress Normalization**. *The stresses* σ_{slow} *and* σ_{fast} *exerted on the dislocation by the zone are normalized to the Peierls-Nabarro stress,* σ_{PN} *(*24.12*)* [2].

[1] Nabutovskii and Shapiro's groundbreaking work laid the foundation for a field-theoretical approach to describe the superconductive process zone near a dislocation.

5. **Estimates of** σ_{slow} **and** $\mathbb{L}_{\text{wake}}$. *The stress* σ_{slow} *(24.11) depends on* $T - T_0$. *In our estimates, we approximated this temperature interval using* $T_{\text{d}} - T_{\text{c}}$ *(22.4), effectively estimating* σ_{slow} *at the bifurcation boundary. Given a hysteresis width of around* 10 K *(see Sect.* A.1.1*), we also estimate* $T_0 - T_{\text{c}}$ *to be on the order of* 10 K, *which is useful for determining* $\mathbb{L}_{\text{wake}}$.
6. **Estimate of** V_{PN}. *After approximating* $T - T_0$ *(as outlined in Clause 5), we estimate the Peierls-Nabarro velocity* $V_{\text{PN}} \sim 10\,\text{cm/s}$. *This provides a natural definition for low velocities,* $V \ll V_{\text{PN}}$. *To estimate* $\sigma_{\text{slow}}/\sigma_{\text{PN}}$, *we assumed a velocity of* $V \sim 1\,\text{cm/s}$. *In contrast, dislocation glide typically occurs at velocities around* $10^{-1}c \sim 10^4\,\text{cm/s}$ [2], much greater than Gordon's velocity V_{G} *(see Sect.* A.1.2). *This allows us to confidently apply the velocity-independent expression (24.18) for* σ_{fast}.
7. **Fast Motion and Friction Stress**. *For fast-moving dislocations, our estimates predict extreme friction stress, which limits the propagation of the dislocation-zone complex. This implies that dislocations cannot propagate at velocities* $V \gg V_{\text{G}}$. In such cases, the dislocation either slows down or detaches from the zone to propagate independently at velocities exceeding V_{d}.

27.2 Estimates for the Dislocation Wall Zone

Here, we estimate the properties of the low-angle grain boundary four angles: 0.1°, 0.5°, 1°, *and* 2°.

***

In the case of the dislocation wall, we provide estimates for the grain boundary angles: 0.1°, 0.5°, 1°, *and* 2°, *for which the simulations have been performed.*

1. *For transitions of the displacive type, we obtained the estimate* $T_{\text{d}} - T_{\text{c}} \sim 10^5\,\text{K}$ *for all angles.*[2] *For the order-disorder transition type,* $T_{\text{d}} - T_{\text{c}} \sim 10^2\,\text{K}$.
2. *The estimate of the natural scale,* R_0 *(25.9), for all angles results in* $R_0 \sim 10\,\text{nm}$.
3. *The size,* R_y, *of the process zone in the* OY *direction for the specified angles is illustrated in Fig.* 25.5. *We find that* $R_y \sim 10\,\text{nm}$, *and it increases as the angle decreases (see Fig.* 25.5*). The size* R_x, *across the zone, is smaller but remains within the same order of magnitude.*
4. *The plasticity factor,* k_{pl} (25.14), *depends on temperature. We estimated it using Tables* A.1 and A.2 *for the displacive transitions, and according to the comments below Table* A.2 *for the order-disorder transitions. The estimates are summarized in the plot shown in Fig.* 27.1.

For the displacive transitions (left ordinate), k_{pl} *(25.14) varies between* 10^{-5} *to* 10^{-3} *when* $T - T_{\text{c}}$ *changes from zero to* 30 K, *where the fitting expression (25.12)*

[2] See item one of the discussion below.

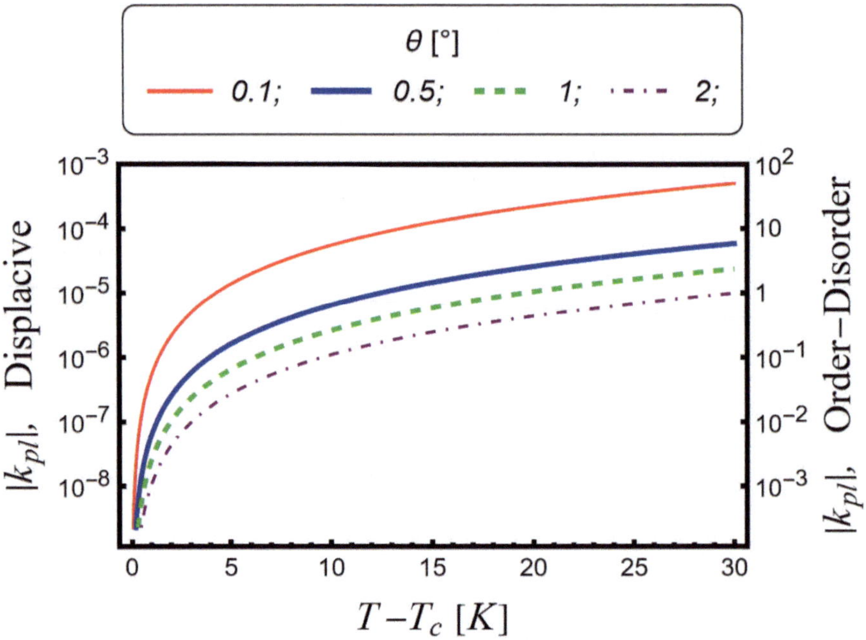

Fig. 27.1 Estimate of the plasticity factor k_{pl} for the dislocation wall. The plot shows k_{pl} in a log-linear format. The left ordinate corresponds to the displacive transitions, while the right one corresponds to the order-disorder ones

remains accurate. In the case of the order-disorder transitions (right ordinate), within the same temperature range, k_{pl} *exceeds* 1, *reaching* 10^2 *at* $\theta = 0.1°$.

***

In summary, *the chain of zones within the dislocation wall displays some distinct characteristics compared to the zones of single dislocations discussed previously.*

1. **High Emergence Temperature**
For transitions of the displacive type, we obtained the estimate $T_{\text{d}} - T_{\text{c}} \sim 10^5$ K *for all angles. Evidently, this value appears quite unphysical, especially since no phase diagram spans such a temperature range. In practice, this large estimate for the temperature interval merely indicates that the chain of the process zones already exists within the dislocation wall immediately below the melting point. However, the zones are highly localized, as discussed in the next item.*
For transitions of the order-disorder type, $T_{\text{d}} - T_{\text{c}} \sim 10^2$ K. *The latter interval corresponds to the typical width of phases in the phase diagram, which is commonly observed also with other defects.*
2. **The Natural Scale and Lateral Zone's Size**
The estimate of the natural scale, $\mathbb{R}_0$ *for all angles results in* $\mathbb{R}_0 \sim 10$ nm. *Recall that in the case of single dislocations or cracks, the zone size near the bifurcation*

is approximately $\mathbb{R}_0$. *However, far from the bifurcation, its size,* $\mathbb{R}_d$ *exceeds* $\mathbb{R}_0$ *by orders of magnitude. In contrast, the sizes of the mature grain boundary zone* $\mathbb{R}_y \sim 10\,\text{nm}$, *and it increases as the angle decreases. The size* $\mathbb{R}_x$, *across the zone, is smaller but remains within the same order of magnitude. Thus, the natural scale,* $\mathbb{R}_0$, *and the mature zone sizes,* $\mathbb{R}_x$ *and* $\mathbb{R}_y$, *are of the same order of magnitude. The latter suggests that the dislocation wall zone only slightly increases in lateral size as the temperature approaches the binodal point. Moreover, as mentioned earlier, although the zone emergence occurs as early as immediately below the melting point, the size of the zone remains small, only slightly larger than the dislocation core, making such zones difficult to observe.*

3. **The Plastic Factor**
The plasticity factor, k_{pl}, *depends on temperature. We estimated it for the displacive and order-disorder transitions. The estimates are summarized in the plot. For the displacive transitions (left ordinate),* k_{pl} *varies between* 10^{-5} *to* 10^{-3} *when* $T - T_c$ *changes from zero to* $30\,\text{K}$. *In the case of the order-disorder transitions (right ordinate), within the same temperature range,* k_{pl} *exceeds* 1, *reaching* 10^2 *at* $\theta = 0.1°$. *It is noteworthy that the plasticity factor,* k_{pl}, *is quite low in the case of displacive transformations but reaches high values, around* 1 *or greater, in the case of order-disorder ones.*

References

1. V.M. Nabutovskiĭ, V.Y. Shapiro, Sov. Phys. JETP **48**(3), 480 (1978)
2. J. Friedel, *Dislocations* (Pergamon Press, Oxford, 1964)

Part V
Anomalous Behavior of Elements of the Dislocation Ensemble Caused by Their Process Zones

In Chap. 26, we discussed the behavior of an isolated, streight, infinite, stress-free dislocation carrying the process zone. We obtained a very general result showing that as soon as $k_{pl} > 1$, the dislocation's line tension becomes negative. The latter translates into its instability arising at $k_{pl} = 1$. In real crystals, however, the dislocations form various structures [1, 2]. Elements of these structures, such as dislocation segments and loops behave differently than such an ideal dislocation. Besides, the real materials exhibit internal stresses $\sim 10^{-5}\mu$ and vacancies [2] curving the dislocation segments. These and some other reasons lead to the variation in the instability point. This Part addresses the manifestation of the plastic instability in elements of the dislocation ensemble.

References

1. J. Friedel, Dislocations (Pergamon Press, Oxford, 1964)
2. J.P. Hirth, J. Lothe, Theory of dislocations (Krieger Publishing Company, Malabar, Florida, US, 1991)

Chapter 28
Elastic Limit of a Crystal Close to a First-Order Transition

Abstract The Frank–Read source is a dislocation segment anchored between two obstacles. When external stress is applied, the segment bends and, upon reaching a critical stress level, generates a closed dislocation loop. This loop then moves away from the obstacles, allowing the source to generate successive loops. The initiation stress of the Frank–Read source defines the crystal's elastic limit—the maximum stress the material can withstand without permanent deformation. This chapter focuses on analyzing the influence of the process zone on the functionality of the Frank–Read source (Korzhenevskii 1986).

In this study we follow the paper [1]. At low Peierls-Nabarro barriers and moderately high temperatures, the elastic limit of a crystal is primarily governed by the minimum stress needed to activate Frank-Read sources [2, 3]. These sources are segments of dislocations pinned at their ends by lattice defects such as Frank dislocation nodes, steps, or impurity atoms.

The relationship between the elastic limit in the absence of a process zone, $\sigma_{\text{el}}^{(0)}$, the dislocation linear tension, $\tau_{\text{disl}}^{(0)}$, and the characteristic size of the Frank-Read source, $\mathbb{L}_{\text{FR}}$, is expressed by:

$$\sigma_{\text{el}}^{(0)} = \frac{\tau_{\text{disl}}^{(0)}}{\mathfrak{g}\mathfrak{b}\mathbb{L}_{\text{FR}}} \tag{28.1}$$

Here, $\mathfrak{g} \sim 1$ is a dimensionless factor [2, 3], and $\tau_{\text{disl}}^{(0)}$ is defined by (1.18). The formation of a process zone reduces the elastic limit.

To examine this, we address a first-order daughter-phase zone within the mother-phase matrix. Using (26.4), we can express $\tau_{\text{disl}} = \tau_{\text{disl}}^{(0)}(1 - k_{\text{pl}}^{(\text{d})})$ and further apply (26.6). The elastic limit in the presence of the daughter-phase process zone, $\sigma_{\text{el}}^{(\text{d})}$, is then:

$$\sigma_{\text{el}}^{(\text{d})} = \frac{\tau_{\text{disl}}^{(0)}}{\mathfrak{g}\mathfrak{b}\mathbb{L}_{\text{FR}}} \left(1 - q_{\text{d}} \frac{T_{\text{d}} - T}{T - T_0}\right) \tag{28.2}$$

As T approaches T_0 from above, the elastic limit $\sigma_{\text{el}}^{(\text{d})}$ decreases. Equation (28.2) allows one to determine the temperature at which it reaches zero, coinciding with $T_{\text{pl}}^{(\text{d})}$, given by (26.8) defining the line (b) on the phase diagram Fig. 26.1. Near this

© The Author(s), under exclusive license to Springer Nature Switzerland AG 2026
A. Boulbitch and A. Korzhenevskii, *Stress-Induced Local Phase Transformations Near Cracks and Dislocations*, Springer Series in Materials Science 356,
https://doi.org/10.1007/978-3-032-00945-6_28

temperature, the reduction in the elastic limit becomes significant, and below $T_{\mathrm{pl}}^{(\mathrm{d})}$, the Frank-Read source begins generating dislocation loops almost without any applied stress. For a mother-phase zone below the bulk binodal, the elastic limit $\sigma_{\mathrm{el}}^{(\mathrm{m})}$ is:

$$\sigma_{\mathrm{el}}^{(\mathrm{m})} = \frac{\tau_{\mathrm{disl}}^{(0)}}{\mathfrak{g}\mathfrak{b}\mathbb{L}_{\mathrm{FR}}} \left[1 - \frac{2A^2(1-2\nu)^2(1+\nu)}{E(1-\nu)\sqrt{\alpha_0 \gamma (T_{\mathrm{up}} - T)}} \right] \tag{28.3}$$

This limit also reaches zero at $T = T_{\mathrm{pl}}^{(\mathrm{m})}$ according to (26.14) representing the line (e) on the phase diagram (Fig. 26.1).

Estimates (26.21) suggest that the temperature range over which the elastic limit decreases is significant.

It is to be noted that the generation of dislocation loops within the region of plastic instability cannot continue indefinitely. Once the available space $\mathbb{L}_{\mathrm{disl}}$ is filled, the stress generated by the newly formed loops hinders the further activity of the Frank-Read source. As a result, while the elastic limit decreases, it cannot reach zero.

***

In summary, *the dislocation line tension governs the material's elastic limit. As the process zone forms, this line tension decreases, reducing the crystal's elastic limit and enhancing its plasticity. Below line (a) on the phase diagram in Fig.* 26.1, *the initiation stress of the Frank-Read source begins to decrease and remains reduced across the entire domain where the daughter phase exists. The Frank-Read source becomes unstable between lines (b) and (e), producing dislocation loops at minimal applied stress. The region on the phase diagram between lines (b) and (e) corresponds to the domain of plastic instability, as defined in Chap.* 26.

References

1. A.L. Korzhenevskii, Sov. Phys. Solid State **28**, 745 (1986)
2. J. Friedel, *Dislocations* (Pergamon Press, Oxford, 1964)
3. J.P. Hirth, J. Lothe, *Theory of Dislocations* (Krieger Publishing Company, Malabar, Florida, 1991)

Chapter 29
Segment of a Dislocation Subjected to Stress Field

Abstract In the model used in this book, only edge dislocations directly form a process zone, while screw dislocations do not. However, screw dislocations contribute indirectly by bending, which introduces an edge component to the Burgers vector. This edge component initiates the process zone, making the bending energetically favorable. For an edge dislocation segment, bending generates a screw component, which in turn reduces the size of the process zone, increases line tension, and limits dislocation deflection. This section presents several examples to illustrate these phenomena (Korzhenevskii and Lisachenko 1988).

29.1 Screw Dislocation Segment

In this section, we focus on a straight segment of the screw dislocation and study its reaction on the externally applied shear stress [1].

***

29.1.1 Segment's Free Energy

We examine the equilibrium configuration of an initially straight segment of an edge dislocation of length $2\mathbb{L}_0$ lying in the slip plane (X, Y), with its Burgers vector $\mathfrak{b}$ along OX. The segment is fixed at its endpoints $X = \pm\mathbb{L}_0, Y = 0$, as shown in Fig. 29.1a.

In the absence of external load, this segment represents a rectilinear screw dislocation (Fig. 29.1a). If the segment bends within the (X, Y)-plane (Fig. 29.1b), its shape can be described by the function $Y = Y(X)$. Since the Burgers vector is always conserved, in the bent state it decomposes into two components—an edge component, $\mathfrak{b}_e$, and a screw component, $\mathfrak{b}_s$:

$$\mathfrak{b}_s = \mathfrak{b}\cos\varphi; \qquad \mathfrak{b}_e = \mathfrak{b}\sin\varphi \tag{29.1}$$

© The Author(s), under exclusive license to Springer Nature Switzerland AG 2026
A. Boulbitch and A. Korzhenevskii, *Stress-Induced Local Phase Transformations Near Cracks and Dislocations*, Springer Series in Materials Science 356,
https://doi.org/10.1007/978-3-032-00945-6_29

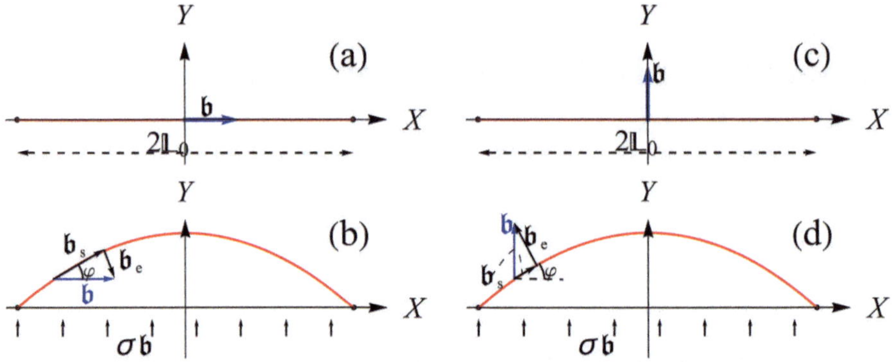

Fig. 29.1 Bending of a dislocation segment. **a** A straight segment of a screw dislocation (red), length $2\mathbb{L}_0$, is pinned by two point obstacles (shown by black disks). Its Burgers vector, $\mathfrak{b}$, is shown by the blue arrow. **b** Bending of this segment in the sliding plane after application of the external Peach-Köhler force $\sigma\mathfrak{b}$. The decomposition of the Burgers vector into its edge and screw components is shown. **c** and **d** Show the same scenario with the initially straight segment of the edge dislocation

where φ is the angle between the tangent vector to the dislocation segment at a given point and the OX direction. Here, $\sin\varphi$ and $\cos\varphi$ are described by:

$$\sin\varphi = \frac{Y'}{\sqrt{1+Y'^2}}; \qquad \cos\varphi = \frac{1}{\sqrt{1+Y'^2}} \tag{29.2}$$

where $Y' \equiv Y'(X) = dY(X)/dX$.

From (29.1), the dislocation's line tension is expressed as a combination of the tensions of its components:

$$\tau = \tau_{\rm s}^{(0)} \cos^2\varphi + \tau_{\rm e}^{(0)}(1-k_{\rm pl})\sin^2\varphi \tag{29.3}$$

where $\tau_{\rm s}^{(0)}$ and $\tau_{\rm e}^{(0)}$ are the line tensions of the screw and edge components, respectively. The factor $(1-k_{\rm pl})$ accounts for the contribution of the edge component's process zone. The line tensions $\tau_{\rm s}^{(0)}$ and $\tau_{\rm e}^{(0)}$ following from the expressions (1.18, 1.19) are given by:

$$\tau_{\rm s}^{(0)} = \frac{\mu\mathfrak{b}^2}{4\pi}\ln\left(\frac{\mathbb{R}_{\rm cr}}{\mathfrak{b}}\right); \qquad \tau_{\rm e}^{(0)} = \frac{\mu\mathfrak{b}^2}{4\pi(1-\nu)}\ln\left(\frac{\mathbb{R}_{\rm cr}}{\mathfrak{b}}\right) \tag{29.4}$$

By considering the angle $\varphi = \varphi(s)$ as a function of the arc length s, the dislocation energy can be obtained by integrating (29.3) over length s of the curved segment:

$$\mathcal{F} = \tau_{\rm s}^{(0)} \int \left[\cos^2\varphi(s) + \frac{1-k_{\rm pl}}{1-\nu}\sin^2\varphi(s)\right] ds \tag{29.5}$$

Converting to Cartesian coordinates (X, Y) and using $ds = \sqrt{1 + Y'^2(X)}dX$ along with the trigonometric relations (29.2) for $\sin\varphi$ and $\cos\varphi$, the energy of the bent segment is:

$$\mathcal{F} = \tau_s^{(0)} \int_{-\mathbb{L}_0}^{\mathbb{L}_0} \left[1 + \frac{1 - k_{\text{pl}}}{1 - \nu} Y'^2(X)\right] \frac{dX}{\sqrt{1 + Y'^2(X)}} \tag{29.6}$$

In subsequent sections, we will refer to this free energy expression for the screw dislocation segment.

29.1.2 Bending of the Segment of the Screw Dislocation Under External Stress

The expressions (29.5) and (29.6) do not account for any external stress applied to the segment. Let us assume that a shear stress, σ_{xz}, is applied to the solid, generating a uniform Peach-Köhler force of $\mathfrak{b}\sigma_{xz}$ per unit segment length. This force performs work, $\mathcal{W} = \mathfrak{b}\sigma_{xz} \int Y(X)dX$, which must be subtracted from the free energy (29.6), leading to a modified free energy $\mathcal{F}' = \mathcal{F} - \mathcal{W}$. The resulting expression is:

$$\mathcal{F}' = \tau_s^{(0)} \int_{-\mathbb{L}_0}^{\mathbb{L}_0} \left\{\left[1 + \frac{1 - k_{\text{pl}}}{1 - \nu} Y'^2(X)\right] \frac{1}{\sqrt{1 + Y'^2(X)}} - \frac{\mathfrak{b}\sigma_{xz}}{\tau_s^{(0)}} Y(X)\right\} dX \tag{29.7}$$

Minimizing this functional allows us to find the equilibrium shape of the dislocation segment. By expanding the functional in terms of $Y'^2(X)$ and keeping the expansion up to second-order terms, we obtain:

$$\mathcal{F}' = \tau_s^{(0)} \int_{-\mathbb{L}_0}^{\mathbb{L}_0} \left\{1 + \frac{1 + \nu - 2k_{\text{pl}}}{2(1 - \nu)} Y'^2(X) - \frac{\mathfrak{b}\sigma_{xz}}{\tau_s^{(0)}} Y(X)\right\} dX \tag{29.8}$$

The variation of this functional with respect to $Y(X)$ gives the following equilibrium equation[1]:

$$Y''(X) = \frac{1 - \nu}{1 + \nu - 2k_{\text{pl}}} \frac{\mathfrak{b}\sigma_{xz}}{\tau_s^{(0)}} \tag{29.9}$$

with boundary conditions $Y(\pm\mathbb{L}_0) = 0$. The solution has a form of a parabola:

$$Y(X) = \frac{1 - \nu}{2(1 + \nu - 2k_{\text{pl}})} \frac{\mathfrak{b}\sigma_{xz}}{\tau_s^{(0)}} \left(\mathbb{L}_0^2 - X^2\right) \tag{29.10}$$

[1] Recall that varying the free energy concerning displacement yields the force balance equation.

The maximum deflection occurs at $X = 0$:

$$Y(0) = \frac{(1-\nu)\mathbb{L}_0^2}{2(1+\nu-2k_{\rm pl})} \frac{\mathfrak{b}\sigma_{xz}}{\tau_{\rm s}^{(0)}} \tag{29.11}$$

At $k_{\rm pl} = k_{\rm pl}^{(*)}$, where

$$k_{\rm pl}^{(*)} = \frac{1+\nu}{2} \tag{29.12}$$

the prefactor in (29.10) diverges, signaling an instability in the dislocation segment. As the system approaches this point, the deflection spontaneously increases. Reaching this point marks the onset of morphological instability in the segment. This result can also be obtained using the de Wit-Köhler line tension anisotropic model.

Note 29.1.1 It is worth noting that the inequality $k_{\rm pl}^{(*)} < 1$ holds. For this reason, the critical condition (29.12) is less strict than the condition (26.5). Therefore, the temperature range defined by the inequality $k_{\rm pl} \geq (1+\nu)/2$ is broader than region IV of the phase diagram shown in Fig. 26.1. Thus, this instability precedes the one defined by the equality $k_{\rm pl}^{(*)} = 1$.

Note 29.1.2 Although the result (29.10) is, for simplicity, obtained for a small deflection, account for nonlinear terms does not restrict the possible deformation of the dislocation. Formally, it suggests infinite deflection of the segment at $k_{\rm pl} = k_{\rm pl}^{(*)}$. However, in practice, the deflection growth is limited once it becomes comparable to the inter-dislocation distance: $Y(0) \sim \mathbb{L}_{\rm disl}$.

29.2 Edge Dislocation Segment

Here, we solve the analogous problem for a straight segment of the edge dislocation.

***

Let us now focus on the edge dislocation segment shown in Fig. 29.1c, d. When this dislocation is bent the decomposition of its Burgers vector into the screw and edge components (as shown in Fig. 29.1d) differs from that given by (29.1):

$$\mathfrak{b}_{\rm e} = \mathfrak{b}\cos\varphi; \qquad \mathfrak{b}_{\rm s} = \mathfrak{b}\sin\varphi \tag{29.13}$$

This leads to the expression for the line tension different from (29.3):

$$\tau = \tau_{\rm s}^{(0)} \sin^2\varphi + \tau_{\rm e}^{(0)}(1-k_{\rm pl})\cos^2\varphi \tag{29.14}$$

and the dislocation free energy in the quadratic approximation has the form:

$$\mathcal{F}' = \frac{\mu \mathfrak{b}^2}{4\pi} \int_{-\mathbb{L}_0}^{\mathbb{L}_0} \left\{ \frac{1-k_{\text{pl}}}{1-\nu} + \left[1 - \frac{1-k_{\text{pl}}}{2(1-\nu)} \right] Y'^2 - \frac{4\pi\sigma_{xz}}{\mathfrak{b}\mu} Y \right\} dX \tag{29.15}$$

It follows the equation of equilibrium:

$$Y''(X) = \frac{1-\nu}{1-2\nu+k_{\text{pl}}} \times \frac{4\pi\sigma_{xz}}{\mathfrak{b}\mu} \tag{29.16}$$

with the same boundary conditions as in the previous section. It has the solution:

$$Y(X) = \frac{2\pi\sigma_{xz}(1-\nu)}{\mathfrak{b}\mu(1-2\nu+k_{\text{pl}})} (\mathbb{L}^2 - X^2) \tag{29.17}$$

Since, on the one hand, $1 - 2\nu > 0$, and on the other hand, $k_{\text{pl}} > 0$ by definition (26.4), the denominator in expression (29.17) never reaches zero. Therefore, unlike the screw segment, the edge dislocation segment subjected to the Peach-Köhler force exhibits no instability.

***

In summary, *the screw dislocation segment subjected to the Peach-Köhler force becomes unstable as* k_{pl} *approaches the critical value* $k_{\text{pl}}^{(*)}$, *which is always lower than the threshold* $k_{\text{pl}} = 1$. *This results in a broader temperature range for the instability of the screw segment compared to the plastic instability domain defined in Sect. 26.4. In contrast, the edge dislocation segment under the Peach-Köhler force undergoes bending, but its amplitude remains bounded, showing no tendency toward unrestrained growth.*

29.3 Dislocation's Zone in an Ultrasonic Stress Field

We examine a segment of either an edge or screw dislocation under the influence of an ultrasonic shear stress field, aiming to describe the spatial and temporal configuration of the process zone. Focusing on the case of small vibration amplitudes, we assess the contribution of the process zone to energy dissipation,[2] *supplementary to the Granato–Lücke mechanism* [2].

***

[2] Boulbitch and Korzhenevskii, first published here.

29.3.1 Regime of the Segment Oscillations and Equation of Motion

We consider a dislocation segment (length L) carrying a process zone and subjected to a shear stress wave $\sigma_{yz} = \sigma_0 e^{i\omega t}$, where σ_0 is the spatially uniform stress amplitude and ω the frequency. Let us first determine the small parameters of the problem.[3]

1. **Temporal scale and time derivative**. Ultrasonic waves exhibit a period $T_{\text{us}} = 2\pi/\omega$, typically ranging from 10^{-4} to 10^{-6} s. This is significantly greater than the characteristic relaxation time of the order parameter, $\kappa \sim 10^{-9}$ to 10^{-13} s. Thus, the order parameter always relaxes to its temporal equilibrium much faster than the stress can change appreciably. This justifies neglecting the $\partial\eta/\partial t$ term in the equation of motion for the order parameter, treating time as a parameter.
2. **Spatial scale and derivative along** Z. The size estimation of the process zone divides into two cases:

   - A process zone in the *vicinity of a soft bifurcation*:
   - One estimates the second derivatives, $\partial^2\eta/\partial X^2$ and $\partial^2\eta/\partial Y^2$ as follows:

$$\frac{\partial^2\eta}{\partial X^2}, \frac{\partial^2\eta}{\partial Y^2} \sim \frac{\eta}{\mathbb{R}_0^2} \tag{29.18}$$

For a straight oscillating dislocation, the only length scale is the wavelength $\lambda \sim 10^{-1}$ to 10 cm. Thus, the derivative along Z is estimated as:

$$\frac{\partial^2\eta}{\partial Z^2} \sim \frac{\eta}{\lambda^2}$$

The lateral zone size is of the order $R_0 \sim 10^{-8}$ to 10^{-6} cm. This yields the estimate $\eta/R_0^2 \gg \eta/\lambda^2$. It follows:

$$\frac{\partial^2\eta}{\partial X^2}, \frac{\partial^2\eta}{\partial Y^2} \gg \frac{\partial^2\eta}{\partial Z^2} \tag{29.19}$$

- *Manifested process zone*: Here, the lateral size is approximately $R_d \sim 10^{-4}$ cm (Table 27.1). As discussed in Sect. 23.1, the derivatives $\partial^2\eta/\partial X^2$ and $\partial^2\eta/\partial Y^2$ are small within the zone and become significant only at the interface, which has a thickness $D \sim 10^{-6}$ cm. Therefore:

1. •

$$\frac{\partial^2\eta}{\partial X^2}, \frac{\partial^2\eta}{\partial Y^2} \sim \frac{\eta}{\mathbb{D}_0^2} \tag{29.20}$$

[3] We follow the reasoning provided by V. Kozub and A. K. Tagantsev in [3], who studied phase transition generation by acoustic waves in dislocation-free crystals.

Due to inequality (29.19) the term $\partial^2\eta/\partial Z^2$ can be omitted in the equation for the order parameter, while the derivatives with respect to X and Y must be retained explicitly.

The resulting equation of state for the order parameter becomes:

$$g\left(\frac{\partial^2\eta}{\partial X^2}+\frac{\partial^2\eta}{\partial Y^2}\right)-\left[\alpha-\frac{(1-2\nu)A\mathfrak{b}_e}{\pi(1-\nu)}\frac{\sin(\theta)}{R}\right]\eta-$$
$$-\beta\eta^3-\gamma\eta^5=0, \tag{29.21}$$

where $\mathfrak{b}_e$ is the edge component of the Burgers vector specifies in the next section, X and Y are independent variables, and $\eta=\eta(X,Y,Z,t)$ is the order parameter, with Z and t treated as parameters, alongside g, A, α, etc.

Note 29.3.1 One may adopt the so-called gradient-free approximation by neglecting $\partial^2\eta/\partial X^2$ and $\partial^2\eta/\partial Y^2$ in (29.21), as previously done in Sect. 14.2.1. The resulting algebraic equation for η is:

$$\left[\alpha-\frac{(1-2\nu)A\mathfrak{b}_e}{\pi(1-\nu)}\frac{\sin(\theta)}{R}\right]\eta+\beta\eta^3+\gamma\eta^5=0 \tag{29.22}$$

The boundary of the zone is then defined as follows.

For second-order zone:

$$\alpha-\frac{(1-2\nu)A\mathfrak{b}_e}{\pi(1-\nu)}\frac{\sin(\theta)}{R}=0$$

For first-order zone:

$$\alpha-\frac{(1-2\nu)A\mathfrak{b}_e}{\pi(1-\nu)}\frac{\sin(\theta)}{R}=\frac{3\beta^2}{16\gamma}$$

Although this solution is simpler, it provides less information about the order parameter. Since we have already constructed the solution to (29.21), such as (21.8), we will use it in the following.

In (29.21) and (29.22), the dependence on Z and t is not immediately obvious. These parameters are implicitly contained in $\mathfrak{b}_e$ (see (29.25) below) and explicitly embedded in R and θ as follows:

$$R=\sqrt{[X-X_0(Z,t)]^2+Y^2};\quad \sin(\theta)=\frac{X-X_0(Z,t)}{\sqrt{[X-X_0(Z,t)]^2+Y^2}} \tag{29.23}$$

The segment oscillates in the XZ-plane, with $X_0(Z,t)$ describing its configuration during oscillation:

$$X_0(Z,t) = \mathfrak{A}\sin(\omega t)\sin\left(\frac{\pi Z}{\mathbb{L}}\right) \tag{29.24}$$

Here, $\mathfrak{A}$ is the oscillation amplitude (see Sect. 1.5 for details). In the long-wavelength limit, $X_0 = X_0(t) \neq X_0(Z,t)$, corresponding to a straight infinite dislocation vibrating uniformly. The solution in this limit is obtained by substituting expressions (29.23) and (29.24) into one of the previously derived solutions.

29.3.2 Shear Wave-Induced Dynamics of the Process Zone Engendered by an Edge Dislocation

The process zone at a dislocation segment, pinned at two points and oscillating as shown in Fig. 29.1c, d, exhibits a more intricate shape dependence than described at the end of the previous section. This complexity arises because each element of the segment acquires a screw component in addition to the edge one, while the total Burgers vector remains conserved. The edge component $\mathfrak{b}_e$ satisfies the following relation:

$$\mathfrak{b}_e = \mathfrak{b}\cos\varphi \tag{29.25}$$

where $\varphi = \varphi(Z,t)$ is the angle between the unit tangent vector to the segment at a given point and the vertical direction, as illustrated in Fig. 29.2.

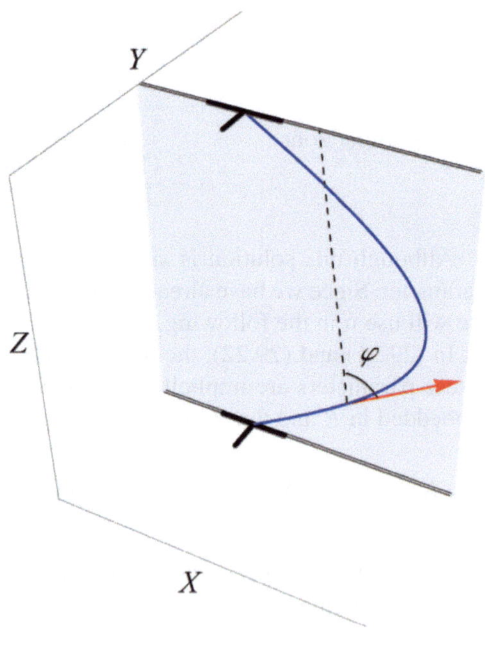

Fig. 29.2 Origin and orientation of the angle φ. The dislocation lies in the (X, Z) plane, which is depicted as a transparent 'glass plate' for clarity. The unit tangent vector to the segment, also confined to the (X, Z) plane, is shown in red in the color version

The expression for $\cos\varphi$ in terms of the segment shape $X_0(Z,t)$ (see (29.24)) is a standard textbook result[4]:

$$\cos(\varphi) = \frac{1}{\sqrt{1 + X_0'^2(Z,t)}}$$

where $X_0'(Z,t) = \partial X_0(Z,t)/\partial Z$ yielding:

$$\cos(\varphi) = \left[1 + \frac{\pi^2\mathfrak{A}^2}{\mathbb{L}^2}\cos^2(\pi z)\sin^2(\omega t)\right]^{-1/2} \tag{29.26}$$

Here, $z = Z/\mathbb{L}$ is the dimensionless length in the vertical direction. Note that the scale along OZ differs from that along OX and OY.

All parameters in the expressions for the stationary order parameter that depend on $\mathfrak{b}$ must be replaced with expressions involving $\mathfrak{b}_e$ (29.25). Specifically:

$$\begin{array}{lll} \mathfrak{b} \to \mathfrak{b}\cos\varphi & B_\mathrm{d} \to B_\mathrm{d}\cos\varphi & \eta_0 \to \eta_0\cos\varphi \\ \mathbb{R}_0 \to \mathbb{R}_0\cos\varphi & r \to r\sec\varphi & a \to a\sec^2\varphi \end{array} \tag{29.27}$$

These substitutions are applied wherever the corresponding parameters appear in the expression for $\eta(x,y,z,t)$. Using one of the previously obtained solutions, such as (21.8) with the parameters appropriately modified, one obtains the corresponding solution for the oscillating dislocation.

Typically, the segment length $\mathbb{L}$ is of the order of the Frank's network mesh size, around 1 μm, while the oscillation amplitude $\mathfrak{A}$ ranges from a few to a few tens of interatomic distances. In this context, the ratio $\mathfrak{A}/\mathbb{L} \ll 1$ is a small parameter. Therefore, it is justified to approximate $\cos\varphi \approx 1$ in all expressions—except for the amplitude of the order parameter:

$$\xi = \sqrt{a_\mathrm{d} - a} \to \sqrt{a_\mathrm{d} - a\sec^2(\varphi)}. \tag{29.28}$$

This exception arises because when $a = a_d - \varepsilon$, with $\varepsilon \ll \mathfrak{A}^2/\mathbb{L}^2$, the quantity $a_d - a\sec^2\varphi$ becomes:

$$a_\mathrm{d} - a\sec^2\varphi \approx \pi^2 a_\mathrm{d}\frac{\mathfrak{A}^2}{\mathbb{L}^2}\cos^2(\pi z)\sin^2(\omega t)$$

(where ε is neglected due to its smallness compared to other terms). In other words, when a is close to a_d, the amplitude is significantly influenced by $\sec^2\varphi$, which cannot be neglected.

4 See, for example [4].

29.3.3 Near-Bifurcation Dynamics of the Process Zone at the Vibrating Edge Dislocation

In the previous section, we established all necessary prerequisites to construct an analytic solution describing the process zone at an oscillating dislocation. Indeed, the arguments presented in Sect. 29.3.1 show that this solution can be readily derived from the known result for a stationary dislocation, such as (21.8).

Based on the rescalings in (5.1) and (6.20), the solution takes the form:

$$\frac{\eta}{\eta_0} \approx (j_2/j_4)^{1/2}\sqrt{a_\mathrm{d} - a\sec^2(\varphi)}e^{-0.37r}\left[1 + 0.62\sqrt{r}\sin(\theta)\right]$$

where $(j_2/j_4)^{1/2} \approx 1.5$, and $\eta_0 = B_\mathrm{d}/\sqrt{g\beta}$, and r, θ are defined as follows:

$$r \approx \sqrt{\left[x - \frac{\mathfrak{A}}{\mathbb{R}_0}\sin(\omega t)\sin(\pi z)\right]^2 + y^2}$$

$$\sin(\theta) \approx \frac{1}{r}\left[x - \frac{\mathfrak{A}}{\mathbb{R}_0}\sin(\omega t)\sin(\pi z)\right] \tag{29.29}$$

resulting in the following solution:

$$\eta(x, y, z, t) \approx 1.5\eta_0\sqrt{a_\mathrm{d} - a\left[1 + \frac{\pi^2\mathfrak{A}^2}{\mathbb{L}^2}\cos^2(\pi z)\sin^2(\omega t)\right]}\times$$

$$\times\, e^{-0.37r}\left[1 + 0.62\sqrt{r}\sin(\theta)\right] \tag{29.30}$$

Figure 29.3 shows the process zone near an edge dislocation at several representative phases of the segment's oscillation. At $\omega t = 0$, as shown in Fig. 29.3a, the dislocation is straight, and its Burgers vector has only an edge component, resulting in a cylindrical distribution of the order parameter.

As ωt increases, the acoustic wave bends the segment in the (x, z) plane. The angle φ reaches its maximum near the pinning points $z = 0$ and $z = 1$. In these regions, the order parameter begins to decrease, and the zone adopts a banana-like shape: thicker at the centre and tapering towards the ends, forming a curved, elongated structure as seen in Fig. 29.3b. Notably, the configurations in Fig. 29.3a, b closely resemble the dislocation zones observed in optical experiments by Belousov and Wolf [5].

With further progression of the oscillation phase, the order parameter continues to diminish. The zone becomes increasingly localized around the midpoint of the segment, where the edge component of the Burgers vector is greatest taking an apple-like form (Fig. 29.3c). At $\omega t = \pi/2$, the entire zone is concentrated near the $z = 0$ plane, forming a pancake-like structure. Its centre shifts from $x = y = 0$ in

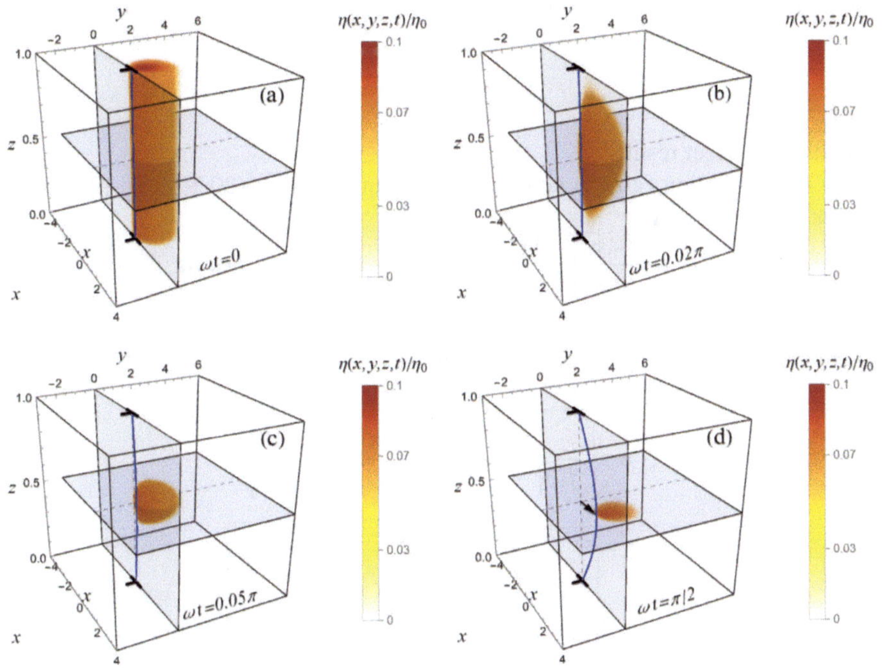

Fig. 29.3 Configurations of the process zone at an oscillating edge dislocation. The dislocation is represented by the solid blue line. Dashed lines along the coordinate axes and the light-blue "glass sheets" lying in the (x, y) and (x, z) planes serve as visual guides. The order parameter distribution is illustrated through a density plot, where the order parameter values are color-coded, with red indicating higher values. The phase of the oscillations is specified at the bottom of each image. In image (d), the deflection of the bent segment is highlighted by a black arrow. Further explanations can be found in the text

the direction indicated by the deflection arrow in Fig. 29.3d. As the oscillation phase continues to increase, the process reverses: pancake → apple → banana → cylinder, and then the cycle repeats.

29.3.4 Energy Dissipation by the Vibrating Zone

29.3.4.1 Dissipation by the Zone of the Single Segment

Any motion of the process zone leads to energy dissipation. In this section, we examine the energy dissipated by the process zone of a vibrating edge dislocation.

The simplest approach employs the theorem proven in Sect. 3.4.3. To recall, this theorem states that, after eliminating the elastic variables, the sole source of energy loss is the evolution of the order parameter. These losses are quantified by the fundamental relation (3.57):

$$\frac{\partial \mathcal{F}}{\partial t} = -2\mathcal{D} \tag{29.31}$$

providing an exact result.

By using relation (20.10) and applying solutions from (29.30) and (29.29), we obtain:

$$\mathcal{D} = \frac{\kappa \mathbb{R}_0^2 \mathbb{L}}{2g\beta} \left[\frac{\mathrm{A}\mathfrak{b}(1-2\nu)}{\pi(1-\nu)} \right]^2 \int\limits_{\Omega} \left(\frac{\partial}{\partial t} \xi \psi \right)^2 dxdydz, \tag{29.32}$$

where the amplitude ξ is specified as follows

$$\xi = \sqrt{\frac{j_2}{j_4} \left\{ a_{\mathrm{d}} - g\alpha \left[\frac{\pi(1-\nu)}{Ab(1-2\nu)} \right]^2 \right\}} \times$$
$$\times \sqrt{\left[1 + \left(\frac{\pi \mathfrak{A}}{\mathbb{L}} \right)^2 \cos^2(\pi z) \sin^2(\omega t) \right]} \tag{29.33}$$

To simplify the calculations, we first express the function $\psi(x, y, z, t)$ in the following form:

$$\psi(x, y, z, t) = \psi(x - \frac{\mathfrak{A}}{\mathbb{R}_0} \sin(\pi z) \sin(\omega t), y, z). \tag{29.34}$$

The term $\mathfrak{A} \sin(\pi z) \sin(\omega t)/\mathbb{R}_0$ in (29.34) is independent of x. Therefore, we perform a variable shift $x' = x - \mathfrak{A} \sin(\pi z) \sin(\omega t)/\mathbb{R}_0$ and integrate over x'.

We then retain only the leading term under the assumption $\mathfrak{A}/\mathbb{L} \ll 1$. In this approximation, we find:

$$\frac{\partial \mathcal{F}}{\partial t} = -\frac{\kappa \mathbb{R}_0^2 \mathbb{L}}{\beta} \left\{ \frac{a_{\mathrm{d}}}{g} \left[\frac{\mathrm{A}\mathfrak{b}(1-2\nu)}{\pi(1-\nu)} \right]^2 - \alpha \right\} \cos(\omega t) \times$$
$$\times \int dz \sin^2(\pi z) \int dx' dy \left(\frac{\partial \psi(x', y)}{\partial x'} \right)^2 \tag{29.35}$$

The integration in (29.35) splits into a straightforward integral over z and another over the (x', y)-plane. For the latter, switching to polar coordinates r and θ, we use:

$$\frac{\partial \psi}{\partial x'} \approx -\frac{e^{-0.37r}}{2\sqrt{r}} \cos(\theta) \left[0.74\sqrt{r} + 0.54(1 + 0.74r) \sin(\theta) \right]$$

This yields the the integral value:

$$\int_{\Omega} \left(\frac{\partial \psi}{\partial x'} \right)^2 dx'\, dy \approx 1.17$$

Finally, the free energy dissipated by the zone of a single dislocation segment over a full period, $2\pi/\omega$, is:

$$\Delta \mathcal{F} \approx -4.5 \frac{\mathfrak{A}^2 \mathbb{L} \kappa \omega}{\beta} \alpha_0 \Delta T \tag{29.36}$$

where $\Delta T = T_d - T$ is the temperature difference between the emergence temperature of the zone, T_d, and the actual temperature T.

29.3.5 Comparison with Granato–Lücke Dissipation

We now compare this result to the analogous expression ΔW in (1.23) from Granato and Lücke, which describes energy dissipation by an undressed dislocation segment:

$$\frac{\Delta \mathcal{F}}{\Delta W} \approx \frac{10 \mathfrak{b}^2 \kappa c_t}{\beta k_B T} \alpha_0 (T_d - T) \tag{29.37}$$

Using the parameters listed in the tables of Annex I, we estimate this ratio. For zones of the displacement type, we find:

$$\frac{\Delta \mathcal{F}}{\Delta W} \sim (T_d - T)/\mathrm{K} \tag{29.38}$$

Thus, if $T_d - T \sim 1\,\mathrm{K}$,the zone's contribution to dissipation is comparable to that of the Granato–Lücke mechanism. However, the zone exists within the temperature interval $T_c < T < T_d$. As seen earlier, the difference $T_d - T_c$ can reach approximately 10^3 K for displacement-type zones (see Table 27.1). Consequently, the zone's contribution may exceed that of Granato–Lücke dissipation by several orders of magnitude.[5] It is also worth noting that, in the case of an order–disorder zone, the larger values of κ and α_0 lead to an increased $\Delta \mathcal{F}/\Delta W$ ratio, making the contribution of the process zone to energy dissipation dominant.

[5] Note that the expression (6.7) for the order parameter used here is valid only near T_d. Far from this point, alternative expressions should be used, as discussed in previous sections. In such cases, the ratio $\Delta F/\Delta W$ increases more slowly with $T_d - T$. A detailed investigation is beyond the scope of this study.

29.3.6 Internal Friction Factor

We now turn to the contribution of the process zone to the internal friction factor. To this end, we estimate the number of 'active' segments within a given volume V. Following [6], we introduce the number of vibrating segments $\mathcal{N}$, along with the total active dislocation length per unit volume L_{act}:

$$L_{\text{act}}=\frac{\mathcal{N}\mathbb{L}}{V} \tag{29.39}$$

as previously defined in Sect. 1.5. This leads to the expression for the total energy dissipated by all segments in the crystal during one oscillation period:

$$\Delta\mathcal{F}_{\text{tot}}\approx -2.26\frac{\mathfrak{A}^2\mathbb{L}\kappa\omega L_{\text{act}}}{\beta}\alpha_0\Delta T\,V \tag{29.40}$$

Using the expression for the ultrasonic wave energy W within the crystal per period (1.26, see Sect. 1.5), and defining the internal friction as:

$$Q^{-1}_{\text{zone}}=\frac{\Delta\mathcal{F}_{\text{tot}}}{W}$$

we get:

$$Q^{-1}_{\text{zone}}\approx 11.9\frac{\kappa\omega L_{\text{act}}\mathbb{L}^4(1-\nu)^2}{\mu\mathfrak{b}^2(1-2\nu)^2\beta}\alpha_0\Delta T \tag{29.41}$$

This internal friction factor can be compared to the value derived by Granato and Lücke (1.28). As expected, the ratio $Q^{-1}_{\text{zone}}/Q^{-1}_{\text{GL}}$ yields the same estimate as $\Delta\mathcal{F}/\Delta W$ (see 29.38).

29.3.7 Screw Dislocation Zone Close to the Bifurcation Point

For a segment of screw dislocation with the configuration shown in Fig. 29.1a, b, the edge component of the Burgers vector $\mathfrak{b}_e$ is given by (29.1):

$$\mathfrak{b}_e=\mathfrak{b}\sin\varphi\equiv\mathfrak{b}\times\frac{X_0'(Z,t)}{\sqrt{1+X_0'^2(Z,t)}}$$

Instead of the substitutions listed in (29.27), the following replacements are introduced for this case:

$$\begin{array}{lll}\mathfrak{b}\to\mathfrak{b}\sin\varphi & B_d\to B_d\sin\varphi & \eta_0\to\eta_0\sin\varphi\\ \mathbb{R}_0\to\mathbb{R}_0\sin\varphi & a\to a\csc^2\varphi & r\to r\csc\varphi\end{array} \tag{29.42}$$

Using these, the ground-state function (21.8) becomes:

$$\psi_{\rm d} \sim \exp(-0.37r) \to \exp\left(-0.37r\frac{\mathbb{L}}{\mathfrak{A}}\right) \tag{29.43}$$

where r is again defined by the first expression in (29.29). From (29.43), we observe that the large parameter $L/\mathfrak{A} \gg 1$ appears in the exponent. This leads to a cut-off over a short distance $\sim R_0\mathfrak{A}/L$, leading to the conclusion: *at small oscillation amplitudes, the process zone at the edge dislocation segment does not form.*

***

In summary, *the application of an ultrasonic wave to a segment of edge dislocation induces complex dynamics in its process zone, manifesting as pulsation. This pulsation contributes to energy dissipation, which can be comparable to or even exceed the dissipation predicted by the Granato–Lücke mechanism. In contrast, when the ultrasonic wave is applied to a segment of screw dislocation that lacks a zone under static conditions, no zone is generated, and dissipation occurs solely via the Granato–Lücke mechanism.*

References

1. A.L. Korzhenevskii, D.A. Lisachenko, Sov. Phys. Solid State **30**, 860 (1988)
2. A. Granato, K.J. Lücke, Appl. Phys. **27**, 583 (1956)
3. V. Kozub, A.K. Tagantsev, Sov. Phys. JETP **62**(1), 126–131 (1985)
4. V.I. Smirnov, *A Course of Higher Mathematics*, vol. 2, Chapter 5, 16 edn. (Pergamon Press, Oxford, 1964)
5. M.V. Belousov, B.E. Volf, JETP Lett. **31**, 450 (1980)
6. J.P. Hirth, J. Lothe, *Theory of Dislocations* (Krieger Publishing Company, Malabar, Florida, 1991)

Chapter 30
Climbing Instability

Abstract Crystals subjected to quenching or irradiation often develop a high concentration of vacancies, resulting in a supersaturated state. These vacancies interact with dislocations by migrating to their cores, where they are absorbed, merging with the half-plane associated with the dislocation's edge component. As a result, the half-plane shortens, effectively driving the dislocation perpendicular to its slip plane in a process known as climb. This motion is driven by the significant energy gain when a vacancy joins the dislocation core, creating an osmotic force that acts on the dislocation normal to its line. This force is so strong that no obstacle can impede the dislocation's climb. A pinned dislocation segment in a crystal with supersaturated vacancies bends and, at a certain level of supersaturation, forms loops. In this chapter, we analyze the effect of the process zone on this phenomenon. Furthermore, in such crystals, prismatic loops are subjected to radial osmotic pressure, which leads to their instability at a critical radius. We examine the contribution of the dislocation's process zone to determining this critical radius. This chapter analyzes the evolution of elements of the dislocation ensemble—Bardeen-Herring sources, and prismatic loops—in a crystal supersaturated with vacancies. This supersaturation can occur due to rapid cooling during quenching or as a result of irradiation. The vacancies accumulate at dislocations, causing them to climb (Korzhenevskii and Lisachenko 1990).

***

30.1 Introduction: Effect of Vacancies on Dislocations

Crystals subjected to quenching or irradiation often develop a high concentration of vacancies, resulting in an oversaturated state. These vacancies interact with dislocations by migrating to the dislocation cores, where they are absorbed, merging with the half-plane associated with the dislocation's edge component. Consequently, the

© The Author(s), under exclusive license to Springer Nature Switzerland AG 2026
A. Boulbitch and A. Korzhenevskii, *Stress-Induced Local Phase Transformations Near Cracks and Dislocations*, Springer Series in Materials Science 356,
https://doi.org/10.1007/978-3-032-00945-6_30

half-plane shortens, effectively propelling the dislocation perpendicular to its slip plane in a process known as climbing. This movement is driven by the significant energy gain when a vacancy joins the dislocation core, creating an osmotic force that acts on the dislocation along the normal to the dislocation line. It is so strong that no obstacle can impede the dislocation's climb. In this section, we provide a brief introduction to the physical description of this phenomenon.

***

The climbing process leads to significant changes in the dislocation structure when the crystal becomes oversaturated with vacancies:

- It is well-known that vacancies oversaturation bends dislocation segments due to the osmotic force, even without external stress.
- This can lead to the formation of dislocation loops. In such cases, the pinned segment is known as *Bardeen-Herring source*.

From a physical perspective, these phenomena are driven by the balance between the linear tension of the dislocation and the osmotic force.[1] The former works to shorten the dislocation, while the latter extends it. This balance is governed by the Helmholtz free energy with a variable number of particles (the Ω-potential):

$$\Omega = W^{(0)} - \mu_{\text{ch}} N \tag{30.1}$$

where $W = \tau \mathbb{L}$ is the dislocation energy (1.14, 1.17), τ is the linear tension of the edge dislocation. Following the approach of the paper [1] we account for the process zone by the factor $(1 - k_{\text{pl}})$:

$$\tau = \tau_{\text{e}}^{(0)} (1 - k_{\text{pl}}) \tag{30.2}$$

and $\tau_{\text{e}}^{(0)}$ is the linear tension of the bare edge dislocation (1.18). For the following analysis, it is convenient to express it in terms of Young's modulus:

$$\tau_{\text{e}}^{(0)} = \frac{E\mathfrak{b}^2}{8\pi(1-\nu^2)} \ln\left(\frac{\mathbb{R}_{\text{cr}}}{\mathfrak{b}}\right) \tag{30.3}$$

Further, $\mathbb{L}$ is the length of the dislocation or its segment, μ_{ch} is the chemical potential of vacancies:

$$\mu_{\text{ch}} = k_{\text{B}} T \ln\left(\frac{c}{c_0}\right) \tag{30.4}$$

Here, k_{B} is Boltzmann's constant, T is the temperature, c and c_0 are nonequilibrium and equilibrium concentrations of vacancies, and N is the number of absorbed vacancies.

[1] For a detailed description of the effect of vacancies on dislocations, see Chap. 17 of Hirth and Lothe's book [2].

If dislocations generate process zones (as assumed here), the crystal should exhibit zone-related effects. The process zone reduces the linear tension of the dislocation's edge component, thereby altering the conditions under which the described structures form.

This chapter aims to explore the influence of the process zone on the dislocation configurations driven by the osmotic force caused by vacancies.

30.2 Deflection of a Plane Dislocation Segment by the Osmotic Force

A pinned dislocation segment in a crystal containing supersaturated vacancies bends and, at a certain supersaturation ratio, c^*/c_0, forms loops. In this section, we analyze the effect of the process zone on this phenomenon.

***

30.2.1 In-Plane Bardeen-Herring Source

Consider a dislocation segment lying in the (X, Y) plane, as shown in Fig. 30.1a. Its associated half-plane lies in the same plane as the dislocation. This plane is visualized as a transparent, light-blue "glass sheet" in the figure.

As soon as supersaturated vacancies appear in the crystal, an osmotic force is exerted on the dislocation line. This force is applied along the outer normal to the dislocation contour, causing the segment to take the shape of a circular arc, visible at an angle φ from the center of the circle, as indicated by black arrows in Fig. 30.1b. The following relations give the segment length $2\mathbb{L}$ and the length of its chord $2\mathbb{L}_0$:

$$\mathbb{L} = \mathbb{R}\varphi \qquad \mathbb{L}_0 = \mathbb{R}\sin\varphi \tag{30.5}$$

The vacancies adsorbed by the dislocation occupy a monoatomic layer with area A and thickness $\mathfrak{b}$, shaded in Fig. 30.1b. Simple geometric calculation yields the area A and the number N of vacancies absorbed therein:

$$A = \mathbb{R}^2\varphi - \mathbb{L}_0\mathbb{R}\cos\varphi; \qquad N = A/\mathfrak{b}^2 \tag{30.6}$$

30.2.2 The Equilibrium State of the In-Plane Bardeen-Herring Source

The Ω-potential of the bent dislocation segment shown in Fig. 30.2b takes the following form:

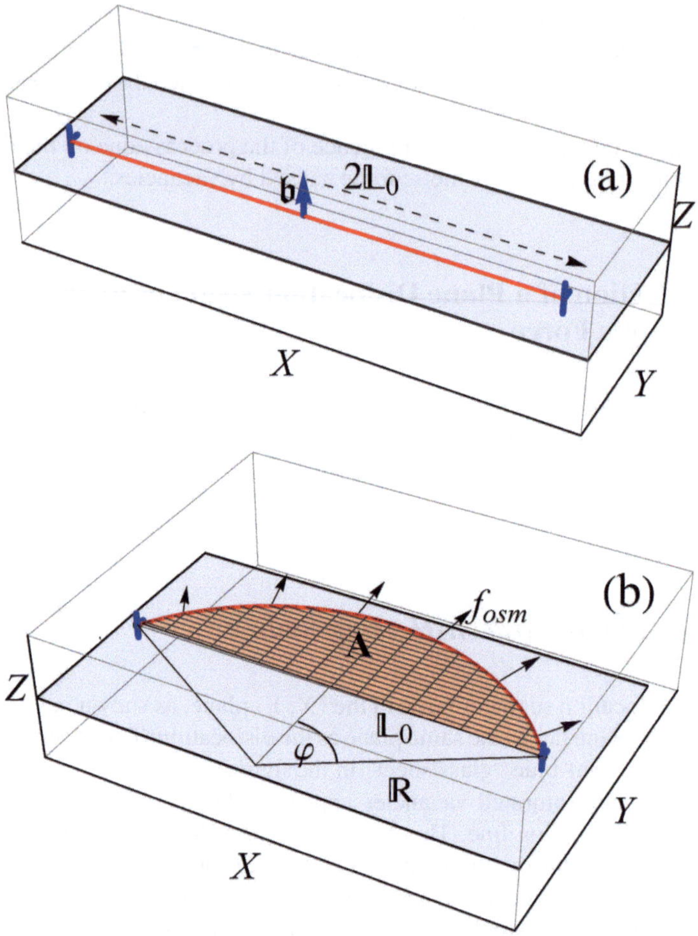

Fig. 30.1 Bardeen-Herring Source. A dislocation segment pinned by two obstacles. The segment (in red) lies within the (X, Y) plane (shown as a glass sheet for clarity). Panel (a) shows the initial position of the straight segment (length $2\mathbb{L}_0$). Pinning defects are located where the dislocation signs are placed in the image. The Burgers vector, $\mathfrak{b}$, is directed along the OZ axis. Panel (b) illustrates the same dislocation forming a circular arc due to the osmotic forces f_{osm}, indicated by arrows. These forces normally act to the segment's contour. The shaded area represents the atomic layer A, which fills with vacancies as the segment bends. It lies between the chord joining the pinning points and the circular arc

$$\Omega = \tau\mathbb{L} - \frac{k_{\mathrm{B}}T}{\mathfrak{b}^2}\ln\left(\frac{c}{c_0}\right)\left(\mathbb{R}^2\varphi - \mathbb{L}_0\mathbb{R}\cos\varphi\right) \tag{30.7}$$

Using the relations from (30.5), we can rewrite the Ω-potential such that it depends only on the unknown parameter φ. It is convenient to divide the Ω-potential by the φ-independent constant product $\tau\mathbb{L}_0$, yielding $\Omega_1 = \Omega/\tau\mathbb{L}_0$. This gives the following expression:

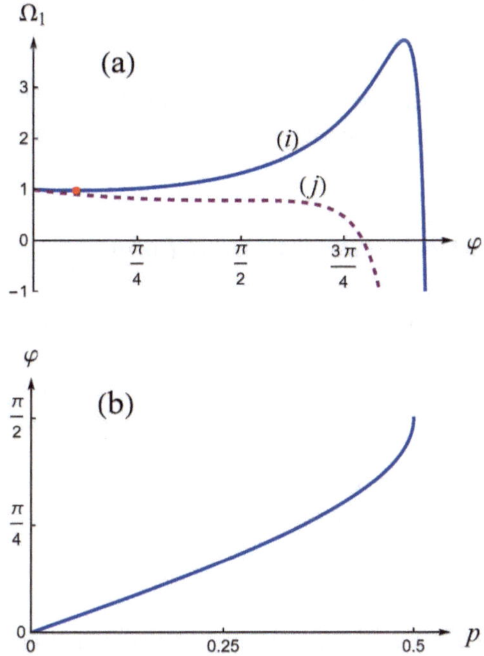

Fig. 30.2 Dependence of the Ω-Potential on Angle φ and the Solution of the Equation of State. **a** The solid blue line (i) shows the behavior of the Ω-potential at $p = 0.16$. The red dot indicates the position of minimum of the Ω-potential. The dashed line (j) displays the run of the Ω_1 at $p = p^*$ at which its minimum vanishes, and the segment becomes unstable. **b** Solution (30.10) of the equation of state

$$\Omega_1 = \frac{\varphi}{\sin\varphi} + p\left[\cot\varphi - \frac{\varphi}{\sin^2(\varphi)}\right] \tag{30.8}$$

where we introduced a dimensionless parameter, p, which absorbs all the constants of the problem:

$$p = \frac{\mathbb{L}_0 k_B T}{\mathfrak{b}^2 \tau_e^{(0)} (1 - k_{pl})} \ln\left(\frac{c}{c_0}\right) \tag{30.9}$$

Figure 30.2a shows the Ω_1-potential as a function of φ.

The potential $\Omega_1 = \Omega_1(\varphi)$ exhibits a minimum, indicated by the red point shown in Fig. 30.2a. We simplify the minimization problem by recognizing that the segment's shape is already known to be part of a circle [2]. This allows us to use the parameterization from (30.5) and directly minimize Ω_1 with respect to φ:

$$\frac{\partial\Omega_1}{\partial\varphi} = 0$$

which we reduced to the equation:

$$\sin\varphi = 2p$$

with the solution:

$$\varphi = \arcsin(2p) \tag{30.10}$$

The solution (30.10) is shown in Fig. 30.2b within the interval $0 < \varphi \leq \pi/2$. When $\varphi = \pi/2$, the segment forms a semi-circle.

30.2.3 Generation of Loops

The minimum of Ω_1 vanishes once the following conditions are satisfied:

$$\begin{cases} \partial\Omega_1/\partial\varphi = 0 \\ \partial^2\Omega_1/\partial\varphi^2 = 0 \end{cases}$$

indicating a transition from a minimum to a saddle point. Numerical solutions for this system give the critical values p^* and φ^*:

$$p^* \approx 0.5; \quad \varphi^* \approx 1.57\text{rad} \approx 90^\circ$$

Once these values are reached, the Bardeen-Herring source starts generating loops.

Using the Eq. (30.9), we determine the critical value of the plastic factor, k_{pl} , above which loop generation occurs:

$$k_{\text{pl}}^* \approx 1 - \frac{2\mathbb{L}_0 k_B T}{\mathfrak{b}^2 \tau_e^{(0)}} \ln\left(\frac{c}{c_0}\right) \tag{30.11}$$

Alternatively, the critical supersaturation value c at a given k_{pl}, above which loops form (i.e., when $c^* \geq c_0$), is:

$$c^* \approx c_0 \exp\left[\frac{0.5\mathfrak{b}^2 \tau_e^{(0)}(1 - k_{\text{pl}})}{\mathbb{L}_0 k_B T}\right] \tag{30.12}$$

30.2.4 Estimates of Parameters of Bardeen-Herring Source

To begin our estimates, we consider Frank's network dislocations as pinning sites. In a "good crystal," the typical mesh size of Frank's network ranges from 10 to 10^2 μm, while in plastically-deformed materials, it reduces to between 10^{-2} to 10^{-1} μm [3]. Furthermore, the supersaturation can reach values as high as $c/c_0 \sim 10^3$. We, further, assume the room temperature, $T = 300$ K.

Our estimates are based on these values, along with additional parameters provided in Tables A.1 and A.2. The results of these estimates are summarized in Fig. 30.3.

Fig. 30.3 Estimates of k_{pl} represented as plots

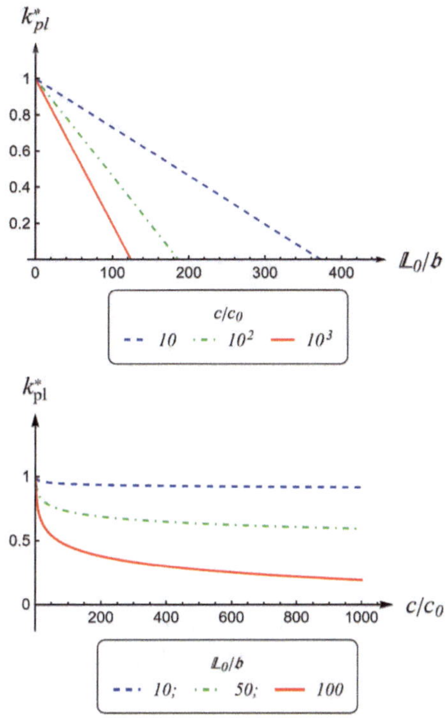

***

In summary, *the estimates suggest that, depending on the parameters, the critical plastic factor* k_{pl} *can fall below unity and even approach zero. This indicates that the Bardeen-Herring source can begin generating loops at* $k_{\text{pl}} < 1$. *As a result, the phase diagram region where the Bardeen-Herring source remains active is considerably broader than the instability domains III, IV, VI, and VII shown in Fig.* 26.1.

30.3 Instability of a Prismatic Dislocation Loop

In crystals with supersaturated concentrations of vacancies, prismatic loops are subjected to radial osmotic pressure, which leads to instability at a critical radius, R^*. *In this section, we analyze the contribution of the dislocation's process zone to determining this critical radius.*

***

Let us consider a circular prismatic dislocation loop lying in the (X, Y)-plane, as shown in Fig. 30.4.

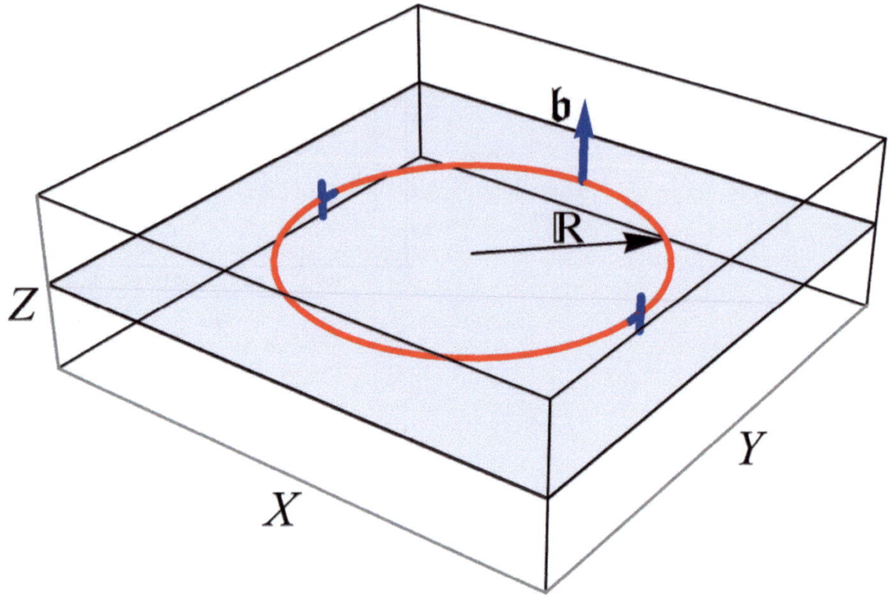

Fig. 30.4 Round prismatic dislocation loop. The loop lies in the (X, Y)-plane, visualized as a "glass sheet" for clarity. The Burgers vector, $\mathfrak{b}$, is perpendicular to the (X, Y)-plane

The expression:

$$W_{\text{loop}} = \frac{\mu \mathfrak{b}^2 \mathbb{R}(1 - k_{\text{pl}})}{1 - \nu} \left[\ln \left(\frac{8\mathbb{R}\alpha_{\text{disl}}}{\mathfrak{b}} \right) \right] \tag{30.13}$$

gives the energy of such a loop, where $\mathbb{R}$ is the loop radius, μ is the shear modulus, α_{disl} is a numerical factor close to 4 [2]. The Ω-potential, which accounts for supersaturated vacancies, is:

$$\Omega = W_{\text{loop}} - k_B T \ln \left(\frac{c}{c_0} \right) \times \frac{\pi \mathbb{R}^2}{\mathfrak{b}^2} \tag{30.14}$$

The notations used here are consistent with those introduced in Sect. 30.1. The Ω-potential, as a function of $\mathbb{R}$, has no minimum but instead displays a single maximum at a critical radius $\mathbb{R}^*$. This means that all loops are inherently unstable: loops with $\mathbb{R} < \mathbb{R}^*$ collapse, while those with $\mathbb{R} > \mathbb{R}^*$grow without limit.[2] The growth process is complex and we do not go into its details. Instead, we focus on determining the critical radius $\mathbb{R}^*$. By differentiating the Ω-potential (30.14) with respect to $\mathbb{R}$ and introducing the following notations:

[2] This conclusion applies only to a single loop. Unlimited growth is constrained by interactions with other defects, if present.

$$q = \frac{2(1 - k_{\rm pl})\mu \mathfrak{b}^3 \alpha_{\rm disl}}{\pi k_{\rm B} T \ln{(c/c_0)}\,(1-\nu)}; \qquad x = \frac{8\mathbb{R}^*}{\mathfrak{b}} \tag{30.15}$$

we obtain the equation:

$$q \ln(x) = x \tag{30.16}$$

This equation admits an exact solution: $x = -qW(-q^{-1})$, where $W(z)$ is the Lambert function (also called the product log function), which solves the equation $z = Wexp(W)$ with respect to W . Recasting in terms of the original variables, we find the critical radius $\mathbb{R}^*$:

$$\frac{\mathbb{R}^*}{\mathfrak{b}} = -\frac{2(1 - k_{\rm pl})\mu \mathfrak{b}^3 \alpha_{\rm disl}}{\pi k_{\rm B} T \ln{(c/c_0)}\,(1-\nu)} W \left[-\frac{\pi k_{\rm B} T \ln{(c/c_0)}\,(1-\nu)}{2(1 - k_{\rm pl})\mu \mathfrak{b}^3 \alpha_{\rm disl}} \right] \tag{30.17}$$

30.3.1 Estimate for the Prismatic Loop's Critical Radius

We consider temperatures of $T = 100$ and 300 K, using additional parameters from Tables A.1 and A.2. Based on these, we estimated the ratio $\mathbb{R}^*/\mathfrak{b}$ within the range $1 < c/c_0 < 10^3$ and $k_{\rm pl} = 0, 0.5$, and 0.9. We found that $\mathbb{R}^*/\mathfrak{b}$ is below unity in all cases, implying that all loops are unstable under supersaturation and exhibit unlimited growth.

***

In summary, *the estimates indicate that prismatic loops of all sizes, $R^* > \mathfrak{b}$, in the presence of supersaturated vacancies, should exhibit growth limited only by interactions with other defects.*

References

1. A.L. Korzhenevskii, D.A. Lisachenko, Sov. Phys. Solid State **32**, 1030 (1990)
2. J.P. Hirth, J. Lothe, *Theory of Dislocations* (Krieger Publishing Company, Malabar, Florida, US, 1991)
3. J. Friedel, *Dislocations* (Pergamon Press, Oxford, 1964)

Chapter 31
Thermofluctuation Motion of Dressed Dislocations

Abstract In this chapter, we analyze the thermofluctuational motion of a single dislocation segment under an external load bypassing a solitary point-like defect in the temperature range near the first-order phase transition. We demonstrate that the formation of the process zone around the dislocation affects the characteristics of this process. In particular, the nature of the motion may shift from being activation-based to non-activation-based. Further, we examine changes in the smooth kink motion associated with the formation of the process zone.

31.1 Overcoming of a Point Defect by the Dressed Dislocation

In this section, we analyze the thermofluctuational motion of a single dislocation segment under an external load bypassing a solitary point-like defect in the temperature range near the first-order phase transition. We demonstrate that the formation of the process zone around the dislocation affects the characteristics of this process. In particular, the nature of the motion may shift from being activation-based to non-activation-based[1].

***

The analysis of how a simple dislocation segment overcomes an elastic point obstacle is fundamental to the thermofluctuation theory of plastic deformation. This problem has been approached using various methods, including the line-tension model, the concept of critical separation angles, concentrated force methods, and small deflection approximations, as explored in studies by De Wit and Köhler [2], Bacon, Foreman [3], and others. Recent efforts have focused on atomistic modeling, with particular emphasis on the statistical aspects of thermofluctuational plasticity. This approach requires extensive numerical calculations. Developing efficient programs for this purpose remains a significant challenge.

[1] Korzhenevskii and Lisachenko 1991.

© The Author(s), under exclusive license to Springer Nature Switzerland AG 2026

A. Boulbitch and A. Korzhenevskii, *Stress-Induced Local Phase Transformations Near Cracks and Dislocations*, Springer Series in Materials Science 356,
https://doi.org/10.1007/978-3-032-00945-6_31

In solids undergoing phase transformations, the formation of zones on dislocations alters their configurations, facilitating the overcoming of barriers. The latter differs significantly from its configuration far from the transition.

31.1.1 Energy of the Bent Dislocation Segment

Consider an edge dislocation segment, fixed at points $X = \pm\mathbb{L}$, $Y = 0$, with the Burgers vector $\mathfrak{b}$ directed along the OY axis. A schematic representation is shown in Fig. 31.1.

The dislocation moves within its slip plane (X, Y) due to an external uniform stress $\sigma_{zy}^{(0)}$, which generates a Peach-Köhler force, F_Y, along the OY axis:

$$F_Y = -t_x \mathfrak{b}_y \left[\sigma_{zy}^{(0)} + \sigma_{zy}^{(\mathrm{p})}(\mathbf{R})\right]$$

Here, $\mathbf{t}$ is the unit vector, tangent to the dislocation, and $\sigma_{zy}^{(\mathrm{p})}(\mathbf{R})$ is the stress generated by the point obstacle (represented as a sphere in Fig. 31.1). The obstacle is located on the OZ axis at $(0, 0, Z_0)$.

The free energy, $\mathcal{F}$, of the dislocation segment, is described by equation (29.6), while the expression for $\mathcal{F}' = \mathcal{F} - \mathcal{W}$, where $\mathcal{W}$ is the work done by the external force, takes the form:

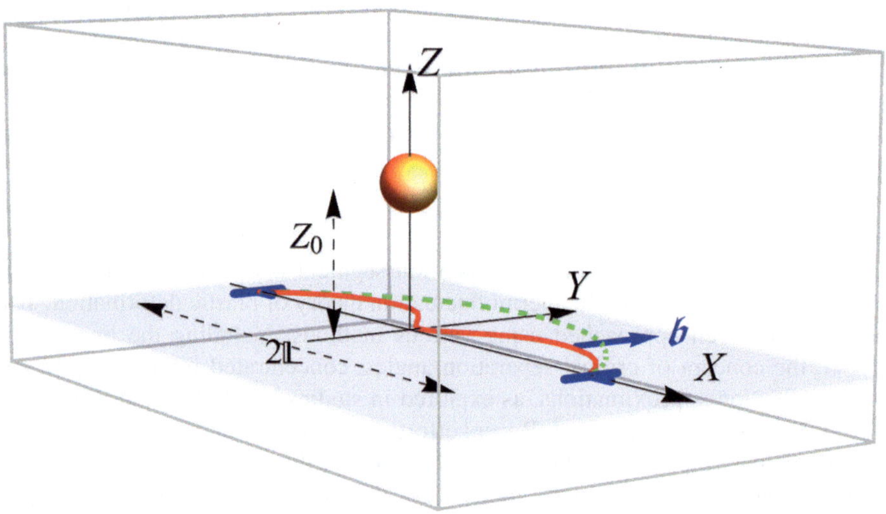

Fig. 31.1 Climbing geometry. The edge dislocation initially lies along the OX axis. For clarity, we depicted a point defect as a sphere outside the (X, Y) plane above the origin. When a Peach-Köhler force along the OY direction is applied, the dislocation bends to the configuration shown by the solid line. Its jump to the configuration depicted by the dashed line represents the climbing process

$$\mathcal{F}' = \tau_s^{(0)} \int_{-\mathbb{L}}^{\mathbb{L}} \left\{ 1 - \frac{(k_{pl} + \nu)\, Y'^2(X)}{(1-\nu)\left[1 + Y'^2(X)\right]} \right\} \times$$

$$\times \sqrt{1 + Y'^2(X)} dX - \int_{-\mathbb{L}}^{\mathbb{L}} dX \int_0^{Y(X)} F_Y(X, Y) dY \qquad (31.1)$$

where the function $Y(X)$ describes the shape of the dislocation segment.

As discussed in other sections, the factor k_{pl}, specified in (26.6, 26.7) and (26.12), accounts for the contribution of the process zone to line tension. For the daughter-phase zone surrounded by the bulk mother phase, the relevant expression is given by formula (26.6), whereas for the mother-phase zone embedded within the daughter-phase matrix, Eq. (26.12) applies. The temperature range, T_{anom}, where anomalies may be observed $T_{pl}^{(m)} \leq T_{anom} \leq T_{pl}^{(d)}$ is defined above by relations (26.21) and follows from the condition $k_{pl}^{(d,m)} = 1$. As discussed in section 29, in the case of a segment, this relation is modified to $k_{pl}^{(d,m)} = (1 + \nu)/2$ (29.12), further broadening the temperature interval.

In an elastically isotropic medium under pure dilatational striction, the process zone generates a significant dilatational strain, described by Eq. (4.9). This strain, $\varepsilon_{\alpha\alpha}(\mathbf{R}) - \varepsilon_{\alpha\alpha}^{(0)}(\mathbf{R}) \sim \eta^2$, decays exponentially with distance from the zone boundary[2]. A dilatational point center does not generate strain in space, except for the point of its immediate location [4]. Thus, phase transformations do not affect the elastic interaction between the dilatational point defect and the dislocation-zone complex. Instead, *it is the perturbation of the line tension by the plasticity factor k_{pl} that causes the anomalies.*

However, for defects generating $\varepsilon_{\alpha\alpha}^{(p)}(\mathbf{R})$, or for striction involving non-dilatational components, the elastic fields of the process zone must be taken into account. This case lies beyond the scope of this book.

The zone radius $\mathbb{R}_d$ near the dislocation reaches values of $\mathbb{R}_d \sim 1\mu$m for the zone of the displacement type and 10 nm for the order-disorder zone (see Table 27.1 at the page 332). In a "good crystal" with a typical dislocation density of $\rho_{disl} \sim 10^6$ cm^{-2} [5], the fraction of the new phase is $\rho_{disl}\mathbb{R}_d^2 \sim 10^{-2}$ for the displacive zone and $\sim 10^{-4}$ for the order-disorder one. Thus, the anomalies should become noticeable at the very beginning of the zone formation process.

The elastic stress field of a defect, $\sigma_{zy}^{(p)}(\mathbf{R}) = 2\mu\varepsilon_{yz}^{(p)}(\mathbf{R})$, is derived from its force dipole tensor $P_{\alpha\beta}$, using the isotropic Green's function $G_{\alpha\beta}(\mathbf{R})$. Details the reader can find in the book of Leibfrid and Broer [6]. The displacement component generated by the point defect, $\mathrm{u}_\alpha^{(p)}$, is:

$$\mathrm{u}_\alpha^{(p)} = P_{\gamma\beta} \frac{\partial G_{\alpha\gamma}(\mathbf{R})}{\partial X_\beta} \qquad (31.2)$$

[2] Examples of this decay can be seen in the spatial behavior of η, as depicted in Figs. 23.1 and 24.3a.

The dipole tensor is spherical for a dilatation center: $P_{\alpha\beta} = P\delta_{\alpha\beta}$, where P represents the defect strength. The strain tensor generated by the defect follows the standard form (3.10), leading to:

$$\varepsilon_{yz}^{(\mathrm{p})}(\mathbf{R}) = \frac{1}{2}\left(\frac{\partial \mathrm{u}_y^{(\mathrm{p})}(\mathbf{R})}{\partial Z} + \frac{\partial \mathrm{u}_z^{(\mathrm{p})}(\mathbf{R})}{\partial Y}\right) \tag{31.3}$$

31.1.2 Point Defect

Outside of its location, the dilation defect generates the following stress field:

$$\sigma_{yz}^{(\mathrm{p})}(\mathbf{R}) = \frac{3P(1-2\nu)YZ}{4\pi(1-\nu)R^5}; \qquad \varepsilon_{\alpha\alpha}^{(\mathrm{p})} = 0 \tag{31.4}$$

We determine the shape of the dislocation segment by solving the Euler equation for the functional (31.1), which takes the form:

$$\frac{y''}{\left(1+y'^2\right)^{3/2}}\left(1 + k_{\mathrm{pl}}\frac{y'^2-2}{1+y'^2}\right) = s_0 + s^{(\mathrm{p})}(\mathbf{R}) \tag{31.5}$$

where we use the following notations: $\mathbf{r} = \mathbf{R}/\mathbb{L}$, $x = X/\mathbb{L}$, $y(x) = Y(X)/\mathbb{L}$, $z = Z_0/\mathbb{L}$, and

$$s_0 = \frac{\sigma_{zy}^{(0)}\mathbb{L}}{\mu \mathfrak{b}}; \qquad s^{(\mathrm{p})}(\mathbf{r}) = D\frac{zy(x)}{r^3}; \qquad D = \frac{3(1-2\nu)P}{4\pi\mu(1-\nu)\mathbb{L}^2}$$

Here, $r = \left(x^2 + y^2(x) + z^2\right)^{1/2}$ is the rescaled distance between the defect and the dislocation at coordinates x and $y(x)$. We are going to analyze Eq. (31.5) numerically.

In the particular case when $k_{\mathrm{pl}} = 0$ and $y'(x) \ll 1$, Eq. (31.5) simplifies to:

$$y'' = s_0 + s^{(\mathrm{p})}(\mathbf{r}) \tag{31.6}$$

This linear case was analyzed by Chernov and Indenbom [7], who showed that under certain conditions, a segment can occupy two positions: stable or metastable (shown as the red and dashed-green curves in Fig. 31.1), between which there is a solution to Eq. (31.6), corresponding to the saddle point. Our simulations of the nonlinear version of Eq. (31.5), shown in Fig. 30.2, agree with these calculations, demonstrating the stable (a), metastable (b), and maximum-energy (c) configurations. The parameters used in the simulations correspond to a dislocation density of $\rho_{\mathrm{disl}} \sim 10^8\ \mathrm{cm}^{-2}$ and defect concentration $n \sim 10^{16}\mathrm{cm}^{-3}$.

The energy differences between these positions, $\mathcal{F}'_{ab} = \mathcal{F}'_c - \mathcal{F}'_a$, $\mathcal{F}'_{ba} = \mathcal{F}'_c - \mathcal{F}'_b$ represent the activation barriers when a dislocation segment jumps between positions (a) and (b) (Fig. 31.2). These barriers determine the thermal fluctuation-driven motion of the dislocation. Figure 31.3 shows how the deflection at the midpoint of the

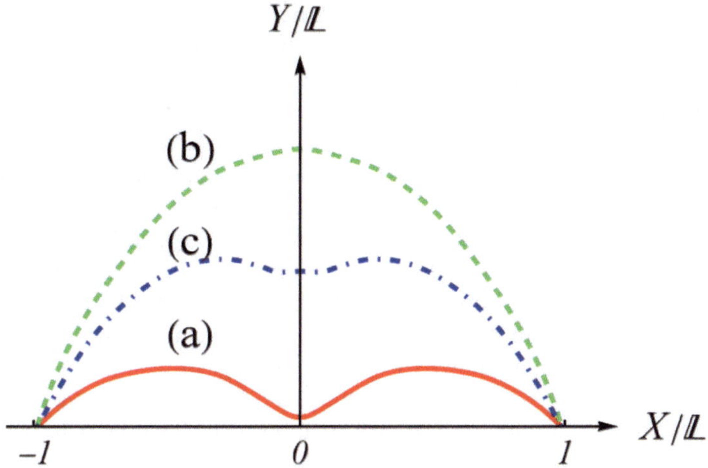

Fig. 31.2 Numerical solution of the Eq. (31.5) shows the stable (**a**), metastable (**b**), and maximum-energy (**c**) configurations of the dislocation in the (X, Y) plane

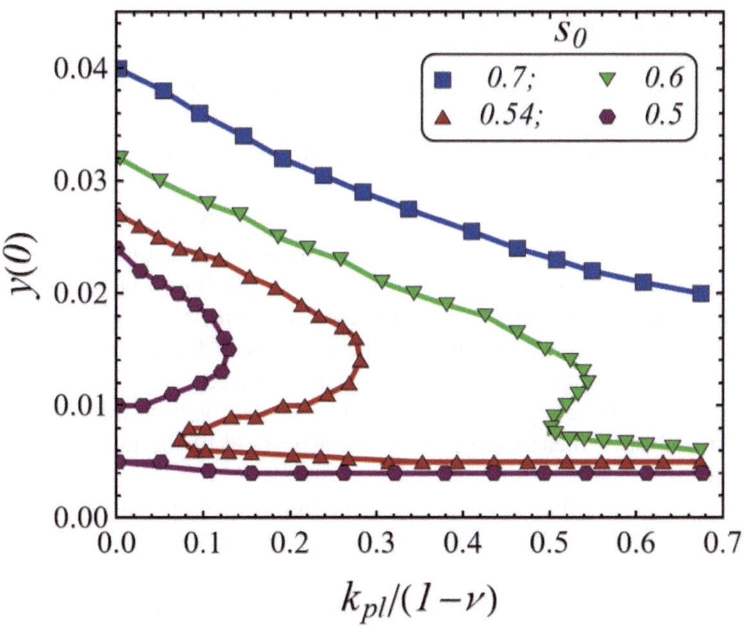

Fig. 31.3 Deflection curve apex of the segment versus k_{pl}. The results are obtained from simulations. Different values of s_0 are shown in the legend

dislocation segment, $y(0)$, depends on the plasticity factor, k_{pl}, at different values of s_0. Our simulations demonstrate that the equilibrium configuration of dislocations can strongly depend on temperature in the presence of a zone, which is controlled by the temperature-dependent plasticity factor $k_{pl} = k_{pl}(T)$, as described in Eq. (26.6).

We can also define the stress range $s_1 < s_0 < s_2$ where the segment exhibits two equilibrium configurations and activated transitions between them. This temperature-dependent range is again determined by $k_{pl}(T)$. Outside this range, where $s_0 < s_1$ or $s_0 > s_2$, only one configuration remains, and the segment's motion becomes activation-free. Figure 31.4 illustrates the stress interval $(s1, s2)$ for activated motion at two different values of z. For $s_0 < s_1$, the segment is pinned by the defect, corresponding to the curve (a) in Fig. 31.2. When $s_0 > s_2$, the dislocation overcomes the defect without activation, as shown by curve (b) in Fig. 31.2.

Within the interval $s_1 < s_0 < s_2$, the segment undergoes thermally activated jumps between the two configurations. The transformation zone affects both the width of this interval and the stress required for hopping, as shown in Fig. 31.4. Translating this into temperature dependence, we find that the activation stress increases with temperature. Furthermore, curve (a) in Fig. 31.4 demonstrates that the hopping domain in the (k_{pl}, s_0) plane is limited.

So far, we have discussed the behavior of an edge dislocation segment. For screw dislocation segments, however, the stress required for hopping decreases with increasing temperature. As a dislocation moves through a network of obstacles, the critical stress varies depending on the orientation of the segment. The same applies to the critical angle of attack on the defect. This phenomenon arises from the influence of the order parameter, which significantly affects the formation of the process zone and, consequently, alters the activation barrier height for different dislocation configurations.

At high $\sigma^{(0)}_{zy}$, the segment may act as a Frank-Read source. After separating from the defect, where $\sigma^{(d)}(r) \ll \sigma^{(0)}$, the critical stress can be estimated for a segment in a non-uniform field. According to Korzhenevskii [8], for the edge dislocation, the critical stress does not change, but for the screw segment, $\sigma \sim 1 - k_{pl}$ (see Sect. 29).

Let us briefly discuss shear-type point defects. These defects generate alternating dilatations. Due to the alternating signs of $\varepsilon^{d}_{\alpha\alpha}$, these defects always produce a process zone, regardless of the signs of A and P. Consequently, both the renormalization of the dislocation's linear energy and the elastic field of the defect zone must be accounted for.

Simulations show that in such cases, the behavior of the dislocation segment becomes more complex. For instance, tetrahedral defects cannot be definitively classified as attractive or repulsive, as the type of their interaction depends on their orientation relative to the dislocation segment. As shear centers generate elastic fields with complex angular dependencies, it can result in the appearance of more than two stable configurations between which thermally activated jumps occur. It is also worth noting that experimental studies of the plastic properties of crystals near first-order phase transitions reveal that anomalies depend not only on the phase transition properties (e.g., order parameter symmetry, jumps of the lattice deformation)

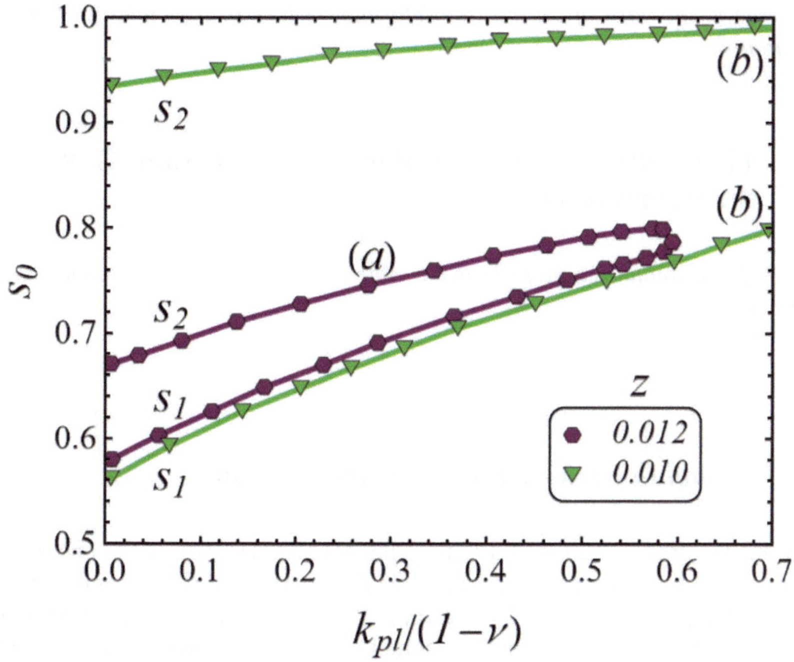

Fig. 31.4 Effect of zone formation on activation stress For $z = 0.012$, activated motion occurs within the "tongue" formed by the purple line (**a**), while for $z = 0.01$, it lies between the two green lines (**b**)

but also on the dislocation structure and the predominant type of point defects. To better understand the origins of mechanical anomalies, it is advisable to compare samples with different defective structures, where the anomalous contributions from the "dressing" effect vary significantly.

The analytical calculation of the friction force on two-dimensional extended defects, caused by dislocation dressing near phase transition points, is feasible. However, assessing the additional pinning contribution requires a significantly more complex mathematical method than that for dressed dislocations interacting with point defects (see, e.g., [9], which addressed twin boundary pinning). Therefore, these solutions are not included in this book. Despite this complexity, the underlying physical picture remains unchanged.

***

In conclusion, *we established that the activation energy considerably changes during overcoming of point defects by the dislocation. Within the simple example we demonstrated the approach of the numerical simulation of its variation. As the dislocation mostly overcomes defects by thermal fluctuations, while its velocity exponentially depends on the barrier height, the zone formation must manifest itself in* *the dramatic increase of the plastic deformation rate in the vicinity of the phase transition.*

31.2 Thermally Activated Motion of a Dislocation Kink in the Peierls Relief

This section examines changes in the smooth kink associated with the formation of the process zone[3].

***

31.2.1 Smooth Kink on the Screw Dislocation

The time required to generate a kink exponentially depends on its energy, $\mathcal{F}$. To determine the temperature dependence $\mathcal{F}(T)$ as the system approaches the phase transition temperature ($T \to T_0$), we apply the continuum approximation, which is valid when the kink is smooth. This condition arises in materials where the linear tension energy of the dislocation significantly exceeds the Peierls potential height $U_{\rm P}$ (where the subscript "P" denotes Peierls), resulting in a kink length L spanning several interatomic distances [5], as schematically illustrated in Fig. 31.5.

Let us consider a straight screw dislocation. In the absence of external stress, it resides in a groove of the periodic Peierls potential, $U_{\rm P}$. The groove is aligned along the X-axis, and the Peierls potential varies periodically with Y, such that $U_{\rm P} = U_{\rm P}(Y)$. The configuration of the dislocation is described by $Y = Y(X)$.

The perturbation in free energy due to kink formation is given by:

$$\Delta\mathcal{F} = \int \Bigg\{ \frac{\tau_{\rm s}^{(0)}}{\sqrt{1+Y'^2}} \left(1 + \frac{1-k_{\rm pl}}{1-\nu} Y'^2\right) +$$

$$+U_{\rm P}(Y)\sqrt{1+Y'^2} - \left[\tau_{\rm s}^{(0)} + U_{\rm P}(Y_0)\right] - \sigma\mathfrak{b}(Y - Y_0)\Bigg\}\, dX \qquad (31.7)$$

Here, $\mathfrak{b}$ is the Burgers vector, $\tau_{\rm s}^{(0)}$ (29.4) represents the energy per unit length of the screw dislocation far from the phase transition point, and $Y' = dY(X)/dX$. The temperature-dependent dimensionless parameter $k_{\rm pl}(T) > 0$ (26.4) accounts for the contribution of the process zone that influences the edge component of the smooth kink.

---

[3] Kozhenevskii and Lisachenko, 1993 [10].

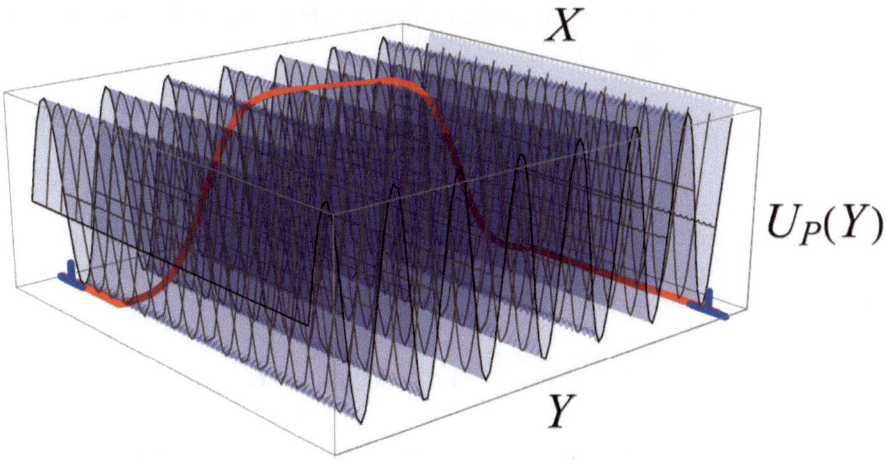

Fig. 31.5 Smooth kink on a screw dislocation in a periodic Peierls potential The Peierls potential is depicted as a transparent blue sinusoidal surface. The screw dislocation, shown as a red line, transitions smoothly from one potential groove to another. The Burgers vector, $\mathfrak{b}$, is oriented along the OX axis

The equilibrium condition:

$$-\frac{\partial U_P(Y)}{\partial Y}_{|Y=Y_0} = \sigma \mathfrak{b} \tag{31.8}$$

determines the position Y_0 which minimizes the potential relief under the applied external stress σ.

In this expression:

- The first term (31.7) represents the surface tension energy.
- The term $U_P(Y)\sqrt{1+Y'^2}$ accounts for the contribution of the Peierls barrier.
- The subtraction of $\tau_s^{(0)} + U_P(Y_0)$, isolates the energy perturbation due to kink formation by removing the energy of a straight dislocation resting in the groove.
- Finally, $\sigma \mathfrak{b}(Y - Y_0)$, represents the work done by the external force acting on the kink.

As noted, we focus on the case where the Peierls potential is much smaller than the dislocation's line tension:

$$U_P \ll \tau_s^{(0)} \tag{31.9}$$

Under this condition, the kink's deflection varies slowly, such that $Y'^2 \ll 1$. Expanding the integrand to second order in Y'^2, we obtain:

$$\left[\tau_s^{(0)} \frac{1+\nu-2k_{pl}}{2(1-\nu)} + \frac{1}{2} U_P(Y)\right] Y'^2 + U_P(Y) - U_P(Y_0) - \sigma \mathfrak{b}(Y - Y_0)$$

Due to inequality (31.9), the second term in square brackets can be neglected. The free energy then simplifies to:

$$\Delta\mathcal{F} = \int \left[\tau_s^{(0)} \frac{1+\nu-2k_{\text{pl}}}{2(1-\nu)} Y'^2 + \Delta U_{\text{P}}(Y) \right] dX, \tag{31.10}$$

where $\Delta U_{\text{P}}(Y) = U_{\text{P}}(Y) - U_{\text{P}}(Y_0) - \sigma\mathfrak{b}(Y - Y_0)$.

Minimizing the functional (31.10) with respect to $Y(X)$ leads to the corresponding differential equation. By drawing an analogy with classical mechanics, we can immediately write down its first integral:

$$\tau_s^{(0)} \frac{1+\nu-2k_{\text{pl}}}{2(1-\nu)} Y'^2 - \Delta U_{\text{P}}(Y) = C, \tag{31.11}$$

where C is an integration constant. For a kink configuration connecting the minima $Y_{\min}$ and maxima $Y_{\max}$ of the Peierls potential, $C = 0$. Substituting this condition into the expression for the activation energy, $\Delta\mathcal{F}$, and noting that $Y'^2 dX = Y' dY$, we obtain:

$$\Delta\mathcal{F} = \sqrt{2\tau_s^{(0)} \frac{1+\nu-2k_{\text{pl}}}{1-\nu}} \int_{Y_{\min}}^{Y_{\max}} \sqrt{\Delta U_{\text{P}}(Y)} dY \tag{31.12}$$

Assuming that the Peierls potential $U_{\text{P}}(Y)$ depends weakly on temperature (except near $T = T_c$), the temperature dependence of the activation energy primarily arises from the expression $1 + \nu - 2k_{\text{pl}}$, arising from the modification of the dislocation's linear tension by the process zone. The activation energy prefactor vanishes when $k_{\text{pl}} = k_{\text{pl}}^{(*)}$, where:

$$k_{\text{pl}}^{(*)} = \frac{1+\nu}{2}$$

This result matches the relation derived in (29.12).

To analyze the mobility of the kink, denoted here as $\mathfrak{M}$, we express it in Arrhenius form:

$$\mathfrak{M} = \mathfrak{M}_0 \exp\left(-\frac{\Delta\mathcal{F}}{k_B T}\right) \tag{31.13}$$

To highlight the role of the process zone in kink mobility, let us consider the ratio of the mobility of the kink dressed by the process zone, $\mathfrak{M}_{\text{PZ}}$, to that of the undressed kink, $\mathfrak{M}_{\text{NZ}}$, of the undressed kink. Here, "PZ" refers to "Process Zone," while "NZ" refers to "No Zone." This mobility ratio takes the form:

$$\frac{\mathfrak{M}_{\text{PZ}}}{\mathfrak{M}_{\text{NZ}}} = \exp\left(-\frac{\Delta\mathcal{F}}{k_B T}\right)$$

where $\Delta\mathcal{F}$ is the difference between the free energies of the dressed and undressed kinks. Specifically, $\Delta\mathcal{F}$ can be expressed as:

$$\Delta\mathcal{F} = k_B T \left(\sqrt{1+\nu-2k_{pl}} - \sqrt{1+\nu}\right)\mathfrak{D}$$

where for brevity, we denoted:

$$\mathfrak{D} = \frac{1}{k_B T}\sqrt{\frac{\mathcal{T}_s^{(0)}}{2(1-\nu)}}\int_{Y_{min}}^{Y_{max}} \sqrt{\Delta U_P(Y)}dY$$

This factor encapsulates all k_{pl}-independent constants. Using these notations, the mobility ratio becomes:

$$\frac{\mathfrak{M}_{PZ}}{\mathfrak{M}_{NZ}} = \exp\left[\left(\sqrt{1+\nu} - \sqrt{1+\nu-2k_{pl}}\right)\mathfrak{D}\right] \tag{31.14}$$

Observing that $\left(\sqrt{1+\nu} - \sqrt{1+\nu-2k_{pl}}\right)\mathfrak{D} > 0$, it can be concluded that $\mathfrak{M}_{PZ}/\mathfrak{M}_{NZ}$ always exceeds unity.

Figure 31.6a depicts the behavior of the mobility exponent, $\ln(\mathfrak{M}_{PZ}/\mathfrak{M}_{NZ})$ for the kink on the screw dislocation as k_{pl} varies between 0 and 1.

***

It is concluded *that the mobility of the dressed kink on a screw dislocation is consistently higher than that of the bare kink. Furthermore, as* k_{pl} *increases, the mobility ratio* $\mathfrak{M}_{PZ}/\mathfrak{M}_{NZ}$ *increases monotonically, reaching its maximum value:*

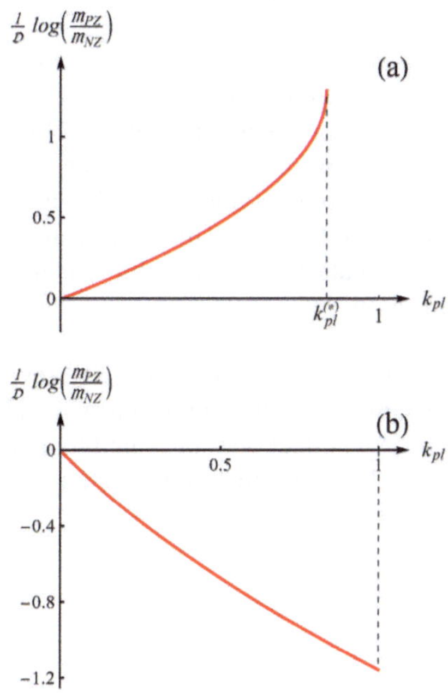

Fig. 31.6 Logarithm of the mobility ratio $\ln(\mathfrak{M}_{PZ}/\mathfrak{M}_{NZ})$, equivalent to $-\Delta\mathcal{F}/\mathfrak{D}$. **a** Shows $-\Delta\mathcal{F}/\mathfrak{D}$ for a kink on a screw dislocation, while **b** shows it for a kink on an edge dislocation. The defnition of $\mathfrak{D}$ is provided in the text

$$\frac{\mathfrak{M}_{\rm PZ}}{\mathfrak{M}_{\rm NZ}} = \exp\left(\mathfrak{D}\sqrt{1+\nu}\right) > 1$$

at $k_{\rm pl} = k_{\rm pl}^{(*)}$. *For* $k_{\rm pl} \geq k_{\rm pl}^{(*)}$, *an instability arises. In this regime, the kink elongates and adopts a meandering shape as its extension becomes energetically favorable. The kink stabilizes once its fragments approach nearby dislocations, as observed in prior examples.*

In contrast, the mobility of the dressed kink on an edge dislocation remains persistently lower than that of the bare kink.

31.3 Some Further Results

The above results do not exhaust our findings. However, some require a complex mathematical framework or lengthy calculations. Below, we list several such results with references, without discussing them in detail.

***

Kinetic Self-Roughning of Dislocations. In the paper [12], it was demonstrated that the dependence of the local friction force on the angle between the Burgers vector and the dislocation line tangent leads to its roughening near the phase transition points. However, as far as we know, the statistical properties of this roughening and its contribution to the dry friction force hindering dislocation motion have not been analyzed to date.

Instability Related to a Descending Branch of the Friction Force. According to the general formula (20.12), the contribution to the viscous force from dislocation dressing is a nonlinear function of velocity. As this contribution vanishes at $V \to V_{\rm d}$, the total viscous friction force on the dislocation—including relatively small contributions from phonon generation and scattering—exhibits a decreasing region with increasing velocity. This leads to an instability of the steady-state dislocation motion. Mathematically, this instability is most clearly reflected in correlation and response functions [11].

Dynamic Strength Anomaly. The review [14] reported a several-fold increase in the dynamic strength of high-purity metals (Ti, Fe, Co, etc.) near phase transitions. This universal effect was attributed to a large friction force on dislocations dressed by a new-phase nucleus, as estimated in [13, 15]. Further study of this effect would be valuable.

References

1. A.L. Korzhenevskii, D.A. Lisachenko, Sov. Phys. Solid State **33** (5), 877 (1991)
2. G. de Wit, J.S. Köhler, Phys. Rev. **116**, 1113 (1959)
3. D.J. Bacon Physica status solidi **23**, 527–538 (1967); A.J.E. Foreman, Phil. Mag. **15**, (137), 1011–1021 (1967)
4. J.D. Eshelby, *Elastic Inclusions and Inhomogeneities* (North Holland Publishing Co., Amsterdam, 1961)
5. J. Friedel, *Dislocations* (Pergamon Press, Oxford, 1964)
6. G. Leibfried, N. Breuer, *Point Defects in Metals* (Springer, 1978)
7. V.M. Chernov, V.L. Indenbom, Fizika Tverdogo Tela **10**, 3331 (1967). (in Russian)
8. A.L. Korzhenevskii, D.A. Lisachenko, Sov. Phys. Solid State **30**, (5), 860 (1988)
9. A.L. Korzhenevskii, D.A. Lisachenko, Sov. Phys. Solid State **34**(4), 568–570 (1992)
10. A. L. Kozhenevskii, D.A. Lisachenko, Izv. Akad. Nauk ser. fiz. **57**, (3), 11–14 (1993) (in Russian)
11. A.L. Korzhenevskii, R. Bausch, R. Schmitz, Phys. Rev. E **63**(5), 056105 (2001)
12. A.L. Korzhenevskii, R. Bausch, R. Schmitz, Phys. Rev. Lett. **83**(22), 4578–4581 (1999)
13. A. Boulbitch, P.E. Pumpyan, Sov. Phys. Crystallogr. **35**(2), 156 (1990)
14. E. Zaretsky, AIP Conf. Proc. **1426**, 1553–1558 (2012); E. Zaretsky, Mech. Solids **49**, 623–634 (2014)
15. A.L. Korzhenevskii, R. Bausch, R. Schmitz, Phys. Rev. B **67**(10), 100103 (2003)

References

Chapter 32
Experiments Reporting Dressed Dislocations

Abstract In this chapter, we supplement the theoretical results presented earlier with experimental data revealing local phase transitions at dislocations. A substantial body of evidence has been gathered from various scattering experiments conducted near phase transition points. Here, however, we focus on the most significant results. Experimental data can be broadly categorized into that obtained through direct imaging and that derived from indirect methods. Direct imaging includes optical and electron microscopic observations of process zones at dislocations in real three-dimensional space. Indirect methods encompass data from experiments involving the scattering of X-rays, neutrons, electrons, and light. A special category of observations includes measurements of superconductivity at dislocations.

***

Optical Observations. Belousov and Wolf observed anomalously strong small-angle light scattering near the first-order phase transition point ($T_0 = 235$K) in NH_4Br crystals [13]. This phase transition exhibits a narrow hysteresis of approximately 0.05K. In the high-temperature phase, the NH^{+4} ions exhibit orientational disorder, with the symmetry group O_h^1 and appearing optically homogeneous. In the low-temperature phase, antiparallel ordering of the ions occurs, resulting in a uniaxial crystal with D_{4h}^7 symmetry group.

When a temperature gradient is applied along one of the three C_4 symmetry axes of the β-phase, the optical axis of the γ-phase aligns with the [001] direction. Observations using a polarizing microscope revealed γ-phase zones elongated along directions close to [100], [010], and [001], forming a non-periodic three-dimensional lattice. As the phase transition point is approached, nuclei appear near the slip planes of dislocations deliberately introduced by a razor. These observations suggest dislocation clusters decorated with process zones.

Vlasko-Vlasov, Dedukh, Nikitenko, and Indenbom studied magnetic orientational transitions in single crystals of gadolinium iron garnet ($Gd_3Fe_5O_{12}$) [14]. Using a polarizing optical microscope, they observed the formation and growth of zones occupied by distinct magnetic phases localized at single dislocation of growth.

© The Author(s), under exclusive license to Springer Nature Switzerland AG 2026
A. Boulbitch and A. Korzhenevskii, *Stress-Induced Local Phase Transformations Near Cracks and Dislocations*, Springer Series in Materials Science 356,
https://doi.org/10.1007/978-3-032-00945-6_32

Paraelectric-Ferroelectric and Paraelectric-Antiferroelectric Transitions. Using in-situ transmission electron microscopy (TEM) and X-ray diffraction, the authors of the paper [1] directly observed that dislocations stabilize the low-temperature anti-ferroelectric (orthorhombic) phase over the paraelectric (cubic) phase upon heating. They also reported that dislocations serve as preferential nucleation sites for the antiferroelectric phase during cooling. This observation aligns with our predictions, particularly the one illustrated in Fig. 20.1. These findings suggest that dislocation stress fields can locally induce or maintain a crystal structure that differs from the bulk.

Transformations Austenite-Martensite, Shape-Memory Alloys. Local α-Fe $\rightarrow$ γ-Fe transitions at dislocations were reported by Nguen and colleagues [2].

Fukuda, Saburi, Doi, and Nenno provided direct optical and electron microscopic evidence of martensitic plate formation of the rhombohedral R-phase at individual dislocations within the B2 bulk austenite phase in a Ti-Ni-Al alloy [15].

Ubarra and co-authors observed the nucleation of low-temperature β_3' and γ_3' martensite plates at dislocations in the high-temperature austenite β-phase of a Cu-Al-Ni shape-memory alloy [3].

Transitions Involving Hydrogen in Metals. Using neutron scattering, Heuser, Udovic, and Ju investigated palladium containing a low concentration of dissolved hydrogen, $PdH_{0.0008}$, characteristic of the α-phase [16]. They observed a new low-energy peak consistent with the signature of the high hydrogen content in the β-Pd hydride phase. This β-phase appeared to be localized at edge dislocations within regions of $1 - 5$ nm.

Ulvestad and co-authors employed Bragg coherent diffractive imaging (3D X-ray microscopy) to study dislocation networks and phase evolution in Pd nanocrystals during hydrogen absorption [4]. Their observations revealed localized $\alpha \rightarrow \beta$ phase transitions. In nanoparticles exceeding 300 nm, the β-phase did not nucleate until a dislocation emerged near the phase boundary. Essentially, the new hydride phase formed preferentially at dislocation sites, suggesting that the strain from the dislocation core facilitated its nucleation.[1]

Phase Separation and Reordering in Solid Solutions. Zolotorev and co-authors reported phase separation at dislocations in the Ni_4Mo alloy [6].

In his study of silicon containing dissolved lithium, Vanderwalker observed that lithium phase separation in the vicinity of dislocations was accompanied by changes in the crystal structure. This indicates the formation of a transformational process zone at the dislocation, driven by stress-induced local phase separation. TEM analysis

[1] Note that the local α-α' phase transformation of hydrogen in metals at dislocations and cracks was discussed by one of us decades ago [5] Although this transition differs in some respects from the α-hydride phase transformation, the two share certain features, and the study [5] predicted behaviour comparable to that reported in the works cited above.

revealed that the Li-rich phase formed small droplets decorating the dislocation lines [7].

Applying the atom probe tomography, Mukherjee, Assali, and Moutanabbir observed the atomic-level details of the behavior of solutes in the vicinity of a dislocation in GeSn. They demonstrated the presence of the dislocations-induced local phase transformation [8].

Indirect Observations. Dislocation pile-ups observed by TEM were found to facilitate the nucleation of α-martensite [17] and to occur during the B2 $\rightarrow$ R phase transformation in NiTi shape memory alloys [9].

Basinski, Duesbery, and Murty investigated the behavior of lithium single crystals subjected to low-temperature deformation. They observed abrupt yield drops and hysteresis in the stress–strain curve, suggesting the formation of domains containing a new phase. Although not directly imaged, X-ray diffraction revealed regions with a crystal structure distinct from the bulk bcc phase of lithium. The authors concluded that the most plausible explanation for the flow-stress anomalies was a martensitic phase transformation localized at dislocations [10].

Local Superconductivity at Dislocations and Twins. The formation of superconductivity localized along low-angle twin boundaries was first discovered by Khaĭkin and Khlyustikov and has been documented in seven metals: Sn, In, Nb, Re, Tl, Al and Pb ([11] and references therein). Additionally, similar phenomena have been observed in the organic metal $\beta - (ET)_2X$, where X represents I_2, IBr_2, or AuI_2 and ET stands for (ethylenedithio) tetrathiofulvalene [12].

References

1. Z. Li, Z. Liu, Z. Fu, T. Hu, Z. Yu, L. Zhang, X. Chen, G. Wang, F. Xu, Microstructures **5**, 2025019 (2025)
2. N.H. Ngu, A.I. Novikov, Yu.A. Skakov and V.A. Solov'ev, Fiz. Met. Metalloved. **64**, 915 (1987)
3. A. Ibarra, D. Caillard, J.M. San Juan, M.L. Nó, Appl. Phys. Lett. **90**, 101907 (2007)
4. A. Ulvestad, M.J. Welland, W. Cha, Y. Liu, J.W. Kim, R. Harder, E. Maxey, J.N. Clark, M.J. Highland, H. You, P. Zapol, S.O. Hruszkewycz, G.B. Stephenson, Nat. Mater. **16**, 565–571 (2017)
5. A. Boulbitch, J. Alloys Compd. **196**, (1–2), 29–36 (1993)
6. S.P. Zolotarev, I.B. Sidorova, Y.A. Skakov, V.A. Solov'ev, Sov. Phys. Solid State **20**, 450 (1978)
7. D.M. Vanderwalker, Phys. Status Solidi (a) **90**
8. S. Mukherjee, S. Assali, O. Moutanabbir, NanoLetters **21**(23), 9882–9888 (2021)
9. H. Vashishtha, D.M. Collins, Acta Mater. **240**, 118331 (2022)
10. Z.S. Basinski, M.S. Duesbery, G.S. Murty, Acta Metall. **29**, 801–805 (1981)
11. I.N. Khlyustikov, A.I. Buzdin, Sov. Phys. Usp. **31**, 409 (1988)
12. Zvarykina et al., Sov. Phys. JETP **67**, 1891 (1988)
13. M.V. Belousov, B.E. Volf, JETP Lett. **31**, 450 (1980)
14. V.K. Vlasko-Vlasov, M.V. Indenbom, Sov. Phys. JETP **59**, 633 (1984); V.K. Vlasko-Vlasov, L. M. Dedukh, V. I. Nikitenko, Sov. Phys. Solid State **23**, 1058 (1981)
15. T. Fukuda, T. Saburi, K. Doi, S. Nenno, Mater. Trans. JIM **33**(3), 271–277 (1992)

16. B.J. Heuser, T.J. Udovic, H. Ju, Phys. Rev. B **78**, 214101 (2008)
17. J. Liu, Ch. Chen, Q. Feng, X. Fang, H. Wang, F. Liu, J. Lu, D. Raabe, Mater. Sci. Eng. **A703**, 236–243 (2017)

Chapter 33
Concluding Remarks

Abstract In this chapter, we first summarize new predictions concerning the process zone—its properties and its impact on individual cracks and dislocations, as well as on the macroscopic properties of solids—arising from the field-theoretical approach. We then provide a list of problems that remain unsolved and can be addressed in future work. By tackling these and many other open questions not listed here, future research can further elucidate the complexities of defect–process zone interactions, their dynamics, and their broader implications for material behavior.

33.1 A Brief Summary of Our Main Qualitative Results

Let us summarize the most important qualitative findings of this study:

Defects and Process Zones. Defects inducing stress concentrations alter the effective transition temperature in their vicinity.[1] Such defects generate process zones—spatial domains localized at the defects that contain phases distinct from those in the bulk of the crystal.

Order Parameter and Zone Properties. The process zone is governed by the order parameter, which represents the degree of freedom associated with the transformation. For displacive transitions, it corresponds to one or several of the optical phonon modes; for order-disorder transitions, it may represent a combination of probabilities for occupying specific lattice positions; or it may involve a combination of these cases. The order parameter qualitatively distinguishes the structure and symmetry of different phases and provides a quantitative estimate of the degree of this distinction. The structure, configuration, motion, and other properties of the process zone are fully described by the spatial distribution and dynamics of the order parameter. The boundary of the process zone corresponds to the kink in the order parameter distri-

[1] Another scenario, involving ferroelastic zones reacting directly to stress concentrations, lies outside the scope of this discussion.

© The Author(s), under exclusive license to Springer Nature Switzerland AG 2026

A. Boulbitch and A. Korzhenevskii, *Stress-Induced Local Phase Transformations Near Cracks and Dislocations*, Springer Series in Materials Science 356,
https://doi.org/10.1007/978-3-032-00945-6_33

bution, which arises naturally as a solution to the time-dependent Ginzburg-Landau equation governing the statics and dynamics of the order parameter.

Daughter-Phase Process Zone in the Phase Diagram. We refer to the original phase as the mother-phase and the phase occupying the process zone as the daughter-phase. A daughter-phase zone represents a localized region of the daughter phase embedded within the matrix of the mother-phase. The daughter-phase process zone localized at the defect does not exist uniformly across the entire phase diagram. Instead, it emerges at specific temperatures (emergence temperatures): either above the binodal for bulk first-order transitions or above the transition temperature for bulk second-order transitions.

Mother-Phase Process Zone in the Phase Diagram. The mother phase zone represents a localized region of the mother phase embedded within the matrix of the daughter-phase in the region of the phase diagram where the bulk daughter-phase is stable. A mother-phase process zone exists below the bulk binodal and is bounded from below by a local binodal, where the mother-phase zone vanishes. Unlike the daughter-phase zone, the mother-phase zone does not completely disappear below the local binodal but instead adopts a beak-shaped configuration in the order parameter distribution.

First-Order Limitations. First-order process zones exist only within a limited region of the phase diagram.

Defect Motion and the Zone Emergence Temperature. The movement of defects, accompanied by their process zones, reduces the temperature at which the zone emerges. At a critical defect velocity, the process zone disappears entirely. Beyond this velocity, the defect propagates without forming a process zone.

Zone Deformation and the Viscous Force. Defect motion deforms the process zone, compressing it in the front and stretching it in the rear. This deformation generates a nonlinear viscous force opposing defect propagation.

Metastable Wake. For first-order zones, motion generates a metastable wake in the elongated rear portion of the zone. Within this wake, the order parameter exists in a nonequilibrium state, leading to its gradual decay. The wake has a finite size determined by the balance between decay and propagation speeds and therefore moves at the same velocity as the defect. In the case of a crack, the wake assumes a droplet-like shape, whereas for a dislocation, it adopts a cucumber-like configuration.

Transformation Toughness. The presence of a process zone at the crack tip significantly increases fracture toughness, a phenomenon known as *transformation toughness*. This transformation toughness arises from surface tension at the zone's interface. Notably, it occurs for a stationary crack and does not require any form of crack propagation. The fracture toughness depends on the zone's position within the phase diagram and the thermodynamic properties of the bulk phase transition.

Crack Velocity Jumps. The propagation of a crack with a process zone at its tip induces a nonlinear viscous force. This force results in a nonmonotonic dependence of crack velocity on the driving force, characterized by rapid increases and decreases that appear as velocity jumps.

Stick-Slip Motion of a Crack. The nonlinear viscous friction force exhibits a nonmonotonic dependence on crack velocity, featuring a descending branch. Under certain conditions, such as those realized during wedging or in the double cantilever beam configuration, this behavior leads to stick-slip crack propagation.

Line Tension of a Dislocation and Plasticity. The process zone at a dislocation has negative energy, reducing the dislocation's line tension. This reduction enhances the plasticity of the solid. The effective dislocation line tension can drop to zero, leading to an onset of plasticity. This transition occurs at two specific temperatures: one lies below the temperature of the daughter-phase zone's emergence but above the bulk binodal, while the other lies below the bulk binodal but above the temperature at which the mother-phase zone vanishes. The dislocations carrying the process zone are unstable between these two temperatures.

Dislocation Bending and Process Zone Reconfiguration. Bending dislocations modify the structure of their process zones. In some cases, this enhances bending, amplifying plasticity and leading to plastic instability in parts of the dislocation ensemble.

33.2 Future Directions

While the above summary highlights key qualitative findings, it does not encompass the full scope of results obtained in our studies. For brevity, we have omitted certain examples, particularly those related to multicomponent order parameters and some aspects concerning dislocations. Below we outline several promising directions for future research.

***

Interaction of Defects Possessing Zones. The interactions between defects—particularly dislocations—that carry process zones remain poorly understood, as does their collective influence on crystal plasticity. Of particular interest is the behavior of process zones associated with dislocation clusters. Several key questions remain open: Can disconnected zones formed by multiple dislocations coalesce into a unified process zone? If so, what positions would the dislocations preferentially occupy with respect to the merged zone—would they align along its boundary, effectively acting as misfit dislocations?

This line of inquiry also includes the possible fusion of zones generated by dislocations with those at crack tips. Particularly compelling is the dynamics of such coalescence processes, which have yet to be systematically studied.

Multicomponent Order Parameters and the Parallelization of Theoretical and Experimental Studies. This book concentrates on systems with a one-component order parameter. Nonetheless, most of the phenomena discussed here extend, either directly or with minor modifications, to multicomponent systems. Some initial steps in this direction were taken in our earlier work [1]. However, such systems are known to exhibit complex behavior, and we expect them to give rise to previously unanticipated effects. Much remains to be explored in this area. Theoretical studies cannot be conducted in a fully general form; they must account for the symmetry of the crystal under study. This applies to the structure of the free energy as a function of the components of the order parameter, as well as to the elastic energy.This, in turn, requires experimental input followed by group-theoretical analysis. Ideally, all coefficients of the Landau potential should be determined experimentally—a task that also requires a concerted contribution from both experimentalists and theorists.

Process Zones Forming under Phase Separation. In solid solutions and alloys, such as AuCu, hydrogen–metal systems, and many others, phase separation often occurs. In such cases, the order parameter is constructed from the concentrations of the constituent species. In the simplest and most common case, one deals a scalar order parameter, and Landau's potential describing the phase separation admits a cubic term. The behavior of cracks and dislocations in solid solutions is important for understanding their mechanical response. Hence, the process zones forming during the phase separation should be addressed. One of us has previously investigated the statics of process zones formed during the $\alpha - \alpha'$ phase separation in hydrogen—metal systems [2]. However, zones in more complex systems—such as those forming during the α'-hydride transition, or in strongly anisotropic alloys—as well as the consequences of a key feature of phase separation dynamics—namely, its diffusive nature—have not yet been analyzed.

***

By addressing these open questions and many others that we did not list here, future research can further unravel the complexities of defect-process zone interactions, their dynamics, and their broader implications for material behavior.

References

1. A. Boulbitch, Y.M. Gufan, A.L. Korzhenevskii, Phys. Rev. E **96**, 013005 (2017)
2. A. Boulbitch, J. Alloys Compd. **196**, (1–2), 29–36 (1993)

Annex A
Parameters Typical for Inorganic Solid

One needs to get an idea of the typical values of fundamental zone properties, such as zone size and formation temperature. These properties depend on material constants whose experimentally measured values may vary. Theoretical estimates of these constants often lack precision and can involve substantial errors. Therefore, we provide estimated values for the material constants of $BaTiO_3$*—a representative inorganic, dielectric, non-metallic material that undergoes displacive phase transitions and has been extensively studied experimentally. These constants are listed in Tables* A.1 *and* A.2*, along with references. If no reference is provided, the parameter was derived using values from other tables and established physical relations.*

***

A.0.1 Generic Parameters

See Table A.1.

A.0.2 Parameters of Ginzburg-Landau Equation

For order-disorder transition estimates, values from Table A.1 are used, except for α_0 and κ. Typically, $\alpha_0 \in [10^{-2}, 10^{-3}]$ K^{-1} in these cases. The kinetic constant κ depends on the transition type. For displacive transitions:

$$\kappa \sim 10^{-12} \text{ to } 10^{-13}\,\mathrm{s} \tag{A.1}$$

Order-disorder transitions show more variability. For example: $\kappa \sim 10^{-7}$ s for $AgNa(NO_2)_2$, $\sim 10^{-11}$ s for KNO_3 or $NaNO_2$, and $\sim 10^{-12}$ s (Rochelle salt, triglycine sulphate, and KD_2PO_4) [9, 10]. We assume $\kappa \sim 10^{-10}$ s for estimates.

© The Editor(s) (if applicable) and The Author(s), under exclusive license to Springer Nature Switzerlandr AG 2026

A. Boulbitch and A. Korzhenevskii, *Stress-Induced Local Phase Transformations Near Cracks and Dislocations*, Springer Series in Materials Science 356,
https://doi.org/10.1007/978-3-032-00945-6

Table A.1 Material constants of $BaTiO_3$

Property	Dimension	Value	References
Elastic modules,* C_{11}; C_{12}; C_{44}	$10^{12} \times \frac{\text{erg}}{\text{cm}^3}$	1.73; 0.82; 1.08;	[5]
Young's modulus, E	$10^{12} \times \frac{\text{erg}}{\text{cm}^3}$	1.7	
Poisson's ratio, ν	Dimensionless	1/3	
Spontaneous strain spur, $\varepsilon_0 = \varepsilon_{\alpha\alpha}$	Dimensionless	−0.028	[6]
Rayleigh velocity, V_R	cm/s	4.8×10^5	
Fracture toughness, K_{IC}	$\text{erg/cm}^{5/2}$	10^7	[1, 2]
Lattice size, $\mathfrak{a}$, Burger's vector, $\mathfrak{b}$	nm	0.4	
Phase interface width, $\mathbb{D}$	nm	14	[3]
Upper spinodal temperature,** T_{up}	K	396.9	[4, 5]
Binodal temperature,**T_0,	K	393	[4, 5]
Curie temperature,** T_c	K	381.3	[4, 5]
gMass density, ρ	g/cm^3	6.02	
Sound speed, $c_t = [E/2\rho(1+\nu)]^{1/2}$	cm/s	3.3×10^5	

* The components of the tensors of elastic modules in a two-index form: $C_{11} = C_{1111}, C_{12} = C_{1122}$, and $C_{44} = C_{1212}$
** The cubic-tetragonal transition

Table A.2 Parameters of Ginzburg-Landau equation for $BaTiO_3$

Property	Dimension	Value	References
α_0	$10^{-5} \times \text{K}^{-1}$	3.7	[5]
β_1	$10^{-13} \times \text{cm}^3/\text{erg}$	−6.17	[5]
γ	$10^{-22} \times \text{cm}^6/\text{erg}^2$	1.65	[5]
g	$10^{-16} \times \text{cm}^2$	3.3	[3]
A	Dimensionless	2.5	[5]
κ*	$10^{-12} \times \text{s}$	3.5	[7, 8]

* The relaxation mode

A.1 Parameters of Phase Transformation Theory

In the following sections, we will provide order-of-magnitude estimates for key parameters related to phase transformations. These parameters are not tied to specific defects or zone properties but are fundamental to understanding phase transition behavior.

***

A.1.1 Hysteresis

One important parameter is the hysteresis, defined as the temperature difference between the upper spinodal and the Curie point (3.33):

$$\Delta T_{\text{hyst}} = T_{\text{up}} - T_{\text{c}} = \frac{\beta^2}{4\gamma\alpha_0}$$

Using the data from Tables A.1 and A.2, we estimate:

$$\Delta T_{\text{hyst}} \sim 10\,\text{K}$$

The temperature difference between the upper spinodal and binodal, $T_{\text{up}} - T_0$, is of the same order of magnitude. Based on Table A.1, we find: $T_0 - T_{\text{c}} = 12\,\text{K}$.

A.1.2 Gordon's Velocity

Figure A.1 shows the temperature dependence of Gordon's velocity within the range $T_0 \leq T \leq T_{\text{up}}$, where T_0 is the binodal and T_{up} is the upper spinodal temperatures. All parameters for calculating Gordon's velocity are derived from the values listed in Tables A.1 and A.2.

Gordon's velocity reaches zero at the binodal point, and near this point, it can be approximated by the linear relation:

$$V_{\text{G}} \approx k(T - T_0)$$

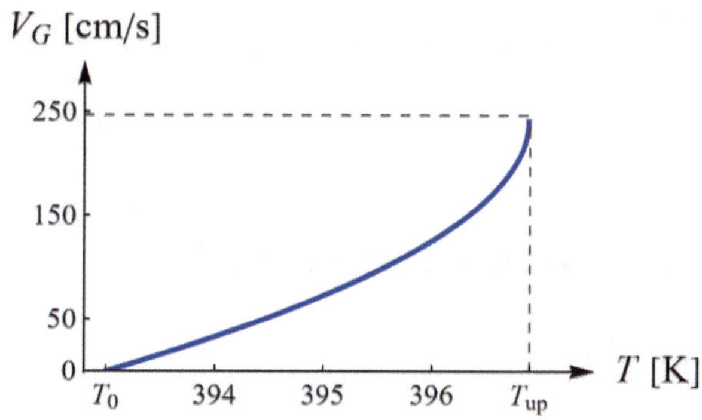

Fig. A.1 Gordon's Velocity versus Temperature. The velocity is calculated for a crystal with parameters from Tables A.1 and A.2, showing its dependence on the ratio $\alpha\gamma/\beta^2$

where the coefficient k is given by:

$$k = \frac{8\alpha_0}{\beta\kappa}\sqrt{\frac{g\gamma}{3}}$$

Using the values from Table A.1, we estimate:

$$k \approx 31\frac{\text{cm}}{\text{s K}}$$

Gordon's velocity reaches its maximum value, $V_{\text{G}}^{(\text{max})}$, at the upper spinodal point $T = T_{\text{up}}$. For the crystal parameters in Tables A.1 and A.2, we estimate:

$$V_{\text{G}}^{(\text{max})} \sim 10^2 \text{ cm/s},$$

For order-disorder transitions, this estimate is much lower, yielding:

$$V_{\text{G}}^{(\text{max})} \sim 10 \text{ cm/s},$$

A.1.3 Surface Tension of the Phase Interface

The surface tension of the phase interface, denoted τ_{s}, can be expressed by the following expression:

$$\tau_{\text{s}} \approx \frac{g[\eta^2]}{\mathbb{D}} \tag{A.2}$$

Here, $[\eta^2]$ is approximately $|\beta_1|/\gamma$, representing the discontinuity of the square of the order parameter at the first-order transition point, and $\mathbb{D}$ is the width of the phase interface. The derivation of this expression is quite lengthy. One can find it in the appendix to the paper [11]. Utilizing values from Tables A.1 and A.2, we estimate the surface tension of the phase interface:

$$\tau_{\text{s}} \approx 1 \text{ erg/cm}^2$$

A.2 Conversion of Some Units

See Table A.3.

Table A.3 Conversion table of units utilized in this book

Parameter name	Off-system unit	SI	CGSE
The line slope, k_0	1 K/kbar	10^{-8} K/Pa = 10 K/*G* Pa	10^{-9} K cm^3/erg
Fracture toughness, K_{IC}	–	1 MPa m$^{1/2}$	10^8 erg/cm$^{5/2}$
Frequency	1 cm^{-1}	4.77 GHz	4.77 GHz

Annex B
Simulations of Formation of a Soft Process Zone

B.1 Relaxational Method

A straightforward numerical solution of the static equations—(10.19), (12.13), and (12.12) for the crack tip zone, or (20.17) and (20.19) for the dislocation zone—with zero Dirichlet boundary conditions returns only the trivial solution $u = 0$ for any value of the control parameter a. This occurs because the function $u(x, y) = 0$ is, indeed, one of the three solutions to each of the aforementioned equations. Consequently, a solver starting from $u = 0$ at the boundary converges to the trivial solution and remains there.

To overcome this, we employed the relaxation method. We introduced an artificial pseudo-time variable τ_p and a new dependent variable $u_{\text{dyn}} = u_{\text{dyn}}(x, y, \tau_p)$, where the subscript "dyn" denotes "dynamic" and "p" stands for "pseudo." Instead of solving the static equation (12.12), we solved the following pseudo-time-dependent equation:

$$\frac{\partial u_{\text{dyn}}}{\partial \tau_{\text{p}}} = -f(u_{\text{dyn}}) \tag{B.1}$$

where $f(u_{\text{dyn}})$ has the following form:

$$f(u_{\text{dyn}}) = -\Delta u_{\text{dyn}} - 2^{1/3}\nu \frac{\partial u_{\text{dyn}}}{\partial x} + \left[a - \frac{\left[\left(x^2 + y^2\right)^{1/2} + x \right]^{1/2}}{\sqrt{2}\left(x^2 + y^2 + \varepsilon\right)^{1/2}} \right] u_{\text{dyn}} + u_{\text{dyn}}^3 \tag{B.2}$$

Here, $u_{\text{dyn}} = u_{\text{dyn}}(x, y, \tau)$, τ is the pseudo-time, $v = V/V_{\text{d}}$. The expression $\left[\left(x^2 + y^2\right)^{1/2} + x \right]^{1/2} / \sqrt{2}\left(x^2 + y^2 + \varepsilon\right)^{1/2}$ corresponds to $\cos(\theta/2)/\sqrt{r^2 + \varepsilon}$ and describes the case of the crack which we take here as the example. The parameter a is defined in (10.5). The regularization parameter $\varepsilon = 0.0001$ prevents divergence at

© The Editor(s) (if applicable) and The Author(s), under exclusive license to Springer Nature Switzerlandr AG 2026

A. Boulbitch and A. Korzhenevskii, *Stress-Induced Local Phase Transformations Near Cracks and Dislocations*, Springer Series in Materials Science 356,
https://doi.org/10.1007/978-3-032-00945-6

$\rho = 0$, ensuring numerical stability. In this form, (B.1) and (B.2) are well-suited for simulation.

The boundary condition is $u_{\text{dyn}}|_{\partial\Omega} = 0$. A nonzero initial condition $u_{\text{init}}(x, y, 0) \neq 0$ serves as the starting point in the functional space.

Equation (B.1) can be interpreted as describing the motion of a point in functional space, where $\partial u_{\text{dyn}}/\partial\tau_{\text{p}}$ acts as velocity of this point, and $f(u_{\text{dyn}})$ is the driving force. The left-hand side represents viscous friction with unit mobility. In other words, (B.1) governs the motion of a massless point under viscous friction in functional space. In these terms, the condition

$$f(u_*) = 0 \tag{B.3}$$

implies that the driving force vanishes at some point u_* in functional space. This point is a fixed point of the nonlinear equation (B.1). On the one hand, (B.3) identifies the fixed point of the dynamic (B.1). On the other hand, it coincides with the stationary equation (12.12).

Since $f(0) = 0$, the trivial solution $u_* \equiv 0$ is a stationary point of (B.1). Physically, this represents the bulk mother phase. Under suitable conditions, a nontrivial solution $u_* \neq 0$ exists, corresponding to the localized daughter phase. Hence, a nontrivial stationary point of (B.1) and (B.2) yields a solution to the static (12.12) describing the process zone.

Moreover, $u_* = 0$ is a repulsive fixed point, whereas $u_* \neq 0$ is attractive. Thus, any initial condition $u_{\text{init}}(x, y, 0) \neq 0$ leads to convergence:

$$u_{\text{dyn}}(x, y, \infty) = u_* \neq 0$$

In practice, we approximate the solution $u(x, y)$ by $u_{\text{dyn}}(x, y, \tau_{\max})$ assuming sufficiently large $\tau_{\max}$ so that convergence is nearly achieved.

This convergence property is key to applying the relaxation method to solve equations like (B.1) and (B.2). Regardless of the specific form of the initial condition, as long as $u_{\text{init}}(x, y, 0) \neq 0$ and $u_{\text{init}}|_{\partial\Omega} = 0$, the solution at large τ approximates the solution $u(x, y)$ of (12.12) with high accuracy.

Further discussion on the application of the relaxation method to problems involving bifurcation of the trivial solution can be found in [12].

B.2 The Simulation Approach

B.2.1 Domain

Simulations were performed using the COMSOL software. Exploiting the problem's symmetry, (10.4) was simulated in the half-plane $y \geq 0$. A semi-circular domain was selected. Since the solution describes a kink whose boundary forms a closed contour, one should choose the domain such that this contour lies entirely within it.

Furthermore, the kink contour should be positioned sufficiently far from the domain's outer boundary to ensure that the boundary condition $u_{\text{dyn}}|_{\partial\Omega} = 0$ does not affect the solution. Through trial and error, the diameter $D = 50$ dimensionless units was found to be sufficiently large for the solution to vanish well away from the boundary.

A no-flux boundary condition was imposed on the symmetry plane $y = 0$, and the condition $u = 0$ was applied to the remainder of the boundary. The initial condition was a smoothed step function, non-zero only near the point $(x_1 = 0, y_1 = 0)$.

Meshing

For the second-order process zone, its scale corresponds to the characteristic size R_{d} (10.25). The rescaling (see Sect. 10.1) ensures that one unit in the rescaled dimensionless coordinates equals the scaling factor $\mathbb{R}_0 \sim \mathbb{R}_{\text{d}}$ in physical space. Thus, the entire process zone is contained within the semi-circular domain of radius 50 dimensionless units.

This simplifies meshing: the mesh size should be approximately one unit or smaller, which already gives a satisfactory resolution. To further improve the mesh, we began with a uniform mesh of size 5 and then applied COMSOL's *adaptive mesh refinement* option. This yields a coarser mesh in regions where $(\nabla u)^2$ is small and a much denser mesh where $(\nabla u)^2$ is large. Details of this procedure are provided in Sect. E.2.5.

B.2.2 Convergence Control

A stable solution of the static equation (10.4) corresponds to a fixed point of the dynamic equation (B.1). The pseudo-time stepping method converges well away from the points $a = 0$ and the bifurcation point $a = a_{\text{d}}$. In practice, we achieved reliable convergence for $a > 0.1$, beyond the global bifurcation threshold. Closer to $a = 0$, equilibrium solutions could not be obtained. Near the bifurcation point $a = a_{\text{d}}$, the method introduced a small regular error, which is discussed in detail below. Aside from this, the method provided good convergence, allowing us to examine the distribution of the rescaled order parameter $u(x, y)$ near the crack tip.

To assess convergence, we used a convergence indicator: a scalar quantity that levels off if the solution stabilizes. This could be the maximum of the solution, $u_{\max}$, or the Hilbert norm

$$\|u\| = \left(\int_\Omega u^2(x, y)\, dx\, dy \right)^{1/2}.$$

Figure B.1 illustrates the evolution of such an indicator over time. This behavior helps identify the minimum simulation time required to achieve sufficient convergence.

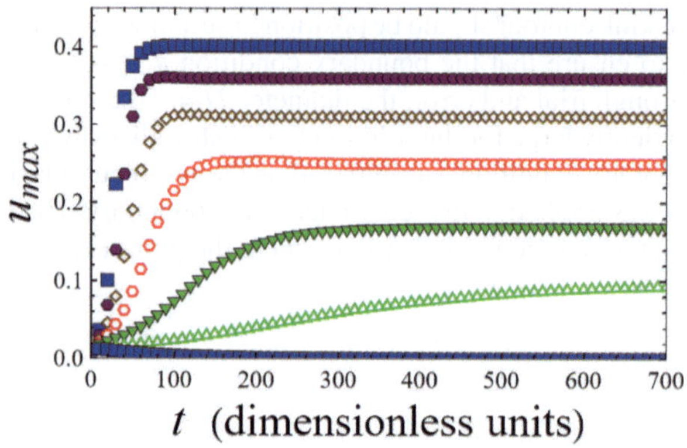

Fig. B.1 Illustration of Convergence Control in Numerical Calculations. With increasing of the the simulation time (in dimensionless units) of the behavior of the amplitude of the rescaled order parameter, $u_{\max}$, is shown at different values of q. Filled disks: $q = 0.3$, filled squares: 0.32, filled diamonds: 0.34, vertex-up triangles: 0.36, vertex-down triangles: 0.38, open circles: 0.39, empty squares: 0.41. Note that in the case of $q = 0.39$ (point-down triangles) the $u_{\max}(t)$ dependence still exhibits a slope and for satisfactory convergence a much longer process was used (not shown here)

The dynamic system was solved using the direct MUMPS solver with BDF time stepping. Convergence to the fixed point was monitored via the behavior of $u_{\max}$, the maximum of $u(x, y, \tau_{\max})$. Through experimentation, we found that 700 pseudo-time steps were typically sufficient for convergence, although in some cases up to 3000 steps were required. Figure B.1 provides an example of such a convergence analysis for simulations where all parameters were held constant except for a, which was varied.

It is evident that far from the bifurcation point, convergence occurs well before 700 pseudo-time steps. Predictably, the scenario changes near the bifurcation (e.g., $a = 0.38$ and 0.39, corresponding to downward triangles and open circles in Fig. B.1). Even in these cases, however, 700 steps typically ensured reliable convergence.

Annex C
Numerical Justification of Free Energy of the Mother-Phase Zone

In the analysis of the rescaled free energy perturbation ΔF (22.16), detailed features in the immediate vicinity of the dislocation were neglected. The results, presented in (22.16), show that ΔF increases logarithmically with the size ratio $\mathbb{R}_{cr}/\mathfrak{b}$. This behavior supports the assumption that the order parameter's structure near the dislocation core can be neglected, despite the slow progression of the logarithmic growth. To validate this approximation, we performed a numerical analysis of the perturbation ΔF.

Figure C.1a shows a semi-logarithmic plot of the absolute value of the free energy perturbation, $|\Delta F|$, as a function of $\mathbb{R}_{cr}/\mathfrak{b}$.

The fitting parameter $k_1 \approx 0.37 \pm 0.20$ characterizes the contribution of the region close to the dislocation to the free energy. Its variation, $\Delta k_1 \approx 0.20$, and the relative deviation $\Delta k_1/|\Delta F|$, shown in Fig. C.1b, falls below 0.1 for $\mathbb{R}_{cr}/\mathfrak{b} \gtrsim 10$, where the crystal size exceeds approximately ten lattice constants—typically the case. This result confirms the negligible influence on the free energy perturbation of the zone located in the immediate vicinity of the dislocation.

© The Editor(s) (if applicable) and The Author(s), under exclusive license to Springer Nature Switzerlandr AG 2026

A. Boulbitch and A. Korzhenevskii, *Stress-Induced Local Phase Transformations Near Cracks and Dislocations*, Springer Series in Materials Science 356,
https://doi.org/10.1007/978-3-032-00945-6

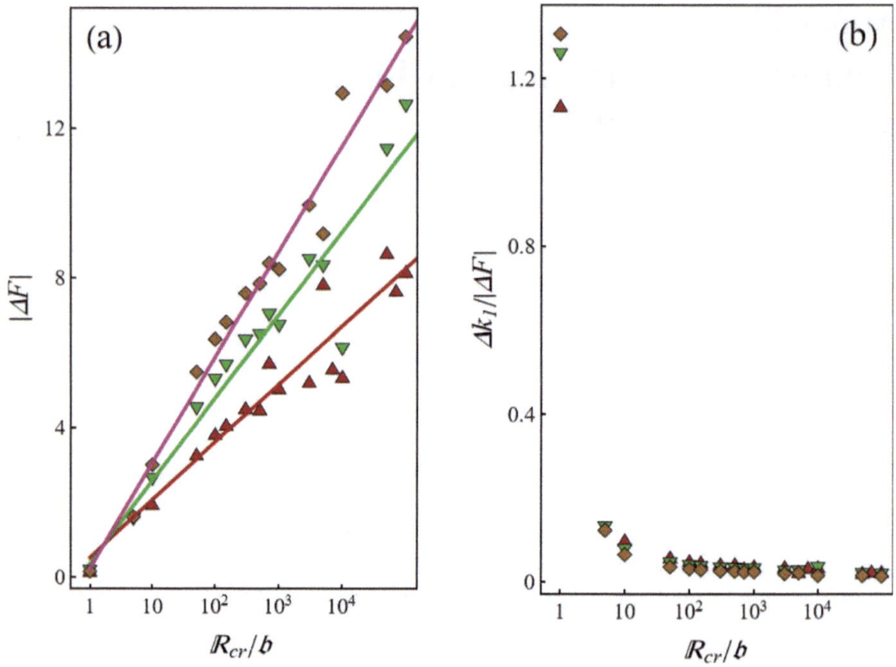

Fig. C.1 Relationship Between Free Energy Perturbation, ΔF, and the Ratio $R_{cr}/\mathfrak{b}$. **a** The plot includes dots representing the evaluated ΔF from simulations with $b = 1$, diamonds for $a = 0.15$, downward-pointing triangles for $a = 0.12$, and upward-pointing triangles for $a = 0$. The linear relations $\Delta F \approx -k_1 - k_2 \ln(R_{cr}/\mathfrak{b})$ fit all the data, with the color of the lines matching the corresponding data points. **b** Shows the variation of the ratio $\Delta k_1/|\Delta F|$ with $R_{cr}/\mathfrak{b}$

Annex D
Simulation of the Crack Velocity Jumps

D.1 Rescaling

We addressed the problem by numerically solving the system (18.3) using the finite element method. We rescaled equation (18.3) to dimensionless variables, applying the transformations:

$$x \to d_1 x_1, \quad y \to d_1 y_1, \quad \eta(x, y) \to d_2 u(x_1, y_1),$$

with

$$d_1 = \frac{g}{\kappa V_R}, \quad d_2 = \frac{\kappa V_R}{\sqrt{g\beta}}.$$

The rescaled equations (18.3) take the following form:

$$\begin{cases} \Delta_1 u + \frac{V}{V_R} \times \frac{\partial u}{\partial x_1} - \left[q_1 - \frac{\nu_c^{3/2}}{\sqrt{2(1-\mathbb{G})}} \times \frac{\cos(\theta/2)}{\sqrt{r_1}} \right] u - u^3 = 0 \\ \frac{V}{V_R} = \frac{\mathbb{G}}{1+q_2(1-\mathbb{G}) \int \left(\frac{\partial u}{\partial x_1} \right)^2 dx_1 dy_1} \end{cases} \tag{D.1}$$

where $\Delta_1 = \partial^2/\partial x_1^2 + \partial^2/\partial y_1^2$, $r_1 = (x_1^2 + y_1^2)^{1/2}$, and the dimensionless control parameters q_1 and q_2 are defined as:

$$q_1 = \frac{\tau_c \nu_c^2}{4}; \quad q_2 = \frac{(\kappa V_R)^3}{g\beta\Gamma} \tag{D.2}$$

Technical details of the simulations are provided below.

© The Editor(s) (if applicable) and The Author(s), under exclusive license to Springer Nature Switzerlandr AG 2026

A. Boulbitch and A. Korzhenevskii, *Stress-Induced Local Phase Transformations Near Cracks and Dislocations*, Springer Series in Materials Science 356,
https://doi.org/10.1007/978-3-032-00945-6

D.2 Technical Details

The simulations were carried out using the COMSOL software. The equations were solved in the half-plane $y \geq 0$. A semi-elliptic domain was defined by trial and error so that the solution vanishes well before reaching the domain boundaries. A no-flux boundary condition was imposed along $y = 0$, and a Dirichlet condition $u = 0$ was applied to the remaining boundaries.

As explained in Sect. B.1, direct simulation of the static equation (D.1) under these boundary conditions results in the trivial solution. To circumvent this, we instead introduced a pseudo-dynamic equation (B.1):

$$f(u_{\text{dyn}}) = -\Delta u_{\text{dyn}} - \frac{V}{V_R} \times \frac{\partial u_{\text{dyn}}}{\partial x} + \left[q_1 - \frac{\nu_c^{3/2}}{\sqrt{2(1-\mathbb{G})}} \times \frac{\cos(\theta/2)}{\sqrt{r}} \right] u_{\text{dyn}} + u_{\text{dyn}}^3 \tag{D.3}$$

which replaces the first equation in (D.1), while the second equation remains unchanged. A smoothed step function was used as the initial condition. As in the previous case, the stable solution of the static equation corresponds to a fixed point of the dynamic one. The integral equation was implemented via the *Model Coupling* feature in COMSOL.

The dynamic system was solved using the direct MUMPS solver with BDF time stepping. Convergence to the fixed point was monitored through $u_{\max}$, the maximum of $u_{\text{dyn}}(x, y)$. As discussed in Sect. B.1, 700 pseudo-time steps were typically sufficient to achieve convergence, although in some cases up to 3000 steps were required.

Annex E
Relaxation Method for the First-Order Zone

E.1 Equation of Motion and Its Interpretation

E.1.1 Equation of Motion

In the case of the first-order zone one uses the relaxation method by simulating (B.1) with the following driving force:

$$f(u_{\text{dyn}}) = -\Delta u_{\text{dyn}} - 2^{2/3} v \times \frac{\partial u_{\text{dyn}}}{\partial x} + \left[a - \frac{\sqrt{\sqrt{x_1^2 + y_1^2} + x_1}}{\sqrt{x_1^2 + y_1^2 + \varepsilon}} \right] u_{\text{dyn}} - b u_{\text{dyn}}^3 + u_{\text{dyn}}^5 \tag{E.1}$$

where $v = V / V_{\text{d}}$. The cut-off ε we included to avoid the infinity during the numeric process. It differs from (B.2) by the coefficient $2^{2/3}$ instead of $2^{1/3}$in front of v, but also by term u_{dyn}^5 and the negative cubic term. These latter variations introduce considerable differences into the numeric process. Below we describe the nature of these differences.

E.1.2 Domain

Using the mirror symmetry of the problem with respect to the $y = 0$ plane we simulated the equation of motion of a zone in the upper half-plane $y \geq 0$. We choose a rectangular domain there, its bottom boundary being at $y = 0$. In all cases obtained in the simulations, the solution represents a plateau located in some region around the point $x = y = 0$ bounded by a kink and rapidly vanishing outside of the plateau/kink region. For this reason, the position of the other boundaries of the simulation domain

© The Editor(s) (if applicable) and The Author(s), under exclusive license to Springer Nature Switzerlandr AG 2026

A. Boulbitch and A. Korzhenevskii, *Stress-Induced Local Phase Transformations Near Cracks and Dislocations*, Springer Series in Materials Science 356,
https://doi.org/10.1007/978-3-032-00945-6

was defined by trial and error, so that the solution vanishes well before the boundaries. By trial we have proved that provided the domain boundaries are at a certain distance from the plateau well exceeding the kink width, their positions do not influence the solution.

We set a symmetry (no-flux) boundary condition at the boundary $y = 0$ and the Dirichlet condition $u|_\Omega = 0$ at the rest of the rectangle boundary.

As the initial condition, we used a smoothed step function, provided by COMSOL and exhibiting a smooth, narrow kink. We modified it such that it is only different from zero within either a cardioid (used for $v < v_G$) or an elongated ellipsoid (used for $v > v_G$). In the latter case, we shifted the center of the ellipsoid somewhat backward from the origin of the coordinates. These initial shapes proved to be close enough to the smooth final solutions and offered a reasonably fast convergence.

E.2 New Features of the Relaxation Process in the First-Order Case

E.2.1 First-Order from the Functional Space Point of View

The free energy F (8.11) constitutes a hypersurface in functional space. Any solution $u_{\text{dyn}}(x, y, \tau_p)$ of (B.1) with (E.1) traces a trajectory on this hypersurface. Let the basis functions of the functional space be denoted $u_0, u_1, u_2, \ldots$. The trajectory $(u_0(\tau), u_1(\tau), u_2(\tau), \ldots)$ progresses from the initial state $(u_0(0), u_1(0), u_2(0), \ldots)$ to the final state $(u_0(\infty), u_1(\infty), u_2(\infty), \ldots)$.

According to the definition, (B.3) minimizes the free energy functional. In this context, the dynamic equation (B.1) may be interpreted as describing the viscous motion of a massless point particle on a hypersurface within a $1 + \infty$-dimensional space $(F', u_0, u_1, u_2, \ldots)$, forming a potential well. The particle moves from an initial position towards the minimum of this well, $u_{\min}$, where it comes to rest. At this equilibrium point, the driving force vanishes (B.3).

Above the bifurcation, this minimum corresponds solely to the mother phase located at $u = 0$. Below the bifurcation, the free energy hypersurface acquires a new configuration, introducing additional minima. Consequently, the particle moves from the initial position to one of these new minima. Given the infinite dimensionality of the functional space, the numerical process may involve multiple dimensions.

If the hypersurface F' possesses several minima, these correspond to fixed points of (B.1), (E.1). The solutions $u(x, y)$ of the static equation (12.12) represent fixed points of the dynamic equation (B.1), attainable in the limit:

$$u(x, y) = \lim_{\tau_p \to \infty} u_{\text{dyn}}(x, y, \tau_p)$$

In summary, for the first-order case, the hypersurface F' presents multiple minima, thus, distinguishing it from the second-order scenario.

E.2.2 Secondary Fixed Points

In the case of the first order zone, the scenario becomes more intricate due to the presence of hysteresis. Within this region, the hypersurface F' possesses at least three minima, corresponding to the mother and daughter phases. This implies that certain trajectories may settle into one of the secondary minima rather than the primary one.

Moreover, the hypersurface F' associated with the first-order potential can exhibit an even more complex structure, featuring several secondary minima unrelated to either the mother or daughter phases.

For instance, one such secondary minimum corresponds to the solution illustrated in Fig. E.1. This solution displays straight segments along its boundary (highlighted by arrows in Fig. E.1) instead of the anticipated curved form.

On the one hand, this suggests the solution lies near a fixed point. On the other hand, the target solution is expected to be smooth. Based on the parameters used in the simulation that produced this contour, we anticipated a roughly cardioid shape. Although the resulting contour is kinked, its size and form suggest proximity to the desired cardioid-like configuration. The contour shown in Fig. E.1 remains static throughout an extended simulation period, exhibiting no visible evolution of the straight boundary segments. This indicates that the numerical process has converged to a fixed point—though not the intended one.

The origin of such a minimum can be understood from energetic considerations. Indeed, the straight boundary segments lead to an increase in the "homogeneous" component of the energy:

$$\int \left\{ \frac{[a - U(x, y)]u^2}{2} - \frac{bu^4}{4} + \frac{u^6}{6} \right\} dx\, dy,$$

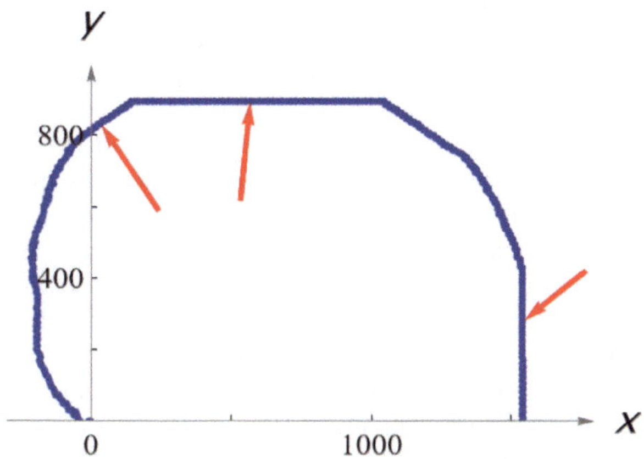

Fig. E.1 Upper Half of the Process Zone Boundary Obtained by Simulation. The boundary exhibits straight portions. The crack is at $x < 0$ with its tip at $x = 0$. The solution in the shown state already exhibits a good convergence which implies that the configuration corresponds to some minimum of the hypersurface. The boundary, however, possesses straight portions indicated by arrows, suggesting that the solution is trapped in a secondary minimum since some parts of the boundary appear slightly closer to the crack tip than they would in a smooth, cardioid-like configuration. At the same time, the straight portions of the boundary contribute to a reduction in the "inhomogeneous" energy component:

$$\int \frac{(\nabla u)^2}{2} \, dx \, dy,$$

This latter energetic gain explains the emergence of additional secondary minima in the functional space F', and thus the appearance of secondary fixed points during the relaxation process.

If a trajectory approaches one of these secondary fixed points in functional space, it may become permanently trapped. Furthermore, the hypersurface may feature saddle points and shallow valleys, which—within limited simulation durations—can be indistinguishable from permanent capture.

E.2.3 Do Secondary Fixed Points Deliver False Solutions?

Whether such secondary fixed points pertain to the problem under study is not a trivial matter. On the one hand, the primary minimum of F' in functional space is most pronounced and indeed corresponds to thermodynamic equilibrium.

On the other hand, the secondary minima reflect physical reality. Indeed, in experiments involving crack loading, process zones invariably evolve over time. This evolution tends to become trapped in similar types of secondary minima. As a result, the boundary of the zone may be observed to consist, wholly or in part, of straight segments.[1] Thus, the relaxation method closely mimics experimental conditions. For this reason, the choice between primary and secondary minima remains open.

In this book, we focus on the general properties of the zone–wake complex. Accordingly, we consider only the fixed points associated with the most pronounced minimum of the free energy F, leaving all secondary fixed points aside. The corresponding solution must exhibit a smooth, continuous boundary—meaning that the zone boundary should not display steps significantly exceeding the mesh size. This serves as the first criterion for selecting an acceptable solution.

Whenever we obtained a stable solution with straight boundary segments flanked by kinks or steps, we revised the initial condition and re-ran the simulation. We aimed to guess an initial condition close to the expected smooth solution. In most cases, after a number of attempts—sometimes considerable—we succeeded in finding an initial shape close to the desired final one, which then led to the required smooth solution.

Furthermore, we systematically monitored the convergence of each solution, using the method described in Sect. B.2.2.

[1] Pinning defects can also lead to the division of the boundary into segments. Here, however, we discuss only the mechanism related to secondary minima.

In general, the closer the simulation approached the binodal, the more challenging it became to find a smooth solution. The most problematic results occurred at $a = 0.2$, where the reliability of solutions could not be confirmed within the velocity range $0 \leq v \leq 0.5$. Consequently, these results are not presented in this book.

E.2.4 Initial Conditions

Let us recall that $u(x, y) \equiv 0$ is always a solution of (12.12). Applying the boundary condition $u|_{\partial\Omega} = 0$ implies that a straightforward simulation of the static equation would return this trivial solution. The latter represents one of the most dominant among the secondary fixed points discussed in Sect. E.2.2. To avoid this outcome, one begins with a seed distribution of the rescaled order parameter $u_{\text{init}}(x, y)$ that ensures the process starts outside the basin of attraction of this trap. The system then evolves step-by-step until the result sufficiently approximates a solution to (12.12). This constitutes the essence of the relaxation method.

Accordingly, the initial condition $u_{\text{init}}(x, y)$ plays a critical role in the relaxation method. The closer $u_{\text{init}}(x, y)$ is to the actual solution, the faster the convergence of the process.

We constructed the seed function $u_{\text{init}}(x, y)$ using the *step(x)* function, a smooth, twice-differentiable version of the Heaviside function $\theta(-x)$, which equals 1 for $x < 0$ and 0 for $x > 0$, with a continuous transition between these values. This function is available within COMSOL and can be used to define a suitable seed. For instance, in simulations of the zone at the tip of a stationary crack, with $b \approx 1$, we used the following seed function:

$$u_{\text{init}}(x, y) = step\left(x^2 + y^2 - R\left(\sqrt{x^2 + y^2} + x\right)\right),$$

as shown in Fig. E.2a, where R is the guessed size of the zone. In this case, the resulting shape should resemble a cardioid, which we expect to approximate the true interface boundary.

Furthermore, since the bifurcation is strong, the solution amplitude is expected to be slightly below unity. Therefore, we also set the seed function's amplitude to 1. This approach effectively avoids the dominant secondary trap at $u(x, y) \equiv 0$.

Let us note that the cardioid shown in Fig. E.2a has a smooth edge, though this is not visible at the scale of the image.

In the search for the lower spinodal, we employed a different seed function. This function was required to have a small amplitude, placing it close to the trivial solution $u(x, y) \equiv 0$. Under these conditions, the numerical process converges to the trivial solution for all points above the lower spinodal. However, below the spinodal, the seed function gradually evolves into the non-trivial solution $u(x, y) \neq 0$.

We constructed such a function using COMSOL's built-in *rect(x)* and *rn(x,y)* functions. The $rect(x)$ function equals zero for $x < 0$ and $x > 1$, and equals one

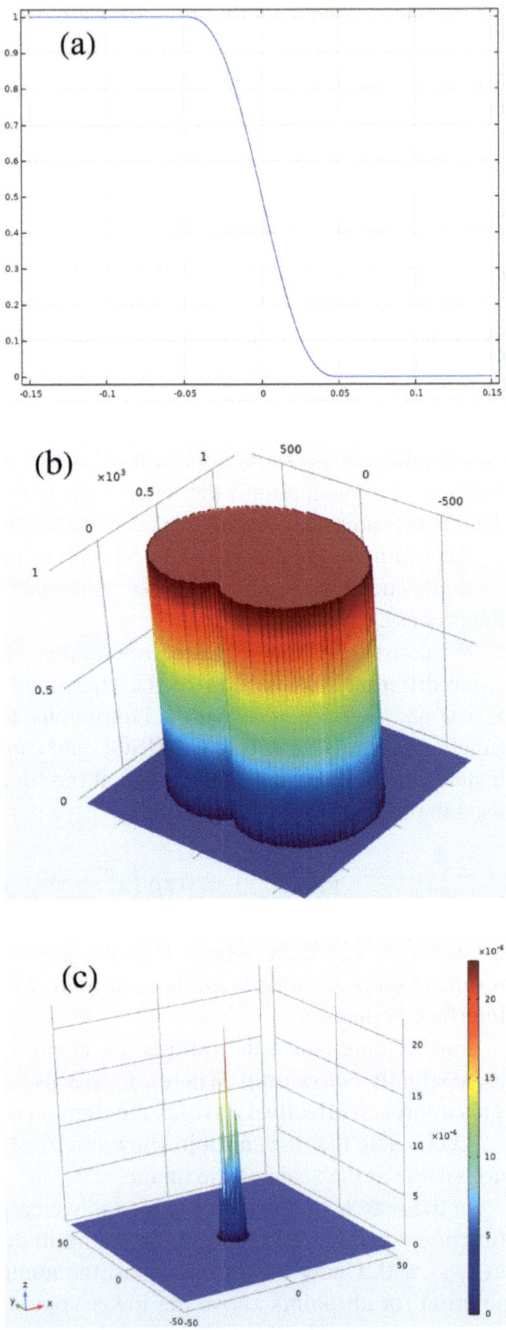

Fig. E.2 Illustration of Seed Functions Utilized in This Study. **a** The step function, $y = step(x)$, provided by COMSOL. **b** The seed function, $u_{\text{init}}(x, y)$ used to simulate the process zone in search for the lower spinodal. **c** The seed function used for dislocations

for $0 < x < 1$, with smooth transitions between these levels. The $rn(x, y)$ function generates random values. The seed function used for the dislocation simulation was defined as:

$$u_{\text{init}}(x, y) = 10^{-5}\,|rn(x, y)| \cdot rect\left(\sqrt{\left(\frac{x}{R}\right)^2 + \left(\frac{y+R}{R}\right)^2}\right)$$

The absolute value ensures that the solution evolves towards the positive minimum of the potential. This function is illustrated in Fig. E.2c.

E.2.5 Meshing

As previously mentioned in Sect. B.2, meshing for the second-order process zone is straightforward.

The situation is quite different for the first-order zone, where multiscaling effects emerge. On one hand, the feature size of the zone–wake complex spans from several tens to several thousands of dimensionless units, as observed in simulations (see Chap. 11). On the other hand, within the zone, the distribution of u in the daughter phase remains shallow almost everywhere, except near the crack tip. Outside the zone, in the mother phase, the absolute value of u is negligibly small. A sharp transition appears only within the narrow interface between the mother and daughter phases, where the order parameter rises steeply—this kink is only a few units wide, as shown in Fig. 13.11.

Thus, to determine the overall shape of the zone–wake complex, one must solve the equation over a domain extending across several thousand units, while simultaneously achieving resolution on the scale of a few units to accurately describe the kink. Complicating matters further, the position of the zone boundary is not known in advance—its determination is part of the solution.

This challenge allows for significant simplification. Since the solution is shallow outside the kink, we used a mesh that is fine in the region of the kink and coarse elsewhere. However, because the kink's location is not known beforehand, a fixed mesh cannot be constructed in advance.

Fortunately, COMSOL provides the ability to generate a dynamic mesh with variable element sizes. This mesh automatically detects the kink region and reconstructs itself accordingly, using the *Adaptive Mesh Refinement* option in conjunction with multiple studies (see below).

We begin with a homogeneous mesh of intermediate resolution. During each study run, the Adaptive Mesh Refinement procedure constructs ten meshes and computes ten corresponding solutions, each using one of these meshes. For each subsequent solution, a new mesh is generated based on the control function $(\nabla u)^2/10$ obtained from the previous solution step. This process allows the mesh to be refined in regions where $(\nabla u)^2$ is large and coarsened where it is small.

Furthermore, COMSOL enables the definition of a new study that uses the previous solution as its initial condition. Following this approach, one starts with a coarse, homogeneous mesh—such as the one shown in Fig. E.3a—and gradually arrives at a highly non-uniform mesh. This refined mesh features fine resolution near the kink

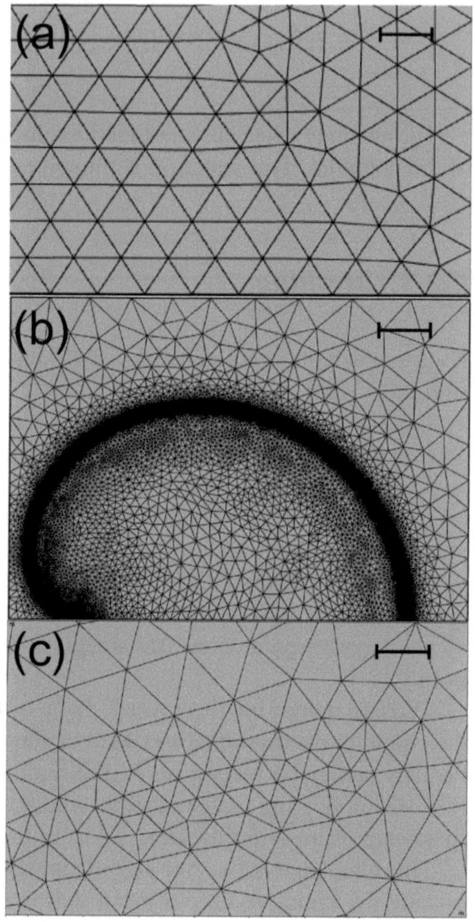

Fig. E.3 Illustration of Mesh Evolution. **a** A fragment of the initial mesh. The bar corresponds to 50 dimensionless units. **b** The final mesh. It exhibits a variable mesh size, which is rough far from the kink. The mesh density dramatically increases in the kink region. The bar length is 1000 units. **c** A blown-up fragment of the same mesh from the middle of the kink. The bar shows 5 units

and a coarser grid elsewhere, as illustrated in Fig. E.3b. A magnified view of the mesh within the kink region is displayed in panel (c) of Fig. E.3.

Depending on the size of the zone, the entire process may involve several to several tens of study steps, producing from a few dozen to several hundred meshes accordingly.

E.2.6 Adaptive Mesh Refinement in COMSOL

Disclamer. The details provided below are valid for COMSOL version 5. We cannot predict any changes that may (or may not) be introduced in future versions. However, the guidance offered here should assist users in locating and configuring the necessary features, even in future releases.

E.2.6.1 Adaptive Mesh Refinement

The *Adaptive Mesh Refinement* feature is deeply buried within COMSOL's interface. For that reason, we provide a step-by-step guide on where and how to locate and configure this option.

To make mesh adaptation to high gradients possible, one must activate the *Adaptive Mesh Refinement* feature in COMSOL by following these steps.

E.2.6.2 Step 1

Within this step, one creates the initial mesh and initiates the mesh refinement process.
a In the `Model Builder` tree, locate the node `...Model`. Right-click this node and select `Add Mesh`. The `Settings Mesh` tab appears. In this tab choose the initial mesh. We recommend selecting `Physics-controlled mesh -> Normal`, which provides a predefined mesh setup, as shown in Fig. E.3a. Depending on the specific problem, `Fine` or `Coarse` options under `Physics-controlled mesh` may also be suitable. Alternatively, one may choose `User-controlled mesh` and specify the mesh parameters manually. This choice generally does not strongly affect the final mesh, though it may slightly influence the number of studies required. In certain challenging cases, adjusting the initial mesh has proven beneficial.
b In the `Model Builder` tree, find the node `Study 1/Step 1: Time Dependent` and select it. In the `Settings: Time Dependent` tab, expand the `Adaptation` section and check the `Adaptive mesh refinement` checkbox. After the simulation, multiple meshes will appear under the `Mesh` node in the `Model Builder`. The first is the initial mesh; the others are refined versions. Any of these meshes can be visualised by clicking on them.

E.2.6.3 Step 2

This step configures the properties of the Adaptive Mesh Refinement for the use in the further studies.
a In the `Model Builder` tree, navigate to `Study 1/Step 1: Time Dependent/Solver Configurations/Solver 1/Solution 2/ Time-Dependent Solver 1`. Click on the `Adaptive Mesh Refinement` node.
b In the `Settings: Adaptive Mesh Refinement` tab, configure the following parameters. While some experimentation may be needed, we typically used:

- `Adaptation Method/General modification`[2]
- `Maximum number of refinements` -> 10

[2] This feature is available starting from the version 5.0 of COMSOL onward.

- `Allow coarsening` -> checked
- `Maximum coarsening factor` -> 3
- `Element selection`/either `Rough global minimum` (with `Element growth factor = 1.3`), or `Fraction of the worst error` (with `Element growth factor` = 0.3)

Optionally, check `Store solution when the new meshes are created`. This enables reuse of any intermediate solution, although memory limitations may require disabling this option.

E.2.6.4 Step 3

This step enables one to re-simulate the problem using one of the newly obtained meshes as the starting mesh, and the last solution as the initial value for the new simulation.

a Click `Add Study` from the main ribbon. This adds a new node `Study 2` to the `Model Builder`. Select `Study 2/Step 1: Time Dependent` and configure the settings.

b Under `Values of Dependent Variables`:

- `Settings` -> `User-controlled`
- `Method` -> `Solution`
- `Study` -> Choose the previous study (e.g., Study 1, Time Dependent)
- `Solution` -> `Refined Mesh Solution...`
- `Time` -> `Last`

c Under `Mesh Selection`, choose the last mesh listed under `Model/Meshes` in the `Model Builder`. This ensures that the new study begins with the most recently refined mesh.

d In the *Adaptation* section, check the checkbox `Adaptive mesh refinement`.

With these settings, COMSOL is prepared for the next simulation cycle.

E.2.6.5 Proceeding Further

Repeat Steps 2 and 3 until the solution quality becomes satisfactory. As previously noted, we applied two criteria for convergence:

1. The zone boundary is smooth.
2. The convergence indicator reaches a horizontal asymptote.

In practice, achieving these conditions may require several studies. However, for larger zones, long wakes, or both, more than 20 studies were sometimes necessary. In total, over 200 meshes and solutions were generated in some cases. Only the final solution was retained as the one representing the zone–wake complex.

Figure E.3b displays the final mesh obtained in such a process, dense at the zone interface and sparse elsewhere. A magnified view of the kink region is shown in Fig. E.3c. Comparing panels (b) and (c), one observes a continuous reduction in mesh size by a factor of approximately 200 when moving across the zone boundary.

Such a mesh dramatically reduces the simulation time of the final study and lowers memory requirements. Without it, memory limitations may prevent storage of even a single solution. Moreover, the refined mesh enables reasonably accurate estimates of the solution's properties, including its integrals, derivatives, and other relevant quantities.

E.2.7 Meshing for Simulation of the Mother-Phase Zone

The simulation of the mother-phase zone is considerably simpler than that of the daughter-phase zone, primarily due to the presence of nonzero boundary conditions. Consequently, the associated meshing is also more straightforward. The order parameter distribution is expected to exhibit significant variation in the vicinity of the crack tip or dislocation. To accommodate this, we partitioned the computational domain into several subdomains surrounding the defect, with the mesh size progressively decreasing from the outer subdomains toward the inner ones. Figure E.4 illustrates such a mesh for the case of a crack tip.

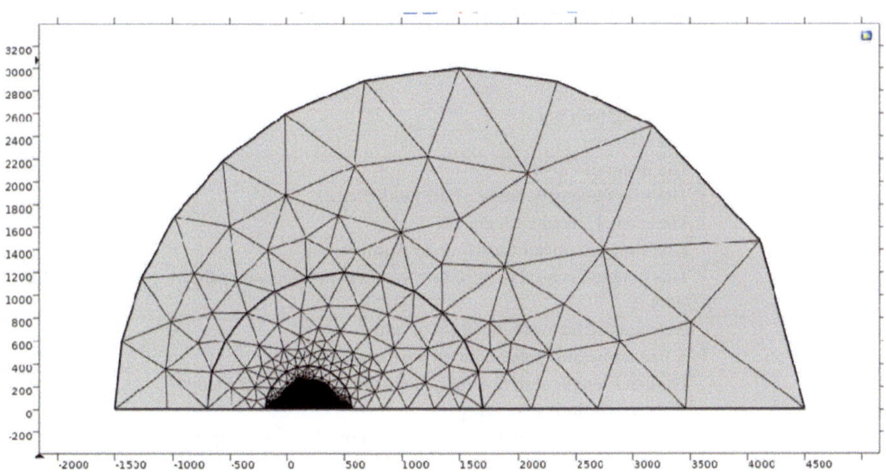

Fig. E.4 Mesh Configuration Around the Crack Tip. The crack tip is located at $x = y = 0$. The domain consists of three subdomains, and the mesh size decreases from the outer domain to the inner one. The Dirichlet boundary condition applies at the semi-circle boundary, while the no-flux Neumann one at the line $y = 0$

E.3 Latin Symbols

Latin symbol	Description	First met
A	The striction coefficient	(3.12)
a	A dimensionless constant of the rescaled Landau potential	(3.21)
a_d	The value of the parameter a at which the the daughter-phase process zone forms	(20.13)
a_b, a_{up}, and a_{low}	The values of the dimensionless parameter a corresponding to the binodal (b), upper (up) and lower (low) spinodals	(3.32) (3.29) (3.30)
B	A factor in the general form of the potential, $U(\mathbf{r})$	(5.3)
B_c	The coefficient of the potential $U(\mathbf{r})$ for the crack	(9.2)
B_d	The coefficient of the potential $U(\mathbf{r})$ for the dislocation	(20.2)
b	A dimensionless constant of the rescaled Landau potential	(8.3)
$b_{cr}^{(d,m)}$	The b value at which the local and bulk binodals intersect	(16.6)
c	The sound velocity	
d	The dimensionless thickness of the process zone interface	(3.68)
E	Young's modulus	(3.11)
${}_2F_1(x, y; z, t)$	A hypergeometric function	Sect. 19.3.3
f	The configurational friction force	(3.58)
$f(\varphi)$	An unknown function of $\sin(\varphi)$ and $\cos(\varphi)$	(5.3)
$f_x + if_y$	A complex force acting within the complex $z = x + iy$ plane	(17.3)
f_{mol}	Molecular force exerted on atoms due to the transition	Sect. 17.3.1
f_v	Configurational friction force acting on the crack tip	Sect. 18.1.1
G	The driving force for the crack propagation	(1.10)
g	A constant of the Landau potential staying in front of $(\nabla\eta)^2$	(3.8)
$\mathbf{H}$	The weight functions vector	(17.15)
h	In the case of the cleavage: the wedge thickness Also the process zone height	(1.7) (14.5)
(h_1, h_2)	Components of the weight functions vector	(17.18)
I_2	The integral	(6.45)
i	The imaginary unit	
J_n $(n = 2, 4, 6)$	The integrals	(6.37)
j_n $(n = 2, 4, 6)$	The integrals	(6.18)
K_I	The stress intensity factor of the mode I crack	(1.2)
K_{IC}	The mode I fracture toughness	(1.8)
k	Rescaling parameter for the order parameter	(5.1)
$k_1, k_2, ...$	This notation is used to introduce variaous intermediate parameters	(22.2), (19.12)
k_g	A geometric factor of the order of unity	(24.4)
$\hat{L}(\alpha)$	A linear operator depending on a parameter α	(2.1)
l	The crack length or semi-length (defined in the text) or the length of the crack during wedging	(1.2) Fig. 19.1
l_{inv}	The dimensionless depth of an invagination of the crack tip process zone	(14.12)
l_{wake}	The wake's dimensionless contact length	(14.23), (24.4)
m	The mass of a crack during wedging	(19.1)
$\hat{N}$	A nonlinear operator	(2.1)
$\mathbf{n}$	A normal unit vector to the zone contour in the weight functions method	(17.19)
$p(x)$	The distribution of a force exerted on the crack boundary	(1.2)
p	Pressure	

Latin symbol	Description	First met
$Q(\mathbf{q})$	The Fourier-transform of $\eta^2(\mathbf{r})$	(4.2)
Q_1, Q_2	Kernels used for calculation ζ_0	(17.10)
$\mathbf{q}, q_\alpha, \mathbf{p}, p_\alpha$	Vectors of the reciprocal space	(4.2)
$\mathbf{R}$	The radius vector	
$\mathbf{r}$	The radius-vector	(1.6)
$\mathbf{r}'$	The radius vector in the co-moving system: $\mathbf{r}' = \mathbf{r} - \mathbf{V}t$	(5.10)
r	The absolute value of the radius-vector	(1.13)
r_c	The correlation radius	
r_0	The distance from the crack tip to the leading zone edge	(14.2)
R_d, R_m	The characteristic dimension of the order parameter distribution in the daughter (d) and mother (m) phases	(6.12)
R_f, R_b	The sizes of the zone leading and trailing parts	(6.39), (6.40)
$\mathbf{S}$	The compliance tensor	
$S(V)$	A combination of the velocity-dependent integrals	(18.5)
S_0	≈ 4.9 is the average value os $S(V)$	(19.6)
T	The temperature (measured in Kelvin)	(3.9)
T_c	The Curie temperature	(3.9)
T_{c0}	The Curie temperature at zero pressure	
T_d	The temperature of the emergence of the daughter-phase zone	(6.10)
T_{up}, T_{low}	The temperatures of the upper and lower spinodals	(3.33)
T_0	The binodal temperature	(3.33)
t	Time	(3.48)
$U_0(\mathbf{r}, \mathbf{t})$	$U_0(\mathbf{r},t) = 2A\varepsilon^{(0)}_{\alpha\alpha}(\mathbf{r},t)$	(4.11)

***

Latin symbol	Description	First met
$\overline{U}$	The potential averaged over the angular coordinates	(6.15)
$U(x, y)$	The rescaled potential	(12.13)
$\overline{U}(\mathbf{r})$	The potential $U(r)$ averaged over the angle	(6.15)
u	Dimensionless (rescaled) order parameter	(3.21)
u_0, u_d	The value of the dimensionless order parameter within the homogeneous bulk daughter phase	(3.28)
$s = s(\mathfrak{z})$	A function defining ζ_0	(17.26)
dS	The element of the zone contour arc	(17.15)
V	The defect velocity	(1.9)
V_d, V_m	The critical velocity value of the daughter-phase (d) and the mother-phase (m) zones	(6.31)
V_G	Gordon's velocity	(3.61)
V_R	The Rayleigh velocity	(1.9)
v	The rescaled (dimensionless) velocity	(3.74)
v_G	Rescaled, dimensionless Gordon's velocity	(3.69)
v_R, v_c	Dimensionless emergence velocity, V_d/V_R, and the ratio, V_d/V_R	(18.8)
W	The elastic energy of the dislocation	(1.14)
x, y, z	Rescaled (dimensionless) spatial coordinates	(1.12)
z	Exponent used in a general form of the potential, $U(\mathbf{r})$	(5.3)
$z = x + iy; Z = X + iY$	Coordinates of the complex plane	(17.3)
$z(\tau)$	In Chap. 19, the dimensionless length of the wedge crack perturbation	(19.11)

E.4 Greek Symbols

Greek symbol	Description	First met
α	The coefficient of the original Landau potential staying in front of η^2	(3.8)
$\sin(\alpha)$	Here α is the semi-angle at the wake's point	(14.16)
α_0	A constant of the Landau potential: $\alpha = \alpha_0(T - T_c)$	(3.9)
$\alpha_{\mathrm{d}}, \alpha_{\mathrm{m}}$	The value of α at the bifurcation with formation of the daughter phase (d) and the mother-phase (m) zone	(6.12)
β_0, β_1 and β	Constant and renormed constants of the original Landau potential staying in front of η^4	(3.8)
Γ	The resistance for the crack propagation	(1.10)
$\Delta\Gamma$	Contribution of the process zone to the resistance to the crack propagation	(18.1)
γ	The constant of the original Landau potential staying in front of η^6	(3.8)
Δ	The Laplace operator	(3.53)
∇	The gradient operator	(3.8)
Δ'	The Laplace operator in terms of the coordinates $\mathbf{r}'$	(6.22)
∇'	The gradient operator in terms of the coordinates $\mathbf{r}'$	(6.22)
δ	Determines the absolute value of the fracture toughness perturbation	(17.33)
$\delta\rho(x, y, z)$	The distribution function perturbation caused by the phase transformation	
$\delta\eta$	The variation of η	(3.52)
ε	Atomic energy or small parameter in various calculations	
$\varepsilon_{\alpha\beta}$ $(\alpha, \beta = 1, 2, 3)$	The strain tensor. The Greek letters $\alpha, \beta, ... = 1, 2, 3$ are the tensor indices	(1.12)
$\varepsilon_{\alpha\alpha}$	The trace of the strain tensor: $\varepsilon_{\alpha\alpha} = \varepsilon_{11} + \varepsilon_{22} + \varepsilon_{33}$ in 3D and $\varepsilon_{\alpha\alpha} = \varepsilon_{11} + \varepsilon_{22}$ in 2D	(1.13)
ε_0	The characteristic value of the spontaneous strain; the notation used in estimates	
η	The order parameter	(2.4)
η_0	The order parameter value in the daughter phase far from the defect also used as the order parameter amplitude	(7.1)
$\eta_{\mathrm{d,m}}$	The order parameter in the zone containing the daughter (d) or mother (m) phases	(6.7)
η_{s}	The value of the order parameter at saturation	Sect. 17.4.1.4
$\Theta(x)$	The Heaviside Θ-function	Sect. 19.2.1
θ	The angle in cylindrical coordinates	(1.13)
κ	The kinetic coefficient of the order parameter	(3.51)
μ	The shear modulus	(1.14), (1.15)
ν	Poisson's ratio	(1.12)
ϑ	The angular variable in the van der Pol-Andronov approach	(19.18)
$\Xi = \Xi(z, z')$	A dimensionless function in the theory of stick-slip crack motion	(19.13)
$\Delta\Xi, \Delta\Xi, ..., \Delta\Xi_3$	Perturbations of the function Ξ	(19.15)
ξ $\xi_{\mathrm{d}}, \xi_{\mathrm{m}}$	The amplitude of the solution of the nonlinear equation, the amplitude of the daughter-phase (d) and mother-phase (m) zones	(2.4), (6.7)

Greek symbol	Description	First met
ρ	The mass density of the crystal. In another case, the dimensionless curvature radius	(3.74)
$\rho(x, y, z)$, $\rho_0(x, y, z)$	The distribution function of atoms in crystal, and that in the high-symmetry phase	
ρ_1	The dimensionless curvature radius of the invagination of the crack tip process zone	(14.8)
ρ_2	The dimensionless rounding radius of the wake's point	(14.15)
ρ_{disl}	Density of dislocations	Sect. 20.2
ρ_{st}	The stationary radius-vector in van der Pol-Andronov method	Sect. 19.4.1.3
Σ	The solid's surface energy (fracture energy)	(1.8)
$\sigma_{\alpha\beta}$ $(\alpha, \beta = 1, 2, 3)$	The stress tensor	(1.1)
σ_0	The characteristic value of the spontaneous stress; the notation used in estimates: $\sigma_0 \sim E\varepsilon_0$	Sect. 17.3.1
σ_f	The stress necessary to overcome the friction due to the dissipation within the zone	(24.7)
τ	Rescaled (dimensionless) time	
τ_{disl}	The line tension of the dislocation	(1.18)
τ_{s}	The surface tension coefficient of the process zone interface	(A.2)
Φ	The free energy density	
$\Phi_{\text{pt}}(\eta)$, $\Phi_{\text{el}}(\varepsilon)$ $\Phi_{\text{int}}(\eta, \varepsilon)$	The free energy densities: the phase transformation, elastic and interaction contributions	
Φ_{tot}	The total free energy density of the crystal	(4.15)
$\Phi(z)$	Kolosov-Mushelishvili Φ-potential	(17.3)
$\varphi_\alpha(\theta)$	The functions yielding the angular dependence of the displacement vector	(1.4)
ϕ	J_4^2/J_2J_6	(8.29)
$\Psi_n(\mathbf{r})$	The eigenfunction in the case of the propagating defect	(6.23)
$\Psi(z)$	Kolosov-Mushelishvili Ψ-potential	(17.3)
$\psi_n(\mathbf{r})$ $(n = 1, 2, ...)$	The eigenfunction of the linear operator, $\hat{L}(\alpha_n)$	(2.3)
$\psi_*(\mathbf{r}) \equiv \psi_{\text{d}}(\mathbf{r}) \equiv \psi_1(\mathbf{r})$	The ground state eigenfunction of the linear operator, $\hat{L}(\alpha_0)$	(2.4)
$d\Omega$	The volume element of the crystal $d\Omega = dxdydz$ in 3D and $d\Omega = dxdy$ in 2D	(2.7)
Ω_{pz}	The area in 2D or volume in 3D of the process zone	(6.17)
ω	The frequency	
ζ_0	The dimensionless factor of the fracture toughness	(17.9)

E.5 Gothic Symbols

Symbol	Description	First met
$\mathfrak{A}$	The amplitude of oscillations of a dislocation segment	
$\mathfrak{a}$	Parameter of the crystal lattice	
$\mathfrak{b}$	Burger's vector	(1.12)
$\mathfrak{C}$, $\mathfrak{C}_1$, $\mathfrak{C}_2$	The contour and its portions in the weight functions method	(17.15)

Symbol	Description	First met
$\mathfrak{D}$	A factor	Sect. 31.2.1
$\mathfrak{g}$	A dimensionless factor in the expression for the Frank-Read source start	(28.1)
$\mathfrak{M}$	The kink mobility	(31.13)
$\mathfrak{M}_{PZ}$, $\mathfrak{M}_{NZ}$	The mobility of a kink with and without the zone	(31.14)
$\mathfrak{q}$	An auxiliary dimensionless parameter	(14.22), (14.23), (19.3)
$\mathfrak{s}$	Phenomenological constant accounting for the energy dissipation by mechanisms different from the process zone	(19.2)
$\mathfrak{u}$	The displacement vector	(1.4)
$\mathfrak{u}_{\alpha}^{(0)}$	The displacement vector generated by the undressed defect	(1.4)
$\mathfrak{u}_{\alpha}(\mathbf{q})$	The Fourier-transform of $\mathfrak{u}_{\alpha}(\mathbf{r})$	(4.2)
$\mathfrak{z}$	$\mathfrak{z} = \alpha\gamma/\beta^2$	(17.27)

E.6 Calligraphic Symbols

Symbol	Description	First met
$\mathcal{A}$	The dimensional coefficient of the free energy	(20.22)
$\mathcal{D}$	The dissipation function	(3.51)
$\mathcal{F}$	The total free energy of the crystal containing a dressed defect	(2.8)
$\mathcal{F}_{\text{eff}}$	The effective free energy of the crystal	(6.43)
$\mathcal{F}_{\text{pt}}$	The generating functional whose variation yields equation of state	(4.14)
$\mathcal{F}_{\text{tot}}$	The free energy of the crystal containing a dressed defect	(4.16)
$\mathcal{F}_{\text{resc}}$	The dimensionless free energy functional	(20.23)
$\mathcal{F}_0$	The free energy of a solid with no process zone	(3.14)
$\mathcal{H}$, $\mathcal{H}_0$	The right-hand Part of the wedging equation	(19.4), (19.5)
$\mathcal{L}$	The crystal Lagrangian	(3.48)
$\mathcal{L}_{\text{tot}}$	The Lagrangian of the crystal containing a dressed defect	(4.17)
$\mathcal{M}$	$\mathcal{M}(x)=x \times \Theta(x)$, where $\Theta(x)$ is the theta-function	(19.6)
$\mathcal{Z}$	Partition function	(3.1)

E.4 Double-Stroke Symbols

Symbol	Description	First met
$\mathbb{A}$, $\mathbb{A}_c$	The dimensionless ratios, α/α_{d} and α/α_{c}	(18.8)
$\mathbb{D}$	The physical, dimensional thickness of the phase (zone's) interface	(A.2)
$\mathbb{G}$	The, so-called, inverse driving force, $1 - \Gamma/G$	(18.9)
$\mathbb{H}$	The width of the process zone in the direction perpendicular ot its propagation	(3.61)
$\mathbb{H}_{\text{arc}}$	The dislocation segmnent projection length	
$\mathbb{L}$	The length of dislocation, its segment, or of a crack in the Oz direction	(1.14)

Symbol	Description	First met
$\mathbb{L}_{disl}$	The average distance between dislocations in a good crystal	Sect. 20.2
hline $\mathbb{L}_{inv}$	The physical depth of the invagination of the crack tip process zone	(17.17)
$\mathbb{L}_{arc}$	The length of a dislocation arc	
$\mathbb{M}$	A combination of parameters	(18.26)
$\mathbb{O}$, $\Delta\mathbb{O}$, $\mathbb{O}_i$	Auxiliary functions used in van der Pol-Andronov method	(19.21)
$\mathbb{R}$	The characteristic size of any process zone in physical coordinates	(17.17)
$\mathbb{R}_0$	The natural spatial scale	(5.1)
$\mathbb{R}_{cr}$	The size of the crystal (sample) in physical coordinates	(1.14)
$\mathbb{R}_{curv}$	The curvature radius of a phase interface in physical coordinates	(3.74)
$\mathbb{R}_{d,m}$	The characteristic size of the process zone containing the daughter (d) or mother-phase (m) in physical coordinates	(8.36)
$\mathbb{R}_1$, $\mathbb{R}_2$, $\mathbb{R}_3$	Denote different contributions to an equation	(19.27)
$\mathbb{R}_{arc}$	The radius of the dislocation arc	
$\mathbb{T}_0$	The natural time-scale	(5.1)
$\mathbb{U}$, $\mathbb{U}_1$,..., $\mathbb{U}_4$, $\mathbb{U}_{app}$	Auxiliary functions used in van der Pol-Andronov method	(19.24)

References

1. R.C. Pohanka et al., J. Am. Ceram. Soc. **61**, 72 (1978)
2. R.C. Pohanka et al., Ferroelectrics **28**, 337 (1980)
3. J. Dec, Ferroelectrics **89**(1), 193 (1989)
4. E.Y. Tonkov, *High Pressure Phase Transformations: A Handbook.* (Gordon and Breach, Amsterdam, 1992), vols. 1–3; E. Y. Tonkov and E. G. Ponyatovsky, *Phase Transformations of Elements Under High Pressure.* (CRC Press, Boca Raton, FL, 2005), vol. 4
5. E.G. Fesenko, V.G. Gavrilyachenko, A.F. Semenchev, in *Domain Structure of Multiaxial Ferroelectric Crystals.* (Rostov University, Rostov-on-Don, 1990) [in Russian]; T.E. Hooper, A.J. Bell, J. Appl. Phys. **127**, 104102 (2020)
6. S.G. Dzhabarov, D.P. Kozlenko, S.E. Kichanov, A.B. Belushkin, B.N. Savenko, R.Z. Mekhtieva, K. Late, Fizika Tverdogo Tela **53**, 2185 (2011) - In Russian
7. J. Hlinka, T. Ostapchuk, D. Nuzhnyy, J. Petzelt, P. Kuzel, C. Kadlec, P. Vanek, I. Ponomareva, L. Bellaiche, Phys. Rev. Lett. **101**, 167402 (2008)
8. J. Petzelt, G.V. Kozlov, A.A. Volkov, Ferroelectrics **73**, 101–123 (1987)
9. V.G. Vaks, *Introduction into Microscopic Theory of Ferroelectrics* (Nauka, Moscow, 1973) - in Russian
10. A.D. Bruce, R.A. Cowley, *Structural Phase Transitions* (Taylor & Francis Ltd., London, 1981)
11. A. Boulbitch, A.L. Korzhenevskii, Phys. Rev. B **96**, 054106 (2017)
12. A. Boulbitch, Mathematica Journal **20**, 1–21 (2018)

MIX
Papier aus verantwortungsvollen Quellen
Paper from responsible sources
FSC® C105338

If you have any concerns about our products,
you can contact us on
ProductSafety@springernature.com

In case Publisher is established outside the EU,
the EU authorized representative is:
Springer Nature Customer Service Center GmbH
Europaplatz 3, 69115 Heidelberg, Germany

Printed by Libri Plureos GmbH
in Hamburg, Germany